T0143014

Computer Analysis of Sequence Data
Part II

Methods in Molecular Biology™

John M. Walker, SERIES EDITOR

Methods in Molecular Biology • 25

Computer Analysis
of Sequence Data
Part II

Edited by

Annette M. Griffin
and Hugh G. Griffin

*Institute of Food Research, Norwich Research Park,
Norwich, England*

Humana Press ✳ Totowa, New Jersey

© 1994 Humana Press Inc.
999 Riverview Drive, Suite 208
Totowa, New Jersey 07512

10 9 8 7 6 5 4 3 2

Library of Congress Cataloging-in-Publication Data

Main entry under title: Methods in molecular biology.

Computer analysis of sequence data / edited by Annette M. Griffin and
 Hugh G. Griffin.
    p. cm. — (Methods in molecular biology ; v. 24–25)
    Includes index.
    ISBN 978-0-89603-246-0 (pt. 1) — ISBN 978-0-89603-276-7 (pt. 2)
    1. Nucleotide sequence—data processing. 2. Amino acid sequence—data processing.
I. Griffin, Annette M. II. Griffin, Hugh G. III. Series: Methods in molecular biology
(Totowa, NJ) ; v. 24–25.
QP620.C65 1994
574.87'328'0285—dc20   93-36758
                                                                        CIP

Preface

DNA sequencing has become increasingly efficient over the years, resulting in an enormous increase in the amount of data generated. In recent years, the focus of sequencing has shifted, from being the endpoint of a project, to being a starting point. This is especially true for such major initiatives as the human genome project, where vast tracts of DNA of unknown function are sequenced. This sheer volume of available data makes advanced computer methods essential to analysis, and a familiarity with computers and sequence analysis software a vital requirement for the researcher involved with DNA sequencing. Even for nonsequencers, a familiarity with sequence analysis software can be important. For instance, gene sequences already present in the databases can be extremely useful in the design of cloning and genetic manipulation experiments.

This two-part work on *Computer Analysis of Sequence Data* is designed to be a practical aid to the researcher who uses computers for the acquisition, storage, or analysis of nucleic acid (and/or protein) sequences. Each chapter is written such that a competent scientist with basic computer literacy can carry out the procedure successfully at the first attempt by simply following the detailed practical instructions that have been described by the author. A Notes section, which is included at the end of each chapter, provides advice on overcoming the common problems and pitfalls sometimes encountered by users of the sequence analysis software. Software packages for both the mainframe and personal computers are described. Many of these packages are capable of carrying out all the basic tasks required by the researcher, such as the translation of a DNA sequence into an amino acid sequence, searching for open reading frames, drawing restriction enzyme maps, manipulation of sequences by addition or deletion of fragments, identification of such specified targets as promoters or inverted repeats, and prediction of such protein characteristics as molecular weight, hydropathy, antigenicity, and secondary structure. Some packages and programs also perform more specialized tasks, such as the assembly of random fragments from a shotgun sequencing project, similarity searches, alignments of similar sequences, and phylogenetic analysis.

By bringing together a wide variety of programs for both mainframe and personal computer, *Computer Analysis of Sequence Data* promises to become an important standard reference work for all those analyzing nucleic acid and protein sequences. As such, the book should prove an invaluable practical tool for the many research workers who, although not wishing to become computer experts, regularly use computers to analyze sequence data.

Annette M. Griffin
Hugh G. Griffin

Contents

Contents
Computer Analysis of Sequence Data, Part I

Contributors

GEOFFREY J. BARTON • *Laboratory of Molecular Biophysics, Oxford, UK*

GRAHAM N. CAMERON • *European Molecular Biology Laboratory Data Library, Heidelberg, Germany*

SUSAN E. DOUGLAS • *Institute for Marine Biosciences, National Research Council, Halifax, Nova Scotia*

MICHAEL GRIBSKOV • *San Diego Supercomputer Center, PO Box 85608, San Diego, CA*

ANNETTE M. GRIFFIN • *Institute of Food Research, Norwich Research Park, Norwich, UK*

HUGH G. GRIFFIN • *Institute of Food Research, Norwich Research Park, Norwich, UK*

JOTUN HEIN • *Institute for Genetics and Ecology, Aarhus University, Aarhus, Denmark*

DESMOND G. HIGGINS • *European Molecular Biology Laboratory, Heidelberg, Germany*

JOHN H. MCVEY • *Haemostasis Research Group, Clinical Research Centre, Harrow, UK*

CARY O'DONNELL • *AFRC Computing Division, Harpenden, UK*

SUE A. OLSON • *International Biotechnologies, Inc., New Haven, CT*

WILLIAM R. PEARSON • *Department of Biochemistry, University of Virginia, Charlottesville, VA*

AIALA REIZER • *Department of Biology, University of California at San Diego, La Jolla, CA*

JONATHAN REIZER • *Department of Biology, University of California at San Diego, La Jolla, CA*

CATHERINE M. RICE • *European Molecular Biology Laboratory Data Library, Heidelberg, Germany*

MILTON H. SAIER, JR. • *Department of Biology, University of California at San Diego, La Jolla, CA*

RODGER STADEN • *MRC Laboratory of Molecular Biology, Cambridge, UK*

MICHAEL ZUKER • *Institute for Biological Sciences, National Research Council of Canada, Ottawa, Ontario, Canada*

CHAPTER 1

Computer Analysis of Sequence Data

Hugh G. Griffin and Annette M. Griffin

1. Introduction

DNA sequencing methodology was developed in the late 1970s *(1,2)* and has become one of the most widely used techniques in molecular biology. The amount of DNA sequencing performed worldwide has increased exponentially each year and this trend will almost certainly continue for the next few years at least. The importance of this technique is underlined by the volume of research funds now being channelled into genome sequencing projects, and by the large sums of money being invested in the development of automated sequencers and sequence analysis systems.

Computers have always been an integral part of DNA sequence analysis. Many large sequencing projects would be virtually impossible without the aid of computers, and the international computer databanks are an essential facility for the storage and retrieval of sequence data. During the 1980s, while the original methods of sequencing were being refined and improved, fast and powerful computer programs were being developed to analyze the copious amounts of data being generated. Indeed, many sequencing projects, such as those utilizing the shotgun approach *(3–6)* depended on a computer for the assembly of the DNA sequence of random fragments. The development of automated sequencers also depended on the improved quality of computer hardware and software that were necessary for the interpretation of data produced by these instruments.

From: *Methods in Molecular Biology, Vol. 25: Computer Analysis of Sequence Data, Part II*
Edited by: A. M. Griffin and H. G. Griffin Copyright©1994 Humana Press Inc., Totowa, NJ

The molecular biologist of today has a vast array of computer programs and packages from which to choose. Not only are these faster and more powerful than earlier versions but are much more "user friendly." Many have easy-to-use help menus and the competent scientist will not need any formal training to enable him/her to perform rapid, in-depth analysis of sequence data. Most of the widely used programs were originally developed for mainframe computers, but now powerful sequence analysis packages are available for IBM and IBM-compatible personal computers (PCs) and for the Apple Macintosh.

Two main factors will influence the scientist's choice of computer and sequence analysis package: (a) what is already in his/her laboratory, and (b) what is affordable to buy. The molecular biologist new to the area of sequence analysis would be well advised to start by becoming familiar with systems already set up in his/her laboratory; additional programs and hardware can be acquired when a need is identified or when funds are available. For the scientist planning to purchase sequencing software the decision on what to buy might be based on available hardware. The mainframe sequence analysis packages are primarily designed for multiuser use and take up a lot of disk space, while the programs developed for personal computers and Macintoshes are ideal for individual (or small group) use. Database searching is perhaps more readily performed on a main frame and the systems manager can make regular updates. A compact disk system is usually employed for database searches on a PC or Macintosh. In an ideal world, the molecular biologist would have access to all available systems and sequence analysis software!

2. Sequence Analysis Tasks

The types of sequence analysis tasks required of a software package by a molecular biologist vary depending on the individual investigator and his/her research interests. However, some of the most common functions required by almost all investigators are as follows:

1. *Storage, retrieval, and editing of user-generated sequence data*. This includes programs to assemble random fragments from shotgun projects *(3–6)*, manipulation of sequences by deletion of segments, joining segments together, and preparation of publication quality output.

2. *Construction of maps.* Most investigators will want to be able to readily produce a restriction enzyme map of their sequenced DNA fragment or of a sequence retrieved from the databanks. Most packages can cope with linear or circular maps and can also search for specified targets other than restriction sites, e.g., promoters, inverted repeats, and consensus sequences.
3. *Translation.* Analysis of a DNA sequence for open reading frames or potential protein coding regions and translation of a DNA sequence into predicted amino-acid sequence in any specified reading frame.
4. *Protein analysis.* Analysis of protein or deduced amino acid sequences. Many options are available, including prediction of molecular weight, hydropathy, antigenicity, and secondary structures.
5. *Similarity searches.* Search for sequences similar to the user specified sequence. These searches can be conducted on a local user generated databank or on the international databanks (EMBL, GenBank, DDBJ, PIR, Swissprot). Searches can be conducted with either DNA or amino-acid sequences.
6. *Alignment with a similar sequence.* This is one of the most popular tasks a scientist asks of a sequence software package. An optimal alignment is achieved between two similar sequences (DNA or amino acid) and the percent identity and/or similarity calculated. Publication quality output is usually required.
7. *Submission and retrieval.* Most investigators will wish to submit their newly generated DNA sequence to the international databanks. Before the establishment of widely accessible international sequence databanks, the only method of disseminating DNA sequence information was by publication in scientific journals. The only way for another scientist to analyze the data was to enter it into his/her computer via the keyboard. As well as being laborious this process inevitably caused inaccuracies. Most sequence data sent to the databanks are submitted in machine readable form (e.g., by e-mail or on disk or tape) thus ensuring accuracy. In addition, there is a facility to submit data as soon as it becomes available but to disallow its release until the information appears in print in the scientific press. This procedure protects the scientist working in a highly competitive area, while permitting the electronic dissemination of scientific data simultaneously with the written version.

Even for the nonsequencer, the databanks can be important. For example, most previously sequenced genes, operons, promoters, and most of the commonly used cloning vectors (e.g., pUC19) can be retrieved from the databanks. This sequence data can be of great use in the design of cloning and genetic manipulation experiments. Also, computer generated codon usage tables can facilitate the design of probes or PCR primers from N-terminal protein sequences, perhaps reducing the amount of degeneracy required. The establishment and operation of the international sequence databanks have contributed immensely to the widespread dissemination and utilization of sequence data.

3. Aim of This Book

Most of the widely available sequence analysis packages for either mainframe or microcomputers are capable of performing the tasks mentioned earlier with varying degrees of ease and efficiency. As no single book of this nature could possibly cover all the available packages, we have chosen six of the most popular for inclusion in this two-volume work. Our choice of packages was mainly random and the fact that a program or package is not mentioned it does not imply that is in any way inferior or less widely used than those that have been included. The Genetics Computer Group (GCG) and STADEN packages were developed for mainframe computers such as the Digital Equipment Corporation VAX running VMS operating system; MicroGenie and PC/Gene were designed for IBM-compatible PCs; DNA Strider and MacVector were written for the Macintosh. It should be noted that versions of these programs may well be available for use with computers other than those for which they were originally developed. For example, versions of the STADEN programs can run on a Macintosh, and on a SUN workstation (using UNIX operating system). In fact, current development is being conducted on the SUN version only. In addition, we have included a number of specialized programs to aid the handling and manipulation of sequence data and to improve the choice of programs capable of carrying out the most popular tasks.

Computer Analysis of Sequence Data is aimed at the competent molecular biologist who nonetheless is a newcomer to the field of DNA sequencing and data analysis. Each contribution is written such that a competent scientist who has only basic computer literacy can carry out the procedure successfully at the first attempt by simply following the detailed practical instructions that have been described by the author. As we all know even the simplest computer programs go wrong from time to time, and for this reason a "Notes" section has been included in most chapters. These notes will indicate any major problems or faults that can occur, the sources of the problems, and how they can be identified and overcome. A comprehensive reference section is also included in most chapters to enable the reader to refer to other publications for more detailed theoretical discussions on the various programs.

4. Mainframes, Personal Computers, and the Macintosh

The two mainframe packages described in *Computer Analysis of Sequence Data* are GCG (in Part I) and STADEN (in Part II). The GCG package is an extensive suite of sequence handling and analysis software. The package was originally built by John Devereux and Paul Haeberli from programs written for Oliver Smithies' laboratory in the University of Wisconsin, Department of Genetics *(7,8)*. Many of the programs within the current package were written by individual scientists (such as William Pearson, Michael Gribskov, and Michael Zuker) and some of these programs are described in *Computer Analysis of Sequence Data* by their original authors, *see* Chapters 22 and 29 in this volume, and Chapter 26 in Part I, as well as being mentioned in the section on GCG (Chapters 2–14 in Part I). However, all programs within the GCG package have been adapted to a certain uniform format that makes them particularly easy to use. All programs "look the same"—if you can run one you can run them all. The output from each program is suitable for input to other programs in the package. Most new users find the interactive interface very easy to use.

The STADEN package is based on the programs of Rodger Staden (Medical Research Council, Laboratory of Molecular Biology, Cambridge, UK). The package is split into a relatively small number of large menu-driven programs. GIP uses a digitizer for entry of DNA sequences from autoradiographs. SAP and DAP consist of programs for assembling and editing gel readings. The STADEN package is widely acknowledged as one of the best in the world for management of random sequencing projects. The programs NIP and PIP provide functions for a comprehensive analysis of individual nucleotide and protein sequences, respectively, whereas SIP can compose and align pairs of either protein or DNA sequences. NIPL, PIPL, and SIPL are programs designed to search sequence libraries. The user interface, which is common to all programs, consists of a set of menus and a uniform way of presenting choices and obtaining input from the user. Help is available by responding to a query with the symbol "?". The X interface, involving pulldown menus, dialog boxes, and buttons, is not available on the VAX version. The STADEN programs are described in Chapters 2–14.

There are now a large number of programs and packages available either commercially or in the public domain that operate on the powerful modern microcomputer. We have chosen two packages, MicroGenie and PC/Gene, which operate on IBM or compatible machines, and two packages, DNA Strider and MacVector, which operate on the Macintosh. People in general, and scientists in particular, tend to have strong views on whether the Macintosh or the IBM-PC is superior. The choice of machine must be left up to the individual. The disadvantages of the use of microcomputers for sequence analysis is their limited memory and computational power. This is not usually a problem for the majority of common sequence analysis tasks although library searching is not as convenient as on a mainframe. In addition, whereas the international databases can be updated on a weekly basis relatively easily on a mainframe, this is not usually as conveniently performed on the microcomputer. An advantage of microcomputers is that they can be dedicated to the task of sequence analysis, thus performing some tasks faster than the mainframe, which is often asked to perform many functions simultaneously. Output from a microcomputer package is usually more readily incorporated into word processing packages and graphics programs for use in producing publication quality hardcopy. Additionally, for the individual investigator or small group, a microcomputer system can be considerably less expensive than setting up a mainframe system.

5. Specialized Programs

Each sequence analysis program or package requires the input sequence to be in a specific format. By format, is meant the style and organization of the computer file containing the sequence, e.g., double or single line spacing, number of characters per block, number of blocks per line, whether characters are numbered or not, and so on.

Unfortunately for scientists who have access to more than one package, a different format is required for each package. For example, STADEN format is incompatible with GCG format, even though many research laboratories operate both packages on the same computer. Therefore we have included in *Computer Analysis of Sequence Data* a chapter on methods for converting sequences between different formats (Chapter 30). READSEQ is a program that can interconvert sequence files between thirteen different formats. In addition, many packages contain programs for converting between sequence formats.

The generation of multiple alignments of three or more related sequences, either DNA or protein, is becoming an increasingly popular function of sequence analysis software. Such alignments can highlight small, highly conserved regions and help identify sequence features such as enzyme active sites or substrate binding domains. Overall levels of homology can be calculated and a consensus sequence produced. Similar programs can be used to explore phylogenetic relationships if the equivalent sequences from many different species are available. Evolutionary trees depicting such relationships can be drawn. Chapters 25–28 describe programs for multiple sequence alignment and construction of phylogenetic trees.

Many programs for sequence analysis are in the public domain. That is to say they are available freely to anyone who requests them, as long as they are not subsequently sold. Chapter 31 explains how to obtain software of this nature via electronic mail.

Finally, all newly generated sequence data should be submitted to the international sequence databanks. Chapter 32 describes how to do this.

References

1. Sanger, F., Nicklen, S., and Coulson, A. R. (1977) DNA sequencing with chain-terminator inhibitors. *Proc. Natl. Acad. Sci. USA* **74,** 5463–5467.
2. Maxam, A. M. and Gilbert, W. (1977) A new method for sequencing DNA. *Proc. Natl. Acad. Sci. USA* **74,** 560–564.
3. Messing, J. and Bankier, A. T. (1989) The use of single-stranded DNA phage in DNA sequencing, in *Nucleic Acids Sequencing: A Practical Approach* (Howe, C. J. and Ward, E. S., eds.), IRL Press, Oxford, UK, pp. 1–36.
4. Anderson, S. (1981) Shotgun DNA sequencing using cloned DNAse 1-generated fragments. *Nucleic Acids Res.* **9,** 3015.
5. Deininger, P. L. (1983) Random subcloning of sonicated DNA: Application to shotgun DNA sequence analysis. *Anal. Biochem.* **129,** 216.

6. Bankier, A. T., Weston, K. M., and Barrell, B. G. (1987) Random cloning and sequencing by the M13/dideoxynucleotide chain termination method, in *Methods in Enzymology*, vol. 155 (Wu, R., ed.), Academic, London, pp. 51–93.
7. Smithies, O., Engels, W. R., Devereux, J. R., Slighton, J. L., and Shen, S. (1981) Base substitutions, length differences and DNA strand asymmetries in the human G-gamma and A-gamma fetal globin gene region. *Cell* **26,** 345–353.
8. Devereux, J., Haeberli, P., and Smithies, O. (1984) A comprehensive set of sequence analysis programs for the VAX. *Nucleic Acids Res.* **12,** 387–395.

CHAPTER 2

Staden: Introduction

Rodger Staden

1. Introduction

In this chapter we give an overview of the chapters on the "Staden Package" of programs. Here the equipment required is described, and the scope of the package and the user interfaces is outlined. The next chapter covers character sets, sequence formats, and sequence library access.

The main programs in the package are as follows:

GIP Gel input program
SAP Sequence assembly program
DAP Sequence assembly program
NIP Nucleotide interpretation program
PIP Protein interpretation program
SIP Similarity investigation program
MEP Motif exploration program
NIPL Nucleotide interpretation program (library)
PIPL Protein interpretation program (library)
SIPL Similarity investigation program (library)
XDAP Sequence assembly program
XNIP Nucleotide interpretation program
XPIP Protein interpretation program
XSIP Similarity investigation program
XMEP Motif exploration program

GIP uses a digitizer for entry of DNA sequences from autoradiographs. SAP, DAP, and XDAP handle everything relating to assem-

From: *Methods in Molecular Biology, Vol. 25: Computer Analysis of Sequence Data, Part II*
Edited by: A. M. Griffin and H. G. Griffin Copyright ©1994 Humana Press Inc., Totowa, NJ

bling and editing gel readings. NIP provides functions for analyzing and interpreting individual nucleotide sequences. PIP provides functions for analyzing and interpreting individual protein sequences. MEP analyzes families of nucleotide sequences to help discover new motifs. NIPL performs pattern searches on nucleotide sequence libraries. PIPL performs pattern searches on protein sequence libraries. SIP provides functions for comparing and aligning pairs of protein or nucleotide sequences. SIPL searches nucleotide and protein sequence libraries for entries similar to probe sequences. The programs whose names begin with a letter X are X11 (*see* Section 3.) versions of the programs. For example XNIP is an X11 version of NIP.

2. Materials

2.1. Versions

The programs run on Apple Macintosh computers, on VAX computers using the VMS operating system, and on SUN workstations (which use the UNIX operating system.) The SUN version should run, with only minor changes, on other machines running UNIX. Currently, the Macintosh version is "frozen" in its April 1990 state, the VAX version is "frozen" in its April 1991 state and all development is being done on the SUN version.

2.1.1. VAX Version

The VAX version will run on any VAX using the VMS operating system. A FORTRAN compiler is required.

2.1.2. UNIX Version

The UNIX version is being used on SUN SPARCstations, SUN SPARCstation 10s DECstation 5000/240s, Silicon Graphics R3000 and R4000, and DEC Alphas. At least 16 Mbytes of memory and 500 Mbyte disk are required. Color monitors are preferable for running the programs that display traces from fluorescent sequencing machines, but monochrome displays are adequate for all other programs. Tape drives are used for archiving and a CD-ROM drive for handling the sequence libraries.

2.1.3. Other UNIX Versions

Users of UNIX machines other than those listed in Section 2.1.1. will require a FORTRAN and ANSI C compilers. When operated directly on the workstation screen all UNIX versions require X11 release 4 or above.

2.1.4. The Macintosh Version

The Macintosh version of the package requires a machine with at least 1 Mbyte of memory and a 20 Mbyte hard disk. It only operates on monochrome screens or color screens set to black/white mode. The package contains only programs SAP, GIP, NIP, PIP and SIP. All further information about this version of the package is contained in Section 7.

2.2. Terminals

The programs can also be operated via a serial port using Tektronix terminals, PCs running MS-Kermit, or Apple Macintoshs running Versaterm Pro. The UNIX versions can also be run from X teminals or microcomputers running X emulators.

2.3. Digitizers

The gel reading input program uses a sonic digitizer called a GRAPHBAR GP7 made by Science Accessories Corp., 200 Watson Blvd., Stratford, CT 06497. When ordering specify that the device should be set to use metric units.

2.4. Sequencing Machines

The programs can handle data produced by the Applied Biosystems Inc. (Foster City, CA), 373A and Pharmacia (Piscataway, NJ) A.L.F fluorescent sequencing machines.

3. User Interfaces

The programs have two user interfaces. The first runs under the terminal emulator xterm and the second runs directly under X. On the VAX, at present only the xterm interface is available, but on UNIX systems, either interface can be used. The xterm version of the package will operate on the workstation screen, X terminals, Tektronix terminals, PCs, or Macintoshes *(see above)*. When run on the workstation screen the programs have separate text and graphics windows, each of which can be moved, resized and iconized, and the text windiow can be scrolled in both directions. The versions that run directly under X can only be used on the workstation screen, X terminals, or using an

X emulator. They produce separate text and graphics windows, an independent, constantly available help window and a separate dialog window. All input is controlled by mouse selection and dialog boxes.

3.1. The xterm and VAX Interface

The user interface is common to all programs. It consists of a set of menus and a uniform way of presenting choices and obtaining input from the user. This section describes: the menu system; how options are selected and other choices made; how values are supplied to the program; how help is obtained, and how to escape from any part of a program. In addition it gives information about saving results in files and the use of graphics for presenting results.

3.1.1. Menus and Option Selection

Each program has several menus and numerous options. Each menu or option has a unique number that is used to identify it. Menu numbers are distinguished from option numbers by being preceded by the letter m (or M, all programs make no distinction between upper- and lower-case letters). With the exception of some parts of program SAP, the menus are not hierachical; rather, the options they each contain are simply lists of related functions and their identifying numbers. Therefore, options can be selected independently of the menu that is currently being shown on the screen, and the menus are simply memory aides. All options and menus are selected by typing their option number when the programs present the prompt "? Menu or option number ="

To select a menu, type its number preceded by the letter M. To select an option, type its number. If users type only "return" they will get menu m0, which is simply a list of menus. If users select an option they will return to the current menu after the function is completed. Where possible, equivalent or identical options have been given the same numbers in all programs, so users quickly learn the numbers for the functions they employ most often.

3.1.2. Execution and Dialog

All inputs requested by the program (apart from file names) have default values. In addition, most of the analytical functions have a default path through which they will pass, so when users select an option, in many cases, the program will immediately perform the operation selected without further dialog. However, if users precede an option number by the letter d (e.g., D17), they will force the program to offer dialog about the selected option before the function operates, hence allowing them to change the value of any of its parameters. In addition, alternative suboptions will be made available.

3.1.3. Help

Help about each option can be obtained by preceding the option number by the symbol ? when users are presented with the prompt "? Menu or option number" (e.g., ?17 gives help on the option 17), but there are two further ways of obtaining help. Whenever the program asks a question, users can respond by typing the symbol ?, and they will receive information about the current option. In addition, option number 1 in all the programs will give help on all of a programs functions.

3.1.4. Quitting

To exit from any point in a program, users type ! for quit. If a menu is on the screen, this will stop the program; otherwise they will be returned to the last menu.

3.1.5. Making Selections

Questions and choices are dealt with in three ways. Where there are choices that are not obvious opposites or there are more than two choices, "radio buttons" and "check boxes" are used.

3.1.5.1. CHOOSING BETWEEN OPPOSITES

Obvious opposites, such as "clear screen" and "keep picture," are presented with only the default shown. For example, in this case, the default is generally "keep picture" so the program will display: "Keep picture (y/n) (y) =" and the picture will be retained if the user types Y or y or only return. If the user types N or n, the picture will be cleared. Anything other than these or ? or ! will cause the question to be asked again.

3.1.5.2. CHOOSING ONE FROM MANY

Radio buttons are used when only one of a number of choices can be made at any one time. The choices are presented arranged one above the other, each choice with a number for its selection, and the default choice marked with an X. For example, when the user is reading a new sequence file the following choices of format are offered.

Select sequence file format

    1 Staden
    2 EMBL
X 3 GenBank
    4 PIR
    5 GCG
? Selection (1–5) (3) =

Any single option can be selected by typing the option number, and the default option (here shown as 3) is also obtained by typing only "return". Again, help can be obtained by typing ? and typing ! will quit.

3.1.5.3. CHOOSING AT LEAST ONE FROM MANY

Check boxes are used when any number of a set of choices can be made (i.e., the choices are not exclusive). Choices are made by typing choice numbers. Each choice can be considered as a switch whose setting is reversed when it is selected. Choices that are currently switched on are marked with an X. The user quits from making selections by typing only "return". For example, in the routine that plots base composition, users can elect to plot the frequencies of any combination of bases, e.g.,. only A, A + T, A + T + G, and so forth. The following check box is offered to the user:

X 1 T
  2 C
X 3 A
  4 G
? Selection (1–4) ( ) =

As shown, this will plot the A + T composition. To switch off T, select 1, to switch on C select 2, and so forth, to quit, having set the bases required, type only "return".

3.1.6. Input of Numerical Values

All input of integer or decimal numbers is presented in a standard way with the allowed range shown in brackets and the default value also in brackets. For example: ? Window (5–31) (11) = In this example, users could type any number between 5 and 31, "return" only, ! or ? *(see above)*. Any other input will cause the program to ask the question again. Typing only "return" gives the default value (here 11).

3.1.7. Input of Character Strings

Character strings are requested using informative prompts of the form: **? Search string =**. Where possible, the prompt will be preceded by a default value as in: **Default search string = atatatata ? Search string =**.

Question mark (?) or ! will get help or quit. Where appropriate, for example, when a whole list of strings has been defined one after the other, typing return only will be a signal to the program that input is complete.

3.2. The X Interface

This interface deals with all the types of interactions described above, but options are selected using pull-down menus and all inputs are via appropriately styled dialog boxes and buttons. Default values are accepted by clicking on an "OK" button or typing return on the keyboard. Values are changed by overtyping the defaults. Quit is available from each dialog via a "CANCEL" button. Help is constantly available via a "HELP" button in the main dialog window. Details, such as requesting dialog when an option is selected, are dealt with using a button labeled "execute with dialog," which toggles to "execute."

3.3. Use of the Bell

The programs use the bell to indicate that a task is completed. When the bell sounds, the programs will wait until return is typed. Users can quit from these points by typing ! but no help is available.

3.4. Printing and Saving Results in Files

A few of the functions in the programs automatically write their textual results to disk files, but for most functions, users can choose whether results appear on the terminal screen or go to a file. This applies to both text and graphical results. For these functions, the normal, or default, place for results to appear is on the screen, and users need to decide before the function is selected if they want to redirect the results to a file. In all programs, the option "Redirect output" gives control over whether results appear on the screen or go to a file. When a program is started, results will be sent to the screen. If the option "Redirect output" is selected, users will be given the choice of redirecting either text or graphics to a file. The program will then ask users to supply a file name. From that point on, all results will be sent to the file until the option is selected again, in which case the "redirection file" will be closed and results will again appear on the screen. If these files contain textual results, they can be looked at from within the programs by using option "List a text file." Once the program is left, users can employ an appropriate system command to print the files. There is no function within the programs to direct files to a printer.

3.5. Use of Feature Tables

One particular use of redirection should be noted. The programs can use EMBL/GenBank feature tables as input for directing translation of DNA to protein, and so forth, but the tables must be stored in separate text files and cannot be read directly from the sequence libraries. The only routines that can read the sequence libraries are those available under "Read a sequence." So to create a text file containing the feature table for a particular library entry, users must redirect text output to disk, and then use the option "Read a sequence" to display the appropriate feature table. The feature table will be written to the file, and then the file can be used for controlling translation and so forth.

3.6. Use of Graphics

The analytical programs, including NIP, PIP, and SIP, present the results of many of their analyses graphically.

3.6.1. The Drawing Board and Plot Positions

The position at which the results for any function appear on the screen is defined relative to a notional users "drawing board" of dimension 10,000 by 10,000. This drawing board fills the screen and results are drawn in windows defined using symbols x0,y0 and xlength,ylength, where x0,y0 is the position of the bottom left-hand corner of the window; xlength is the width of the window, and ylength the height of the window. The window positions for each option are read from a file when a program is started. If required, individual users can have their own set of plot positions, and the positions can be redefined from within the programs using the option "Reposition plots."

3.6.2. The Plot Interval

For those analyses that draw continuous lines to represent results (for example, a plot of base composition) the user is asked to supply the "Plot interval." All the analyses produce a value for every point along the sequence, but often it is unnecessary to actually plot the values for all the points. The plot interval is simply the distance between the points shown on the screen. If the user selects a plot interval of 1, every point will be plotted; a plot interval of 3 will show every third point.

3.6.3. The Window Length

The word "window" is used in a further way by the programs. Most of the functions that analyze the content of a sequence (the simplest such routine plots the base composition) perform their calculations over a segment of the sequence of a certain length, display the result, then move on by 1 position, and recalculate. The fixed size of segment over which a calculation is performed is called a "window," and the segment size is the "window length." Many analytical functions request "? Window length =", or more frequently " ? Odd window length =". An odd number is used so that when a result is displayed for a particular window position it is derived from an equal number of points either side of the windows' midpoint.

3.6.4. Use of the Crosshair

All programs that produce graphical output provide a function for using a crosshair to examine the plots. After the crosshair function is selected, the cross will appear in the graphics window, and can be steered around using the mouse or directional keys. Special keyboard characters hit while the function is in operation produce the following results. For all programs, the letter s (for sequence) will show the local sequence around the crosshair position. For the sequence comparison programs that show a dot matrix, the two sequences will be displayed above one another. For the sequencing project management programs, all the aligned sequences in the contig will be displayed. For the sequence comparison programs, the letter m (for matrix) will show a matrix in which all identical characters for a window around the crosshair are marked. The punctuation symbol , will show the local position in sequence units, but leave the crosshair on the screen, whereas the space bar and any other nonspecial character will show the local position and exit the crosshair function. Further special characters are defined in the chapter on managing sequencing projects.

3.6.5. Drawing Scales on Plots

All the programs have a function "Draw a ruler," which will allow users to add scales to the axes of graphical plots. The scale can be positioned anywhere on the plot.

3.6.6. Saving Graphics

Many terminals are not capable of dumping their screen contents to a file for subsequent printing. One convenient way of obtaining hard copy of graphical results is to use a microcomputer as a terminal. On the Macintosh the terminal emulator versa termPro. This allows graphics to be saved as Macintosh files that can be annotated and printed using Macdraw and other painting programs. Alternatively, graphics can be redirected to a file and printed using a laser printer (*see* Section 3.4.).

3.7. The Active Region

All the analytical programs use an "active region" for most of their functions. This is simply the current section of the sequence over which the analysis will be applied. When a sequence is first read in, the active region will be set to its whole length, but the user can restrict the scope of analytical functions by use of an option called "Define active region." However, some functions, such as "List the sequence,"are always given access to the whole sequence and will allow the user to define a limited range after they have been selected.

3.8. Files of File Names

A useful device that is employed by many of the programs is that of "files of file names." If a program needs to perform the same operation in turn on each of 20 files, the user should not have to type in 20 file names. Instead the user types in the name of a single file that contains the names of the other 20 files. This single file is a file of file names. They are used, for example, to process batches of gel readings or to compare a sequence against a library of motifs.

4. Character Sets

There are two types of character sets employed by the programs: those for finished sequences and those used during sequencing projects.

4.1. Character Sets for Finished Sequences

The analytical programs will operate with upper-case or lower-case sequence characters. For nucleic acids, T and U are equivalent. For proteins, the standard one-letter codes are used. The analytical programs also use IUB symbols for redundancy in back-translations and for sequence searches. The symbols are shown in Table 1.

4.2. Symbols Used in Gel Readings

The information stored about a sequence reading has to show the original sequence, recording any doubts about its interpretation, and also, where possible, allow the changes made during editing to be indicated. Lower-case characters are used by the sequence project management programs for recording readings, and upper-case symbols are used when changes are made during editing. Alternatively, the reverse convention can be used. Any other characters in a sequence are treated as dash (–) characters. The symbols are shown in Table 2.

5. Sequence Formats

The data formats for the programs that deal with sequencing projects are described in the chapter on managing sequencing projects. All analytical programs can read sequences stored in several formats. The author distinguish between two sources of input namely: "sequence libraries" and "personal files."

5.1. Personal Sequence Files

The programs can read sequences from files in PIR, EMBL, GenBank, GCG, FASTA, and Staden formats. Staden format means text files with records of up to 80 characters; all spaces are removed; lines with ";" in the first position are treated as comments and will be displayed when the file is read but not included in the sequence; if the first line of data contains a 20-character header of the form ←abcdefghij→ it too will not be included in the processed sequence. This last facility allows the programs to read consensus sequences created by the sequence project management programs. Files in PIR or FASTA format can contain any number of entries (which the user selects by entry name), but all other formats are expected to contain only one sequence. If they contain more only the first will be read.

Table 1
The NC-IUB Characters Used by the Analytical Programs

A,C,G,T		
R	(A,G)	'puRine'
Y	(T,C)	'pYrimidine'
W	(A,T)	'Weak'
S	(C,G)	'Strong'
M	(A,C)	'aMino'
K	(G,T)	'Keto'
H	(A,T,C)	'not G'
B	(G,C,T)	'not A'
V	(G,A,C)	'not T'
D	(G,A,T)	'not C'
N	(G,A,C,T)	'aNy'

5.2. Sequence Libraries

Users may not appreciate the fact that because the sequence libraries are so large, programs need to use indexes to provide rapid retrieval of individual entries. An index is a list of entry names and pairs of offsets. For each entry name, the offsets define the position at which its sequence and annotation start in the large file. The index, which is in any case relatively small, is arranged so that it can be searched quickly—for example, the EMBL cdrom index is sorted alphabetically. When the user supplies an entry name, the program rapidly finds it in the index file and then uses the associated offsets to locate the entry in the larger sequence files.

The sequence libraries are stored in different ways on the VAX and the UNIX machines. On the VAX, we adopted the widely used PIR format and indexing method, and for UNIX, the author used the EMBL cdrom format and indexes.

5.2.1. Sequence Libraries for the VAX Version

On the VAX, all libraries are stored in PIR format, and except for the facility to select entries by accession number, the same functions are provided as those for UNIX. Note that this means that most libraries need reformatting after they have been read from the distribution media. Because, for each entry, the sequence and its annotation are stored separately, the reformatting process consumes significant computer resources. These reformatting programs are avail-

Table 2
The Symbols Used to Record Gel Readings

Symbol	Meaning		Symbol	Meaning	
c	Definitely	c			
t	"	t			
a	"	a			
g	"	g			
1	Probably	c			
2	"	t			
3	"	a			
4	"	g			
d	"	c	Possibly	cc	
v	"	t	"	tt	
b	"	a	"	aa	
h	"	g	"	gg	
k	"	c	"	c–	
l	"	t	"	t–	
m	"	a	"	a–	
n	"	g	"	g–	
r	a or g				
y	c or t				
5	a or c				
6	g or t				
7	a or t				
8	g or c				
–	a or g or c or t				
A	a set by auto edit or corrected by user				
C	c set by auto edit or corrected by user				
G	g set by auto edit or corrected by user				
T	t set by auto edit or corrected by user				
*	padding character placed by auto assembler				
else = –					

able from PIR, and no further information is given here. The programs that search whole libraries of sequences also expect the libraries to be in PIR format.

5.2.2. Sequence Libraries for the UNIX Version

On the UNIX machines, we use the EMBL cdrom as the primary source of sequence data and has chosen their indexing method for all libraries. These indexes leave the sequence libraries in their distribution format and simply provide offsets to the original files. The cdrom provides the EMBL nucleic acid sequence library and the SWISSPROT protein sequence library. Currently, it also includes indexes for entry names, author names, free text, and accession numbers, and has an additional "title" file that, for each entry, consists of entry name, entry length and an 80-character description of the entry. These indexes allow rapid retrieval of entries by name or accession number, and the author and free text indexes can searched for key words. The files can be left on the cdrom or transfered to a hard disk. The programs that search whole libraries of sequences expect the libraries to be in cdrom format or PIR format.

The author has written programs for producing EMBL cdrom-type indexes for other sequence libraries. These allow the use of the PIR protein libraries in CODATA format and between-release updates of the EMBL nucleotide library. Others may wish to use them to produce indexes for libraries, such as GenBank. In addition to the author's programs, the scripts that produce the indexes also use the UNIX sort program. No further details are given here.

5.2.2.1. LIBRARY DESCRIPTION FILES

The following information is only relevent to those installing the sequence libraries on a SUN. To make the sequence library handling as flexible as possible, several levels of files are used. As stated above, at present the author only deals with the EMBL and SWISSPROT libraries as distributed on cdrom, the PIR protein library in CODATA format and GenBank. By including a "library-type" flag in the library description file, the possibility of using alternative formats is left open.

The libraries are described at three levels:

1. A list of libraries and their types, which points to;
2. The files, which name the libraries' individual files and their file types; and
3. The libraries' individual files. The files used are described below.

Level 1: The top-level file is a list of available libraries containing: the library type, the name of the file containing the names of each library's individual files, and the prompt to appear on the user's screen.

For example:

File name: SEQUENCELIBRARIES
File contents:

    A EMBLLIBDESCRP EMBL nucleotide library! in cdrom format
    A SWISSLIBDESCRP SWISSPROT protein library! in cdrom format
    B PIRLIBDESCRP PIR protein library! in CODATA format

The first two libraries are of type A. The logical names are EMBLLIBDESCRP and SWISSLIBDESCRP, and the prompts are "EMBL nucleotide library" and "SWISSPROT protein library". The third library is of type B with logical name PIRLIBDESCRP. Space is used as a delimiter, and anything to the right of a ! is a comment.

Level 2: The file containing the names of the libraries' individual files contains flags to define the file types and the path or logical names of the files. Current file types are:

A Division-lookup,
B Entryname-index
C Accession-target
D Accession-hits
E Brief-directory
F Keyword-target
G Keyword-hits
H Author-target
I Author-hits

For example:

File name: EMBLLIBDESCRP
File contents:

    A STADTABL/EMBLdiv.lkp
    B/cdrom/indices/embl/entrynam.idx
    C/cdrom/indices/embl/acnum.trg
    D/cdrom/indices/embl/acnum.hit
    E/cdrom/indices/embl/brief.idx
    F/cdrom/indices/embl/keyword.trg
    G/cdrom/indices/embl/keyword.hit
    H/cdrom/indices/embl/author.trg
    I/cdrom/indices/embl/author.hit

Level 3: The individual library files. The contents of all files below Division-lookup are exactly as they appear on the cdrom. The Division-lookup file is rewritten so the directory structure and file names can be chosen locally. Its format is I6,1x,A.

For example:

File name: STADTABL/EMBLdiv.lkp
Contents:
```
1/cdrom/embl/fun.dat
2/cdrom/embl/inv.dat
3/cdrom/embl/mam.dat
4/cdrom/embl/org.dat
5/cdrom/embl/phg.dat
6/cdrom/embl/pln.dat
7/cdrom/embl/pri.dat
8/cdrom/embl/pro.dat
9/cdrom/embl/rod.dat
10/cdrom/embl/syn.dat
11/cdrom/embl/una.dat
12/cdrom/embl/vrl.dat
13/cdrom/embl/vrt.dat
```

The files which define all the programs and standard data files used by the package—staden.login and staden.profile, define the file SEQUENCELIBRARIES, that contains the list of available libraries. As should be clear from the description above, the three levels need to be created (actually modified from the contents of the distribution tape), and all names can be changed locally or set to be the same as those on the cdrom.

6. Conventions Used in Text

Obviously, the programs can perform many more operations than there is space to describe but, in the selection of uses shown, the author has tried to give some feel for the programs' scope. For this reason, and the need to conform as closely as possible to the format of the book, specific paths have been chosen through the programs, rather than attempting to describe all routes. For some sections, such as that on the facilities available for editing contigs, this has not been possible and we have instead, a description is given of how the major commands are used. It should also be noted that the user interactions described in the Methods sections are those that would be required if all the options were selected in the "Execute with dialog" mode. In practice, many of the options would normally be used without any dialog being required.

Section 3., this chapter, outlined the different modes of obtaining input from users. Throughout the specific chapters, the following conventions have been adopted to indicate which mode of input is being employed. When a program requests numerical or string input the term "Define" has been used, as in Define "Minimum search score." When a program requests that a choice is made between several options, as in the case of radio buttons or check boxes, the term "Select" has been used. When a program offers a choice between two options in the form of a yes or no answer, as in "Hide translation," the terms "Accept" or "Reject" have been used. When the digitizer program uses the stylus for input, the term "Hit" has been used.

Because it is difficult to produce figures including pull-down menus and dialog boxes, almost all examples containing user input are taken from the xterm interface. However, the actual wording of the prompts is the same for both interfaces.

The programs contain routines for drawing scales on plots and for simple annotation, but in general, such embellishment is not done automatically by the programs. This is because the programs are designed so that many plots can be superimposed, and it is better for the user to decide explicitly to add scales and annotation. More elaborate annotation can be added by saving the graphics output to files that can be handled by, say Macintosh, painting and drawing programs. None of the examples of graphical results shown in the following chapters have added scales: All are exactly as drawn by the programs.

7. Note

Although all the programs in the Macintosh version of the package work, the conversion to this machine was never finished. The package does not provide access to the sequence libraries, handling only simple text files containing sequences, or those generated by the assembly program SAP. The user interface, although using pull-down menus and dialog boxes for all interactions, is not as "Mac like" as many would expect. However, many people find this version very useful, and for others, the digitizer program alone makes the package worth having. Data input from a digitizer is a task suited to a machine like the Macintosh, and the data files can be transferred to a larger machine for assembly and other analysis. With the exception of sequence library access, all the options available in the 1990 VAX version are contained in the package *(1)*. No further details specific to the Macintosh version are given.

Note Added in Proof

The redirection option has been extended to provide Postscript output for graphical results.

Reference

1. Staden, R. (1990) An improved sequence handling package that runs on the Apple Macintosh. *Comput. Applic. Biosc.* **4,** 387–393.

CHAPTER 3

Staden: Sequence Input, Editing, and Sequence Library Use

Rodger Staden

1. Introduction

In this will describe sequence input and editing, as well as the use of sequence libraries.

1.1. Introduction to Sequence Input and Editing

The package contains facilities for input of sequence data from the keyboard, sonic digitizers, and ABI (Foster City, CA) 373A and Pharmacia (Piscataway, NJ) A.L.F fluorescent sequencing machines. Editing of single sequences can be performed using system editors such as EDT on the VAX and EMACS on a UNIX machine. Editing of sequence alignments is discussed in the chapter on managing sequencing projects.

1.2. Introduction to Keyboard Input

The program SAP contains an option to enter sequence at the keyboard. It also creates a file of file names and will list the sequences. Users may choose any four keys to represent the characters A, C, G, and T. For example, four adjacent keys in the same order as the lanes on a gel could be used. The program translates these symbols to A, C, G, and T, and any other characters are left unchanged. No line of input should be longer than 80 characters. Terminate input with the symbol @.

From: *Methods in Molecular Biology, Vol. 25: Computer Analysis of Sequence Data, Part II*
Edited by: A. M. Griffin and H. G. Griffin Copyright ©1994 Humana Press Inc., Totowa, NJ

1.3. Introduction to Input from Digitizer

Digitizers provide a convenient way of entering sequences from films into a computer. The digitizer, which is connected directly to the computer, operates on a light box, and is controlled by a program named GIP *(1)*. The film to be read is taped firmly to the surface of the light box, and the user defines the lane order and the centers of the four lanes to be read. These positions are defined at the point where reading will commence, and the program adjusts their values as the film is read. The user reads the sequence and transfers it to the computer by hitting the centers of the bands progressing up the film. Any number of sets of lanes and films can be read in a single run of the program. Each sequence is stored in a separate file, and a file of file names is also written. The program also uses a menu, which is a series of reserved areas of the light box surface, for entering commands and uncertainty codes. When the pen is pressed in these areas, the program responds accordingly. Each time the pen tip is depressed in the digitizing area, the program sounds the bell on the terminal to indicate to the user that a point has been recorded. As the sequence is read, the program displays it on the screen.

1.4. Introduction to Editing Single Sequences

The editing method used by the programs is designed to give users access to an editor with which they are familiar—i.e., the one on their machine, say EDT on a VAX or EMACS on a UNIX system—and yet to allow them to edit a sequence which contains all the landmarks they need in order to know where they are. Users an create a file containing a simple listing of the sequence (single stranded) with numbering, using "list the sequence," and then edit it with their system editor, using the numbering to know where they are within the sequence. When the edits are complete, they exit from the editor, and the program "analyzes" the edited file to extract only the sequence characters. Similarly, a file containing a three-phase translation, or a file containing a sequence plus its three-phase translation, plus its restriction sites marked above the sequence (*see* Fig. 1), can be edited. In order to be able to "analyze" such complicated listings and correctly extract the sequence the following simple rule is used: All lines in the file that contain a character that is not A, C, T, G, or U are

```
HapII
HpaII
MspI        `MseI
 .          .HincII
 .          .HindII
 .          .HpaI    DsaV
 .          ..       EcoRII
 .          ..       TspAI
 .          ..       . ApyI
 .          ..       . BstNI
 .          ..       . MvaI
 .          ..       . ScrFI              MaeIII
 .          ..     . .                  . BsrI  MseI
ccggttagactgttaacaacaaccaggtttttctactgatataactggttacatttaacgc
        10        20        30        40        50        60
    P  V  R  L  L  T  T  T  R  F  S  T  D  I  T  G  Y  I  *  R
    R  L  D  C  *  Q  Q  P  G  F  L  L  I  *  L  V  T  F  N  A
    G  *  T  V  N  N  N  Q  V  F  Y  *  Y  N  W  L  H  L  T  P
```

Fig. 1. The first page width of a sequence display that can be edited by the program.

deleted. It is obviously important to be aware of this rule and its implications. For protein sequences only a simple listing, i.e., the sequence plus numbering, can be used.

1.5. Introduction to Using the Sequence Libraries

The installation of the sequence libraries is described in the Chapter 2. Direct access to the libraries is provided by all programs that need such a facility: It is not performed by separate programs. The facilities currently offered in NIP, PIP, SIP, NIPL, PIPL, and SIPL include the following:

Get a sequence by knowing its entry name
Get a sequence's annotation by knowing its entry name
Get an entry name by knowing its accession number
Search the key word index on key words
Search the author index

The facilities currently offered in NIPL, PIPL, and SIPL include:

Search whole library
Search only a list of entry names
Search all but a list of entry names

2. Methods

2.1. Sequence Input from Keyboard

1. Select "Type in gel readings."
2. Accept "Use special keys for A, C, T, G."
3. Define the keys in turn.
4. Define "File file names." This is a file of file names so the readings can be processed as a batch.
5. Define the sequence by typing it in using the selected keys. Finish by typing an @ symbol.
6. Define "File name for this gel reading." This is the name for the sequence just entered.
7. Accept "Type in another reading." This cycles round to step 5. If rejected the next step follows.
8. Accept "List gel readings." The batch of readings entered will each be listed, one after the other, headed by their file names, on the screen.

2.2. Sequence Input from Digitizer

1. Tape the autoradiograph down securely on the light box.
2. Start the program (GIP).
3. Define "File of file names."
4. Using the digitizer pen, hit the digitizer menu ORIGIN, program menu ORIGIN, program menu START. After the bell has sounded the program will give the default lane order.
5. If correct, hit CONFIRM; otherwise hit RESET. To reset the lane order, hit the A, C, G, T boxes in the menu in left-to-right order.
6. Hit START, and then hit in left-to-right order, at a height level with the first band to be read, the start positions for the next four lanes. The program will report the mean lane separations and asks for confirmation that they are correct.
7. Hit START.
8. Hit the bands on the film in sequence order. If necessary, use the uncertainty codes in the program menu. Continue until the sequence is finished.
9. Hit STOP.
10. Define "Name for this reading."
11. Accept "Read another sequence." Otherwise the program will stop.

2.3. Sequence Input from the Pharmacia A.L.F.

After processing and base calling on the PC, the data for all 10 clones is contained in a single file, and the user names each using local conventions. Then this single file is transferred to the SUN using PC-NFS. This program allows SUN directories to be mounted as if they were DOS disks, and data can be transferred by use of the DOS copy command. On the SUN, to prepare for processing by program XDAP, the 10 clones are split into 10 separate files each with the names given on the PC. In addition, a file of file names is written. Then the reads for the individual clones need to be examined to clip off the vector sequence and the poor data at the 5' end. *See* Note 2.

2.4. Sequence Input from the ABI 373.

After processing and base calling on the Macintosh the data for each clone are contained in two files: One is simply the sequence, but the main file contains the raw data, trace data, and sequence. For processing, the author does not use the sequence file, since everything needed can be extracted from the main file. The user names each file using local conventions, and then the folder is transferred to the SUN using TOPS. This program allows SUN directories to be mounted as if they were on the Macintosh, and data can be transferred by simply dragging folders on the Macintosh screen. On the SUN, to prepare for processing by program XDAP, a file of file names is written, and the reads for the individual clones are examined to clip off the vector sequence and the poor data at the 5' end. *See* Note 2.

2.5. Editing a Nucleic Acid Sequence Using Restriction Sites and a Translation and Base Numbering as Landmarks

1. Select NIP.
2. Read in the sequence to be edited.
3. Direct output to disk, say creating file edit.seq.
4. Use the restriction enzyme site search routine (*see* Chapter 6) to create a file showing "Names above the sequence," as in Fig. 1.
5. Close the redirection file.
6. Select "Edit the sequence."
7. Define "Name of file to edit." This is the file containing the sequence listing, say edit.seq.
   The system editor will start up.
8. Edit the sequence.
9. Exit from the editor.
10. Accept "Make edited sequence active." The edited sequence will replace the original sequence.

2.6. Searching the Brief Index
of a Sequence Library

1. Select "Read new sequence."
2. Select "Sequence library." The alternative is "Personal file," and, if taken, would be followed by questions about which of the formats "Staden, EMBL, GenBank, PIR, or GCG" it was stored in.
3. Select, say, "EMBL nucleotide library."
4. Select "Search titles for key words."
5. Define "Key words." Type up to five key words separated by spaces.
6. The search will start, and for each match, the program will display the contents of the matching line, which includes the entry name, its length and an 80-character description. After every 20 matches, the program will ring the bell, and the user can escape by typing "!" (*see* Note 4).

2.7. Using Accession Numbers to Retrieve Data
from a Sequence Library

1. Select "Read new sequence."
2. Select "Sequence library."
3. Select, say, "EMBL nucleotide library."
4. Select "Get entry names from accession numbers."
5. Define "Accession number."
6. The program will display the entry names corresponding to the accession number. The last entry name found will become the default entry name.

2.8. Displaying the Annotations for an Entry
in a Sequence Library

1. Select "Read new sequence."
2. Select "Sequence library."
3. Select, say, "EMBL nucleotide library."
4. Select "Get annotations."
5. Define "Entry name." The program will display the annotation for the entry. After every 20 lines the program will ring the bell, and the user can escape by typing "!".

2.9. Reading a Sequence from a Sequence Library

1. Select "Read new sequence."
2. Select "Sequence library."
3. Select, say, "EMBL nucleotide library."
4. Select "Get a sequence."
5. Define "Entry name." The program will make the sequence the active sequence and display its base composition.

2.10. Worked Example
of Sequence Library Access

The worked example in Fig. 2 shows a search for the key word p53, followed by a search of the brief index for the word alpha, followed by search on accession number v00636, followed by "Get annotations" for entry lambda, and finally "Get a sequence" for entry lambda (*see* Note 4).

3. Notes

1. The program menu for GIP is simply a set of boxes drawn on the digitizing surface, each of which contains a command or uncertainty code. Right-handed users will find it is best to position the menu to the right of the digitizing area, but in practice, as long as its top edge is parallel to the digitizer box, it can be put anywhere in the active region. As well as the codes a,cg,t,1,2,3,4,b,d,h,v,r,y,x,-,5,6,7,8 the following commands are included in the menu: DELETE removes the last character from the sequence; RESET allows the lane centers to be redefined; START means begin the next stage of the procedure; STOP means stop the current stage in the procedure; CONFIRM means confirm that the last command or set of coordinates are correct.

   The digitizing device also has a menu of its own. This lies in a two inch wide strip immediately in front of the digitizing box. Pen positions within this two inch strip are interpreted as commands to the digitizer and are not sent to the GIP program. In general, the only time users will need to use the device menu is when they tell GIP where the program menu lies in the digitizing area. This is done by first hitting ORIGIN in the device menu and then hitting the bottom left-hand corner of the program menu. If the bell does not sound after hitting START, try hitting METRIC in the device menu (the program uses metric units, and some digitizers are set to default to use inches, hitting metric switches between the two).

   The user should try to hit the bands as near as possible to the center of the lanes because the program tracks the lanes up the film using the pen positions. If the lane centers get too close, the program stops responding to the pen positions of bands and, hence, does not ring the bell. If this occurs, users must hit the reset box in the menu, and the program will request them to redefine the lane centers at the current reading position. Then they can continue reading. As a further safeguard, the program will only respond to pen positions either in the menu or very close to the current reading position.

```
Select sequence source
X  1 Personal file
   2 Sequence library
?  Selection  (1-2) (1) =2
Select a library
X  1 EMBL nucleotide library
   2 SWISSPROT protein library
   3 PIR protein library
?  Selection  (1-3) (1) =
Library is in EMBL format with indexes
 Select a task
X  1 Get a sequence
   2 Get annotations
   3 Get entry names from accession numbers
   4 Search titles for keywords
   5 Search keyword index for keywords
?  Selection  (1-5) (1) =5
Search for keywords
?  Keyword=p53
P53 hits  11

HSORIP53    1
HSP53G      2
XLP53R      3
HSP53G      4
PP1LREP     5
CAP53       6
HSP53A      7
HSP53MUT    8
MMP53A      9
MMP53B      10
MMP53C      11
 Select a task
X  1 Get a sequence
   2 Get annotations
   3 Get entry names from accession numbers
   4 Search titles for keywords
   5 Search keyword index for keywords
?  Selection  (1-5) (1) =4
Search for keywords
?  Keywords=alpha
Searching for alpha
AAGHA          623 a.anguilla mrna for glycoprotein hormone alpha subunit precu
AAMALI        3338 a.aegypti mali gene encoding alpha 1-4 glucosidase, complete
AAMALIA       1659 a.aegypti maltase-like i (mali) gene encoding alpha-1,4-gluc
AAMALIB       1832 a.aegypti maltase-like i (mali) mrna encoding alpha-1,4-gluc
ACA13GT        371 alouatta caraya alpha-1,3gt gene, 3' flank.
ADHBADA1       102 duck alpha-d-globin gene, exon 1.
ADHBADA2      1145 duck alpha-a-globin gene and 5' flank
ADHBADWP       513 duck (white pekin) alpha ii (minor) globin mrna, complete co
AEACOXABC     5279 a.eutrophus protein x (acox), acetoin:dcpip oxidoreductase-a
AGA13GT        371 ateles geoffroyi alpha-1,3gt gene, 3' flank.
AGAAAGFP       282 c.tetragonoloba alpha-amylase/alpha-galactosidase fusion pro
AGAABL         138 b.subtilis alpha-amylase signal peptide gene e.coli beta-lac
AGAFAMYA        57 synthetic b.stearothermophilus alpha amylase/s.cerevisiae ma
AGAFAMYB        57 synthetic b.stearothermophilus alpha amylase/s.cerevisiae ma
AGAFAMYC        57 synthetic b.stearothermophilus alpha amylase/s.cerevisiae ma
AGAFCOXA        98 synthetic alpha-factor/cox iv fusion gene signal peptide.
AGAGABA       7876 synthetic gossypium hirsutum (cotton) alpha globulin a and b
```

Fig. 2. A worked example of sequence library use (*see* Note 4).

```
AGAMYLS    120 synthetic alpha-amylase gene, 5' end.
AGANPS      95 synthetic gene (jcnf-1) encoding alpha-factor pro-region/han
!
 Select a task
 X  1 Get a sequence
    2 Get annotations
    3 Get entry names from accession numbers
    4 Search titles for keywords
    5 Search keyword index for keywords
 ? Selection  (1-5) (1) =3
 ? Accession number=v00636
Entry name LAMBDA
 Select a task
 X  1 Get a sequence
    2 Get annotations
    3 Get entry names from accession numbers
    4 Search titles for keywords
    5 Search keyword index for keywords
 ? Selection   (1-5) (1) =2
 Default Entry name=LAMBDA
 ? Entry name=
ID    LAMBDA      standard; DNA; PHG; 48502 BP.
XX
AC    V00636; J02459; M17233; X00906;
XX
DT    03-JUL-1991 (Rel. 28, Last updated, Version 3)
DT    09-JUN-1982 (Rel. 1, Created)
XX
DE    Genome of the bacteriophage lambda (Styloviridae).
XX
KW    circular; coat protein; DNA binding protein; genome;
KW    origin of replication.
XX
OS    Bacteriophage lambda
OC    Viridae; ds-DNA nonenveloped viruses; Siphoviridae.
XX
RN    [1]
RP    1-48502
RA    Sanger F., Coulson A.R., Hong G.F., Hill D.F., Petersen G.B.;
RT    "Nucleotide sequence of bacteriophage lambda DNA";
RL    J. Mol. Biol. 162:729-773(1982).
XX
!
 Select a task
 X  1 Get a sequence
    2 Get annotations
    3 Get entry names from accession numbers
    4 Search titles for keywords
    5 Search keyword index for keywords
 ? Selection   (1-5) (1) =
 Default Entry name=LAMBDA
 ? Entry name=
DE   Genome of the bacteriophage lambda (Styloviridae).
 Sequence length= 48502
 Sequence composition
         T          C          A          G          -
      11988.     11360.     12336.     12818.        0.
        24.7%      23.4%      25.4%      26.4%      0.0%
```

Fig. 2 *(continued)*.

2. Few details are given about preparing the data from fluorescent sequencing machines for processing by XDAP since the methods used have been changed significantly since this chapter was written.
3. All of the operations described for the EMBL nucleotide library can be performed in exactly the same way for the SWISSPROT and PIR protein libraries. Using the author's own indexing programs, the GenBank library could also be prepared for access. A search of the key word index is also available for the libraries. On the VAX, EMBL, GenBank, SWISSPROT, and PIR can all be processed.

Note Added in Proof

The procedure to search through the brief index of a sequence library has been replaced by an author name search.

Reference

1. Staden, R. (1984) A computer program to enter DNA gel reading data into a computer. *Nucleic Acids Res.* **12,** 499–503.

CHAPTER 4

Staden: Managing Sequence Projects

Rodger Staden

1. Introduction

Data input, assembly, checking, and editing are the major tasks of
sequence project management. Data input is described in a previous
chapter and everything else is covered here. The programs can deal
with data derived from autoradiographs and from sequencing
machines, such as the Applied Biosystems (Foster City, CA) 373A
and the Pharmacia (Piscataway, NJ) A.L.F.

Two alternative programs for managing sequencing projects are
described. They contain the same assembly and vector screening routines, but they differ in their editing methods. One program SAP (*see*
refs. *[1,2]*) can be operated from simple terminals and emulators, but
the other XDAP *(3)*, requires an X terminal or emulator. XDAP contains a superior editor plus the facility to annotate sequences and display the colored traces for data derived from fluorescent sequencing
machines. Those using autoradiographs will find that SAP is adequate,
but XDAP is essential for users of fluorescent sequencing machines.
Readers should note that several of the methods for displaying contigs
described below are probably of value only to those unable to use the
screen-based contig editor in XDAP.

Fluorescent sequencing machines provide machine-readable data.
This means, given appropriate software, that while making editing
decisions, the user can see displayed on the screen the colored traces
used to derive the sequence. However, data from these machines
require some extra processing. First the machines tend to produce
long sequences with poor quality at their 3' ends, so we have to decide

From: *Methods in Molecular Biology, Vol. 25: Computer Analysis of Sequence Data, Part II*
Edited by: A. M. Griffin and H. G. Griffin Copyright ©1994 Humana Press Inc., Totowa, NJ

how much of the data to use. Second, the sequencing machine does not recognize the primer region (as the user would) so some way of removing it from the data are necessary. The poor quality data from both ends of the sequences and the vector sequences are identified noninteractively by programs CLIP-SEQS and VEP. Alternatively, these tasks can be performed interactively using program TED *(4)*. We term data from the 3' end of a reading, which is not employed in the assembly process "unused" sequence. Note that these data are not lost, but simply ignored until such time as they can be useful for locating joins between contigs or for double-stranding regions of the sequence.

The method described here uses a database to store all the data for each sequencing project. The individual sequence readings derived from autoradiographs or from sequencing machines are initially stored in separate files, but the program copies them into the database during the assembly process. For normal operation, the program handles batches of readings—for example, 24 from a film or machine run. Batch processing is achieved by use of files of file names.

Depending on the strategy employed and the stage of the project, the following operations may be performed.

1. Start a project database.
2. Select primers and templates.
3. Obtain readings.
4. Put individual readings into the computer, and write a file of file names. For data derived from fluorescent sequencing machines, choose which data from the 3' end of the reading should not be used for the assembly process.
5. Screen the batch against any vectors that may be present, excising any vector sequence found and passing to the next step the names of those readings that contain some nonvector sequence.
6. Screen the batch against any restriction sites whose presence would indicate a problem, passing those that do not match on to the next step.
7. Compare each reading in the batch with the current contents of the project database, adding it to the contigs it overlaps, joining contigs or starting new contigs.
8. Check the number of contigs and the quality of the consensus sequence, and plan further experiments. Try to join contigs by searching for overlaps between their ends. (This is particularly useful for those using data

from fluorescent sequencing machines, where although the 3' end of the sequence is not good enough for automatic assembly, it can be valuable for finding overlaps between contigs.)

9. Edit the contigs to resolve disagreements.
10. Produce a consensus sequence.
11. Analyze the consensus sequence, possibly discovering further errors.

Subsets of these operations will be cycled through repeatedly. A pure shotgun strategy would continue using steps 3–7; a pure primer walking strategy would also include step 2. A number of the steps require almost no user intervention; however, checking quality and final editing decisions are still interactive procedures. The program contains several options, such as displays of the overlapping readings in a contig, to help indicate not only the poorly determined regions, but also which clones could be resequenced to resolve ambiguities, or those that can usefully be extended or sequenced in the reverse direction, to cover difficult regions. It is best to use a command procedure or script for handling steps 5–7.

For our projects, we have a script which users employ by typing "assemble filename", where file name is the file of file names for the current batch of readings. This script calls all the necessary options in SAP or DAP (*see* Section 3.) in order to make a backup of the database, screen against any vectors, assemble readings, and print a report. In the text below describes how these operations are performed interactively.

2. Methods

2.1. Starting a Project Database

The assembled data for each project is stored in a database. At the beginning of a project, it is necessary to create an empty database using program SAP or XDAP.

1. Select "Open database."
2. Select "Start new database."
3. Define the database name. Database names can have from 1 to 12 letters, and must not include full stop (.).
4. Accept "Database is for DNA."
5. Define "Database size." This is an initial size and, if necessary, can be increased later using "Copy database." Roughly speaking, it is the number of readings expected to be needed to complete the project.

6. Define "Maximum reading length." This is the length of the longest reading that will be added to the database. The minimum is 512 bases, and the maximum 4096.

The program should confirm that "copy 0" of the database has been started.

2.2. Screening Against Restriction Enzyme Recognition Sequences

For some strategies, it is necessary to compare readings against any restriction enzyme recognition sequences that may have been used during cloning and which should not be present in the data. The function operates on single readings or processes batches accessed through files of file names. The algorithm looks for exact matches to recognition sequences. The recognition sequences should be stored in a simple text file with one recognition sequence per record.

1. Accept "Use file of filenames."
2. Define "File of gel reading names." The input file of file names.
3. Define "File for names of sequences that pass." This is a file of file names for those readings that do not contain the recognition sequences. After the run, it will contain the names of all the files in the batch that do not match any of the restriction enzyme recognition sequences. Hence, it can be used for further processing of the batch.
4. Define "File name of recognition sequences," which is the name of the file of recognition sequences.

2.3. Screening Against Vector Sequences

For most strategies, it is necessary to compare readings against any vector sequences that may have been picked up during cloning. The package contains two routines for screening against vectors. The original function simply reports any matches between the readings and the vector sequences, and only passes on those that do not match. This function should now only be used to screen for any other sequences that should be excluded from the database, because the newer one (program name VEP for vector excising program) is capable of both finding the vector sequences and editing them out automatically.

2.3.1. Clipping Off Vector Sequences

There are two types of vector that may need to be screened out of gel readings: the sequencing vector and for cases where, for example, whole cosmids have been shotgunned, the cloning vector. The two tasks are different. When screening out the sequencing vector, we may expect to find data to exclude, both from the primer region and from the other side of the cloning site (when, for example, the insert is short). When screening out cosmid vector, the researcher may find that either the 5' end, or the 3' end, or the whole of the sequence is vector. Also, for the cosmid search, it is necessary to compare both strands of the sequence. The program (VEP) works slightly differently for each of the two cases. Having read the vector sequence from a file, the program asks for the "Position of the cloning site." A value of zero signifies that the search will be for the cosmid vector. A nonzero value signifies that the search is for the sequencing vector, and so in this case, the program then asks for the "Relative position of the primer site." A negative relative position signifies that a reverse primer is being used; otherwise, a forward primer is assumed.

The program screens a batch of readings using a file of file names and creates a new file of file names that contains the names of all those sequences that include some nonvector sequence. For each sequence that contains some vector, it writes out a new copy of the file in which the vector portion is identified.

The search, which uses a hashing algorithm, is very rapid. Users specify a "Word length," the "Number of diagonals to combine," and a "Minimum score." The word length is the minimum number of consecutive bases that will count as a match. The algorithm treats the problem like a dot matrix comparison and finds the diagonal with the highest score. Then, it adds the scores for the adjacent "Minimum number of diagonals to combined." If the combined score is at least "Minimum score," the sequence is marked to indicate that it contains vector. The score represents the proportion of a diagonal that contains matching words, so the maximum score for any diagonal is 1.0.

1. Define "Input file of file names." This is the file containing the names of all the readings to be screened.
2. Define "File name of vector sequence."
3. Define "Position of cloning site." This is the base number, relative to the beginning of the vector sequence, that is on the 3' side of the insert site. For example, for ml3mpl8 the *Sma*I site is at 6250. A zero value signifies that the search is for cosmid vector.
4. Define "Relative position of 3' end of primer site". This is the position, relative to the cloning site, of the first base that could be included in the

sequence. For ml3mpl8, the 17-mer Sequencing Primer, and the *Sma*I site, the position is 41.

5. Define "Word length." Only words of this length will be counted as matches.

6. Define "Number of diagonals to combine." The scores for this number of diagonals around the highest scoring diagonal will be combined to give the total score.

7. Define "Cutoff score." For a match, at least this proportion of the total length of the summed diagonals must contain identical words.

8. Define "Output file of passed file names." This is the name of the file to contain the names of the readings to pass on to the assembly program.

Processing will commence and finishes with a summary stating the number of files processed, the number completely vectored, the number partly vector, and the number free of vector.

2.3.2. Screening for "Vectors"

This function is contained in both SAP and XDAP and operates on single readings or processes batches accessed through files of file names. The algorithm looks for exact matches of length "minimum match length" and displays the overlapping sequences.

1. Accept "Use file of filenames."

2. Define "File of gel reading names." This is the input file of file names.

3. Define "File for names of sequences that pass." This is a file of file names for those readings that do not contain the vector sequence. After the run, it will contain the names of all the files in the batch that do not match the vector sequence. Hence, it can be used for further processing of the batch.

4. Define "File name of vector sequence." This is the name of the file containing the vector sequence.

2.4. Entering Readings
into the Project Database (Assembly)

Readings are entered into the database using the auto assemble function. This function compares each reading and its complement with a consensus of all the readings already stored in the database. If it finds any overlaps, it aligns the overlapping sequences by inserting padding characters and then adds the new reading to the database. Readings that overlap are added to existing contigs, and readings that do not overlap any data in the database start new contigs. If a new reading overlaps two contigs, they are joined. Any readings that appear to overlap, but that cannot be aligned sufficiently well are not entered and have their names written to a file of failed gel reading names. Note that it is possible that a reading may align well with two contigs (indicating a possible join) but that after it has been added to one of the contigs, the two contigs do not align sufficiently well. In this case, although the reading has been entered into the database, its name will also be added to the file of failed readings. Alignments using more than the maximum number of paddings characters, or exceeding the maximum mismatch may be displayed, but the readings will not be entered into the database. A typical run of the assembly routine is shown in Fig. 1.

1. Accept "Permit entry."
2. Accept "Use file of file names."
3. Define "File of gel reading names." This is the name of the input file of file names, probably passed on from "Screen against vector."
4. Define "File for names of failures." This is a file to contain the names of the readings that the program fails to enter or for which joins are not made.
5. Select "Perform normal shotgun assembly."
6. Accept "Permit joins."
7. Define "Minimum initial match." Only possible overlaps containing exact matches of at least this number of consecutive identical characters will be considered for alignment.
8. Define "Maximum number of pads per reading." This is the maximum number of padding characters permitted in any new reading during the alignment procedure.
9. Define "Maximum number of pads per reading in contig." This is the maximum number of padding characters permitted in the contig in order to align any new reading.
10. Define "Maximum percent mismatch after alignment."

2.5. Searching for Internal Joins

The purpose of this function is to use data already in the database to find possible joins between contigs. Although most joins will be made automatically during assembly, due to poor alignments, some may not have been done. The function is particularly useful for sequences from fluorescent sequencing machines because it may be possible to find potential joins within the unused data from the 3' ends of readings (*see* Note 14).

```
Automatic sequence assembler
Database is logically consistent
? (y/n) (y) Permit entry
? (y/n) (y) Use file of file names
? File of gel reading names=demo.nam
? File for names of failures=demo.fail
Select entry mode
X 1 Perform normal shotgun assembly
  2 Put all sequences in one contig
  3 Put all sequences in new contigs
? Selection  (1-3) (1) =
? (y/n) (y) Permit joins
? Minimum initial match (12-4097) (15) =
? Maximum pads per gel (0-25) (8) =
? Maximum pads per gel in contig (0-25) (8) =
? Maximum percent mismatch after alignment (0.00-15.00) (8.00) =


Results skipped to save space

>>>>>>>>>>>>>>>>>>>>>>>>>>>>>>>>>>>>>>>>>>>>>>>>>>>>>>>>>>>>>>>>>>>>>>>>>>>>>>>
Processing  4 in batch
 Gel reading name=hinw.009
 Gel reading length=   292
 Working
 Contig    1 position   263 matches strand 1 at position   14
 Contig    2 position     1 matches strand 1 at position  156
Total matches found  2
Trying to align with contig  1
 Padding in contig=     1 and in gel=    0
 Percentage mismatch after alignment =  2.9
 Best alignment found
         251         261         271         281
         aattacagcg tt,cctattg acgggcgcat ccac
         ********* ** ** **** ********* ****
         aattacagcg ttcccvattg acgggcgcat ccac
         1          11         21         31
Trying to align with contig  2
 Padding in contig=     0 and in gel=    2
 Percentage mismatch after alignment =  1.4
 Best alignment found
         1          11         21         31         41         51
         tgcacgacat cgagtatgag agttatatcc cgggcgcgct ctgcttgtac atggacctca
         ********* ********* ********* ********* ********* *********
         tgcacgacat cgagtatgag agttatatcc cgggcgcgct ctgcttgtac atggacctca
         156        166        176        186        196        206
         61         71         81         91         101        111
         tgtacctctt tgtctccgtg ctctacttca tgccctccga gcccggcagc gcccacactg
         ********* ********* ********* ********* ***** ** * *********
         tgtacctctt tgtctccgtg ctctacttca tgccctccga gcccg,ca,c gcccacactg
         216        226        236        246        256        266
         121        131
         ctcagacgac ggtcgctgc
         ********* *********
         ctcagacgac ggtcgctgc
         276        286
Overlap between contigs      2 and.   1
Length of overlap between the contigs=  -122
Entering the new gel reading into contig    1
This gel reading has been given the number    4
Working
Trying to align the two contigs
 Padding in contig=     2 and in gel=    0
```

Fig. 1. Part of a typical run of "Auto assemble."

```
Percentage mismatch after alignment =  1.5
Best alignment found
      406         416         426         436         446         456
     tgcacgacat cgagtatgag agttatatcc cgggcgcgct ctgcttgtac atggacctca
     ********** ********** ********** ********** ********** **********
     tgcacgacat cgagtatgag agttatatcc cgggcgcgct ctgcttgtac atggacctca
      1          11          21          31          41          51
      466         476         486         496         506         516
     tgtacctctt tgtctccgtg ctctacttca tgccctccga gcccg,ca,c gcccacactg
     ********** ********** ********** ********** ***** ** * **********
     tgtacctctt tgtctccgtg ctctacttca tgccctccga gcccggcagc gcccacactg
      61          71          81          91          101         111
      526         536
     ctcagacgac ggtcgct
     ********** *******
     ctcagacgac ggtcgct
      121         131
Editing contig    1
Completing the join between contigs      1 and       2

     (Results for other readings skipped to save space)

          Batch finished
          9 sequences processed
          9 sequences entered into database
          2 joins made
```

Fig. 1 *(continued)*.

The program strategy is as follows. Take the first contig and calculate its consensus. If unused data is being employed, examine all readings that are in the complementary orientation, and sufficiently near to the contigs left end to see if they have sufficiently good unused sequence that, if present, would protrude from the left end of the contig. If found add the longest such sequence to the left end of the consensus. Do the same for the right end by examining readings that are in their original orientation. Repeat the consensus calculations and extensions for all contigs hence producing an extended consensus for the whole database. If unused data is not being employed, simply calculate the consensus for the whole database. Now look for possible joins by processing the extended consensus in the following way. Take the last, for example, 500 bases (termed the "probe length" by the program) of the rightmost consensus, compare it in both orientations with the extended consensus of all the other contigs. Display any sufficiently good alignments. Repeat with the left end of the rightmost contig. Do the same for the ends of all the contigs, always comparing only with the contigs to their left, so that the same matches do not appear twice.

Good unused data is defined by sliding a window of "Window size for good data scan" bases outwards along the sequence and stopping when greater than "Maximum number of dashes in scan window" appear in the window. Note that it is advisable to have some sort of cutoff, because if the user simply takes all the data it might be so full of rubbish that the user will not find any good matches. An initial run employing no unused data is also recommended. Sufficiently good alignments are defined by criteria equivalent to those used in auto assemble. However, here only display alignments that pass all tests are displayed.

All numbering is relative to base number one in the contig: Matches to the left (i.e., in the unused data) have negative positions; matches off the right end of the contig (i.e., in the unused data) have positions greater than the contig length. The convention for reporting the orientations of overlaps is as follows: If neither contig needs to be complemented, the positions are as shown. If the program says "contig x in the – sense," then the positions shown assume contig x has been complemented. For example, in the results given in Fig. 2 the positions for the first overlap are as reported, but those for the second assume that the contig in the minus sense (i.e., 443) has been complemented.

1. Select "Find internal joins."
2. Define "Minimum initial match." Only matches containing this number of consecutive identical characters will be found.
3. Define "Maximum pads per sequence." Only alignments containing less than or equal this number of padding characters in each sequence will be found.
4. Define "Maximum percent mismatch after alignment." Only alignments with at least this level is similarity will be found. Particularly when poor data from the 3' ends of sequences derived from fluorescent sequencing machines is used, it is important to allow for a high degree of mismatch—around 75%.
5. Define "Probe length." This is the size of sequence from each end of each contig, that is compared with the total length of all other contigs.
6. Accept "Employ unused data." This means, where available, add the unused data from the 3' ends of sequences, to the ends of the contigs.
7. Define "Window size for good data scan." To decide how much of the unused data should be added to the end of a contig the program scans outward, counting the numbers of dashes (–) over a window of the size defined here.

```
Possible join between contig    445 in the + sense and contig    405
Percentage mismatch after alignment =  4.9
          412         422         432         442         452         462
    405   TTTCCCGACT  GGAAAGCGGG  CAGTGAGCGC  AACGCAATTA  ATGTGAG,TT  AGCTCACTCA
          ********** * ********** ***** *** ********** ********** **********
    445   -TTCCCGACT  G,AAAGCGGG  TAGTGA,CGC  AACGCAATTA  ATGTGAG-TT  AGCTCACTCA
         -127        -117        -107         -97         -87         -77
          472         482         492         502         512
    405   TTAGGCACCC  CAGGCTTTAC  ACTTTATGCT  TCCGGCTCGT  AT
          ********** ********** ********** ********** **
    445   TTAGGCACCC  CAGGCTTTAC  ACTTTATGCT  TCCGGCTCGT  AT
          -67         -57         -47         -37         -27
Possible join between contig    443 in the - sense and contig    423
Percentage mismatch after alignment = 10.4
           64          74          84          94         104         114
    423   ATCGAAGAAA  GAAAAGGAGG  AGAAGATGAT  TTTAAAAATG  AAACG-CGAT  GTCAGATGGG
          **** ***** ********** ********** ****** ** ***** **** **********
    443   ATCG,AGAAA  GAAAAGGAGG  AGAAGATGAT  TTTAAA,,TG  AAACGACGAT  GTCAGATGG,
         3610        3620        3630        3640        3650        3660
          124         134         144         154         164
    423   TTG-ATGAAG  TAGAAGTAGG  AG-AGGTGGA  AGAGAAGAGA  GTGGGA
          *** ****** ********** ** ******* *** ***** ** **
    443   TTGGATGAAG  TAGAAGTAGG  AGGAGGTGGA  ,GAG,AGAGA  GTTGG-
         3670        3680        3690        3700        3710
```

Fig. 2. Typical output from "Find internal joins."

8. Define "Number of dashes in scan window." If the program finds this many dashes in the scan window, it will add no more of the unused data to the end of the contig.

2.6. Editing in XDAP

The XDAP editor is mouse-driven, and can insert, delete, and change readings in contigs. It has facilities to display the traces for data from fluorescent sequencing machines and for annotation of readings. In addition. it allows the poor quality data from the ends of readings to be viewed and, if required, added to the sequences (*see* Note 15).

A typical view of the editor is shown in Fig. 3. This includes the edit window showing an 80-character section of a contig (position 3899–3978). Each reading is numbered and named in the left-hand panel, minus signs indicating those in their reverse orientation. Underneath is their consensus. Some of the sequence letters are lighter than the majority showing that they are "unused." One segment (3933 to 3949) is shaded, which signifies that it has been annotated. The editing cursor is at position 3921. Above this window are the main but-

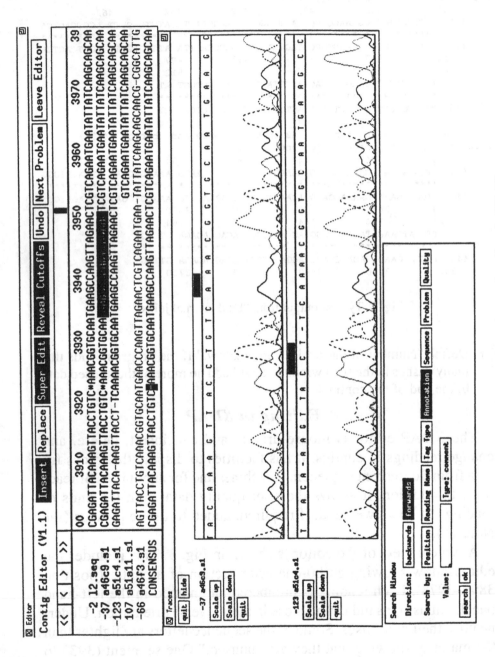

Fig. 3. A typical display from the contig editor in XDAP.

tons the user employs to direct the editing process. Below the edit window is a panel showing the traces for readings 37 and 123. Notice they are centered on the cursor position. Here the traces are shown in four different line styles, but on a color screen, they each have different colors. In the bottom of the figure is the search window. These features are described in the relevant sections below.

2.6.1. Scrolling Through the Contig

The editor allows scrolling from one end of a contig to the other using the scroll bar and scroll buttons, and also the arrow keys. Action of mouse button presses when the mouse pointer is in the scroll bar:

Middle mouse button	Set editor position
Left mouse button	Scroll forward one screenful
Right mouse button	Scroll backward one screenful

The four scroll buttons operate as follows:

"<<"	Scroll left half a screenful
"<"	Scroll left one character
">"	Scroll right one character
">>"	Scroll right half a screenful

The Editor cursor can be positioned anywhere in the edit window by moving the mouse pointer over the character of interest, then pressing the left mouse button. The Editor cursor can also be moved by using the direction arrow keys.

2.6.2. Editing Operations

The editor operates in two main edit modes—Replace and Insert. Replace allows a character to be replaced by another. Insert allows characters to be inserted into a reading. Characters are entered by typing them from the keyboard. Only valid characters are permitted. Characters can be deleted by positioning the cursor one character to their right then pressing the delete key. Normally Insert and Delete apply to the consensus line of the contig only. This restraint can be overridden by using the "Super Edit" mode of operation, although it should be employed with caution, as misuse may corrupt alignments.

Edits can also be performed on the consensus, though they are restricted to insertion and deletion of padding characters ("*"). These edits also have special meanings. A deletion will delete all characters at the position to the left of the cursor in the contig, and move the relative positions of all sequences starting to the right of the cursor position left one character. An insertion will insert the character typed ("*") into all gel reading sequences at the cursor's position in the contig, and move the relative positions of all sequences starting to the right of the cursor position right one character.

2.6.3. Use of Buttons

The effect of the last edit can be undone by pressing the "Undo" button at the top of the editor window. Pressing it n times will undo the last n edits.

The cursor will automatically be positioned at the next problem when the "Find Next Problem" button is selected. The next problem is where the consensus shows either a disagreement ("–") or a pad ("*") character.

The edits to the contig can be saved by pressing the "Leave Editor" button and replying "Yes" to the prompt to "Save changes?" Since no changes are made to the working copy of the database until this point, it is possible to abort the editor if the edit session ends up in an unsatisfactory state.

2.6.4. Displaying Traces for Readings from Fluorescent Sequencing Machines

The original trace data from which the gel reading sequences were derived can be seen by double clicking (two quick clicks) with the middle mouse button on the area of interest. The trace will be displayed with the point clicked at the center of the trace viewport. All traces that are displayed are maintained in one window, which will display a maximum of four traces. When four traces are already being displayed and a new one is requested, the one at the top of the window is removed and the new one is added to the bottom. Traces can be removed individually by using the "quit" button in the panel next to the trace.

2.6.5. Extending Reads with the Unused Data

Sequence data from fluorescent sequencing machines are normally clipped to remove the primer region, and the poor-quality data from the 3' end are marked to be ignored during assembly. Only the sequence used during assembly is made visible in the XDAP editor. However, the unused data is copied into the database and can be viewed from within the editor. Also, the position of this "cutoff" can be altered. To display the unused sequences, press the "Display Cutoff" button at the top of the editor window. The cutoff sequence appears in gray. This sequence can be incorporated into the editable sequence by moving the cutoff position. This is done by positioning the cursor at the end of the sequence, and using Meta-Left-Arrow and Meta-Right-Arrow to adjust the point of cutoff. The Meta key is a diamond on the Sun keyboard (*see* Note 16).

2.6.6. Using the Pop-Up Menu

A pop-up menu is revealed by depressing the "Control" key on the keyboard and at the same time pressing the left mouse button. The menu has the following functions:

Find Next Problem
Save Contig
Create Tag
Edit Tag
Delete Tag
Search

"Find Next Problem" and "Save Contig" are described above. Operations on tags are described in Sections 3.6.7.–3.6.10. and then search is outlined.

2.6.7. Annotating Readings

Parts of a sequence can be annotated to record the positions of primers used for walking, or to mark sites, such as compressions, that have caused problems during sequencing. The annotations are termed "tags." Each tag has a type, such as "primer," a position, a length, and a comment. Each type has an associated color that will be shown on the display. First, the segment to tag is selected, then it is annotated. The consensus sequence cannot be annotated.

2.6.8. Creating a New Annotation

Use the left mouse button to position the start of the selection. While this button is being held down, move the mouse to the other end of the segment. The selection can be extended further using the right mouse button. To create the annotation, invoke the pop-up menu, and select the "Create Tag" function. A small "tag editor" will appear that allows users to select the type of the annotation from a pull-down menu, and specify a comment if desired. To select a new type, pull down the Type menu, and select the entry desired. To enter a comment, simply type into the text window in the tag editor. The annotation is created when the "Leave" button on the tag editor is pressed and is displayed in the color defined in the tag database file (TAGDB).

2.6.9. Editing an Existing Annotation

Position the cursor with the left mouse button on the tag, and select the "Edit Tag" off the pop-up menu. This invokes the tag editor, and changes to the type and comment of the annotation can be made. The tag is updated when the "Leave" button is pressed.

2.6.10. Deleting an Annotation

To delete an existing annotation, position the cursor with the left mouse button on the tag, and select the "Delete Tag" off the pop-up menu.

2.6.11. Searching

Selecting "Search" brings up a window that can remain present during normal editor operation. The window allows the user to select the direction of search, the type of search, and a value to search on. The value is entered into a value text window, then pressing the "search" button performs the search. If successful, the cursor is positioned accordingly. An audible tone indicates failure. Pressing the "ok" button removes the search window. The search window is automatically removed when the contig editor is exited. There are seven different search modes.

2.6.11.1. SEARCH BY POSITION

This positions the cursor at the numeric position specified in the value text window, e.g., a value of "1234" causes the cursor to be placed at base number 1234 in the contig. Positioning within a reading is achieved by prefixing the number with the "@" character, e.g., "@123" positions the cursor at base 123 of the sequence in which the cursor lies. Relative positions can be specified by prefixing the number with a plus or minus character; e.g., "+1234" will advance the cursor 1234 bases. If possible, the cursor is positioned within the same sequence. The direction buttons have no effect on the operation of "search by position."

2.6.11.2. Search by Reading Name

This positions the cursor at the left end of the gel reading specified in the value text window. If the value is prefixed with a slash, it is assumed to be a gel reading name. Otherwise it is assumed to be a gel reading number, e.g., "123" positions the cursor at the left end of gel reading number 123, and "/al6al2.sl" positions at the start of reading al6al2.sl. If the value was "/al6" the cursor is positioned at the first reading that starts with "al6." The direction buttons have no effect on the operation of "search by reading name."

2.6.11.3. Search by Tag Type

This positions the cursor at the start of the next tag that has the same type as specified by the type value menu. To change the type, select from the menu that pops up when the mouse is clicked on the button labeled "Type:." This search can be performed either forward or backward from the current cursor position. To find all tags, use "search by annotation," with a null text value string.

2.6.11.4. Search by Annotation

This positions the cursor at the start of the next tag which has a comment containing the string specified in the value text window. The search performed is a regular expression search, and certain characters have special meanings. Be careful when your value string contains ".", "*", "[", "^", or "$". The search can be performed either forward or backward from the current cursor position.

2.6.11.5. Search by Sequence

This positions the cursor at the start of the next piece of sequence that matches the value specified in the text value window. The search is for an exact match, which means that the case of the value string is important. The search is performed on the gel readings themselves, rather than the consensus sequence. The search can be performed either forward or backward from the current cursor position.

2.6.11.6. Search by Problem

This positions the cursor at the next place in the consensus sequence which is not "A", "C", "G", or "T". The search can be performed either forward or backward from the current cursor position.

2.6.11.7. SEARCH BY QUALITY

This positions the cursor at the next place in the consensus sequence where the consensus for each strand is ot "A", "C", "G", or "T", or where the two strands disagree. The search can be performed either forward or backward from the current cursor position.

2.7. Joining Contigs Interactively Using XDAP

The operation of the join editor in XDAP is very similar to the one for single contigs described above. It allows the user to align the ends of the two contigs by editing each contig separately. First, specify which two contigs are to be joined. The program checks that the two contig numbers are different. (It will not allow circles to be formed!) The Join Editor consists of two Contig Editors in between which is sandwiched a disagreement box. This disagreement box uses exclamation marks to denote mismatches between the two consensuses. A typical example is shown in Fig. 4. Here we see in the top window the right end of one contig and in the bottom window the left end of another. The left end of the overlap is correctly aligned, as indicated by an absence of exclamation marks, but the top contig has an extra character at position 558, which is spoiling the alignment over the next segment. Notice that the "lock" button is highlighted denoting that the user has asked for the two contigs to scroll together.

The best strategy for joining is to identify the exact position of overlap. This is defined as the position in the left contig that the leftmost character of the right contig overlaps. The overlap must be of at least one character. Use the scroll bar and the scroll buttons ("<<", "<", ">", and ">>") for positioning the relative positions of the two contigs. The join position can be fixed by pressing the "lock" button at the top of the Join Editor. Locking allows the two contigs to be scrolled as one when using the scroll bar and buttons, the left ends always in the same position relative to each other. Once locked, it is best to proceed to the right along the contigs, inserting padding characters ("*") into the consensuses to minimize the disagreements. It is important that the user aligns the two contigs throughout the whole region of overlap before completing the join because it is only at this stage that the two contigs can be edited independently. If a join is completed leaving a region of mismatch the consensus will consist of dashes and the assembly function will fail to find overlaps in the bad

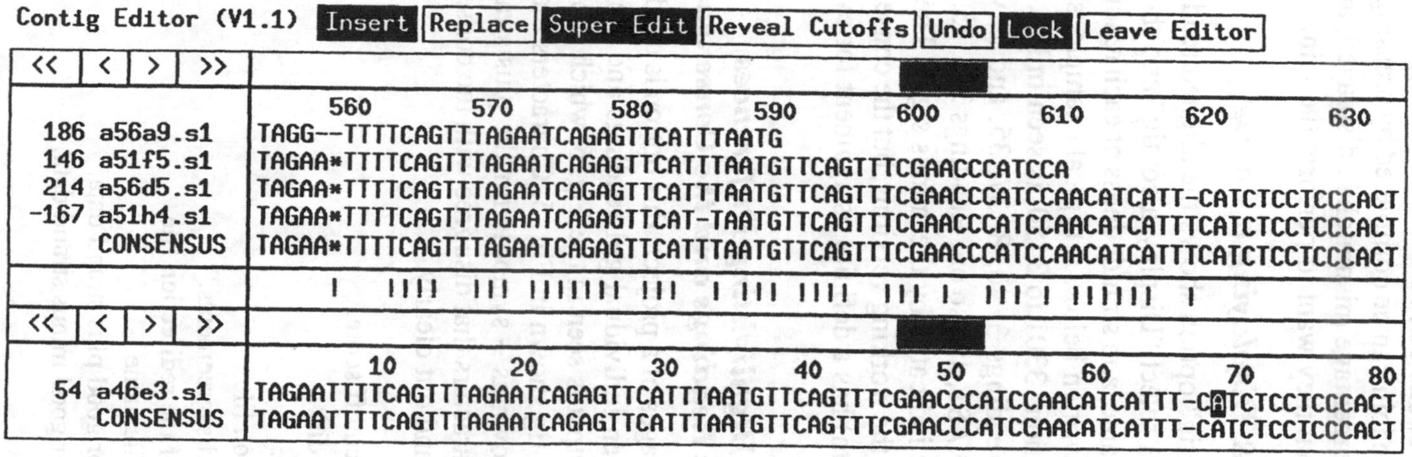

Fig. 4. A typical display from the join editor in XDAP.

section. Misaligned sections can be corrected using the "super edit" mode of the editor. The join can be completed by pressing the "Leave Editor" button. The percentage mismatch is displayed, and users are required to confirm that they want to perform the join.

2.8. Displaying a Contig

The "Display a contig" option shows the aligned readings for any part of a contig. Users select "Display a contig" and then select the contig. The number, name, and strandedness of each reading is shown and the consensus is written below. A typical example, showing part of a contig from positions 3301 to 3450, is seen in Fig. 5. Overlapping this region are readings 3, 40, 8, 37, 35, and 2, with archive names L3.SEQ, A21A7.S1, and so on. Readings 3, 8, 35, and 2 are in reverse orientation as indicated by the minus signs. There are a few padding characters in the working versions, but the consensus (shown below each page width) has a definite assignment for every position except 3376.

2.9. Highlighting Differences Between Readings and the Consensus

During the latter stages of a project, this option is used to highlight disagreements between individual gel readings and their consensus sequences. Typical output is seen in the Fig. 6, which shows the result for the section of contig shown in Fig. 5. Characters that agree with the consensus are shown as + symbols for the plus strand and – for the minus strand. Characters that disagree with the consensus are left unchanged and so stand out clearly.

1. Set the consensus cutoff score.
2. Redirect output to disk.
3. Display the contig.
4. Close the redirection file.
5. Select "Highlight disagreements."
6. Define the name of the redirection file.
7. Define an output file name.
8. Select a symbol for good plus strand data.
9. Select a symbol for good minus strand data.
10. Print the file.

```
                    3310      3320      3330      3340      3350
-3   L3.SEQ      atggttacgccagactatcaaatatgctgcttgaggcttattcgggcgca
40   A21A7.S1    atggttacgccagactatcaaatatgctgcttgaggcttattcgggcgca
-8   A16A2.S1    atggttacgccagactatcaaatatgctgcttgaggcttattcgggcgca
37   A21A2.S1    atggttacgccagactatcaaatatgctgcttgaggcttattcgggcgca
     CONSENSUS   atggttacgccagactatcaaatatgctgcttgaggcttattcgggcgca

                    3360      3370      3380      3390      3400
-3   L3.SEQ      gatctgaccaagcgacag*tttaaa*gtgctgcttgccatt*ctgcgt*a
40   A21A7.S1    gatctgaccaagcgacag*gttaaagttgctgctt
-8   A16A2.S1    gatctgaccaagcgacag*tttaaa*gtgctgcttgccatt*ctgcgt*a
37   A21A2.S1    ga-ctgaccaagcgacag*tttaaa*gtgctgcttgccatt*ctgcgt*a
-35  A16D12.S1               gttttaaa-gtgctgcttgccatttctgcgtaa
-2   L2.SEQ                                      t*ctgcgt*a
     CONSENSUS   gatctgaccaagcgacag*tttaaa-gtgctgcttgccatt*ctgcgt*a

                    3410      3420      3430      3440      3450
-3   L3.SEQ      aaacctatgggt*ggaataaaccaatggacagaatcaccgattctcaact
-8   A16A2.S1    aaacctatgggt*ggaataaaccaatggacagaatcaccgattctcaact
37   A21A2.S1    aaacctatgggtgggaataaaccaatggacagaatcaccgattctcaact
-35  A16D12.S1   aaacctatgggt*ggaataaaccaatggacagaatcaccgattctcaact
-2   L2.SEQ      aaacctatgggt*ggaataaaccaatggacagaatcaccgattctcaact
     CONSENSUS   aaacctatgggt*ggaataaaccaatggacagaatcaccgattctcaact
```

Fig. 5. Typical output from "Display contig."

```
                    3310      3320      3330      3340      3350
-3   L3.SEQ      --------------------------------------------------
40   A21A7.S1    ++++++++++++++++++++++++++++++++++++++++++++++++++
-8   A16A2.S1    --------------------------------------------------
37   A21A2.S1    ++++++++++++++++++++++++++++++++++++++++++++++++++
                 atggttacgccagactatcaaatatgctgcttgaggcttattcgggcgca

                    3360      3370      3380      3390      3400
-3   L3.SEQ      -------------------------*------------------
40   A21A7.S1    +++++++++++++++++++g+++++gt++++++++
-8   A16A2.S1    -------------------------*------------
37   A21A2.S1    ++-+++++++++++++++++++++*+++++++++++++++++++++++++
-35  A16D12.S1               -t--------------------t------a-
-2   L2.SEQ                                     ----------
                 gatctgaccaagcgacag*tttaaa-gtgctgcttgccatt*ctgcgt*a

                    3410      3420      3430      3440      3450
-3   L3.SEQ      --------------------------------------------------
-8   A16A2.S1    --------------------------------------------------
37   A21A2.S1    +++++++++++++g+++++++++++++++++++++++++++++++++++++
-35  A16D12.S1   --------------------------------------------------
-2   L2.SEQ      --------------------------------------------------
                 aaacctatgggt*ggaataaaccaatggacagaatcaccgattctcaact
```

Fig. 6. Typical output from "Highlight disagreements," showing the results for the section of contig displayed in Fig. 5.

2.10. Examining the "Quality" of a Contig

This function reports on the proportion of the consensus that is "well determined' and will display a sequence of symbols that indicate the quality of the consensus at each position or produce a graphical display. Each strand of the contig is analyzed separately using the consensus algorithm, and a position is declared "well determined" if it is assigned one of the symbols a, c, g, t. The current consensus calculation cutoff score is used.

A summary showing the percentage of the consensus that falls into each category of quality is shown. The analysis divides the data into five categories, assigning each a code as shown in Fig. 7. Code 0 means well determined on both strands and they agree, 1 means well determined on the plus strand only, 2 means well determined on the minus strand only, 3 means not well determined on either strand, and 4 means well determined on both strands but they disagree.

If the user chooses to have the data displayed graphically the following scheme is used. A rectangular box is drawn, so that the x coordinate represents the length of the contig. The box is notionally divided vertically into five possible levels which are given the y values: -2, -1, 0, 1, 2. The quality codes assigned to each base position are plotted as rectangles. Each rectangle represents a region in which the quality codes are identical, so a single base having a different code from its immediate neighbors will appear as a very narrow rectangle. Obviously, a single line at the midheight shows a perfect sequence. Figure 8 shows the result for the section of contig shown in Fig. 5.

2.11. Using Graphical Displays
to Examine Contigs

The programs contain three graphical displays to aid the examination of contigs. The first simply gives an overview of all the contigs in the database and provides, with the use of a crosshair, a mechanism for the other two displays to select contigs. One of these displays produces a schematic representation of each of the readings in a contig. The lines in the display show the relative positions of each reading and also their sense. The plot is divided vertically into two sections by a line that is identified by an asterisk drawn at each end. All lines that lie above this line represent readings that are in their

Strands OK			Quality code	Y cordinates		
+	-	and the same	0	0	to	0
+			1	0	to	1
-			2	-1	to	0
neither			3	-1	to	1
+	-	but different	4	-2	to	2

Fig. 7. The codes and coordinates used by the "Quality plot."

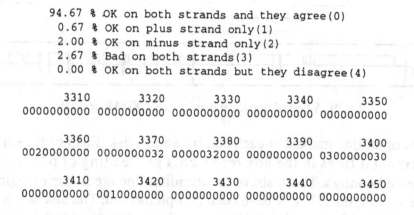

```
94.67 % OK on both strands and they agree(0)
 0.67 % OK on plus strand only(1)
 2.00 % OK on minus strand only(2)
 2.67 % Bad on both strands(3)
 0.00 % OK on both strands but they disagree(4)

     3310       3320       3330       3340       3350
0000000000 0000000000 0000000000 0000000000 0000000000

     3360       3370       3380       3390       3400
0020000000 0000000032 0000032000 0000000000 0300000030

     3410       3420       3430       3440       3450
0000000000 0010000000 0000000000 0000000000 0000000000
```

Fig. 8. Listed output from "Examine Quality" showing the results for the section of contig displayed in Fig. 5.

original sense; all lines below show readings that are in the complementary sense. The final graphical display is of the "quality" of the data as described above.

When these graphical displays are visible users may employ a crosshair, moved by mouse or keyboard commands, to examine the data in more detail. The crosshair is positioned and ,hen keyboard characters S, Q, N, or Z are typed, the program will show the local aligned sequences in a text window, produce the quality plot, give the names of the nearest readings, or zoom into the display.

A typical display of all three plots is shown in Fig. 9. The top rectangle shows a separate line for contigs of each of the projects. The right-hand one is bisected by a vertical line, indicating that it has been selected by the user. The next rectangle below is divided by a

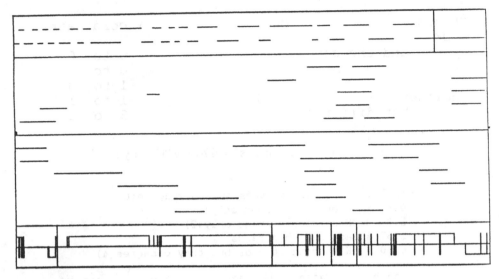

Fig. 9. A typical graphical display from XDAP or SAP.

horizontal line marked at each end by an asterisk. Each of the other horizontal lines in the box represents gel reading of one of the selected contigs. Those above the dividing line are in their original orientation; those below have been complemented. The box below is also divided by a horizontal line and shows the "quality" for each base in the contig. Rectangular areas marked above the central line show sections that only have a good consensus on the minus strand, and rectangles below show good sections from the other strand. Places where the vertical lines reach the top and bottom of the box show disagreements between the two strands. Places with only the midline have a good consensus on both strands.

2.12. Screen Editing Contigs in SAP

When using SAP, the best way for users to edit a whole contig interactively is to use their preferred external editor on the standard display of a contig. When the screen edit function is selected, SAP writes a text file containing a display of the contig and passes it to an external editor—such as EDT on the VAX or emacs on a UNIX system. The user modifies the file using the editor, and when the editor is exited, SAP moves the changed contig back into the project database.

1. Select "Screen edit."
2. Select the contig to edit.

3. Define a temporary file for use by the editor. After a slight pause, the editor will start, and the first page of the contig will appear on the screen.
4. Edit the contig using the editor's standard commands.
5. Exit from the editor.
6. Accept "Put contig back into the database."

2.13. Automatic Editing of Contigs in SAP

This function automatically changes characters in gel readings to make them agree with the consensus sequence. At first sight, this may seem like an unethical procedure, but as is explained in Section 3., it is quite legitimate and saves a great deal of time. Figure 10 shows the effect of using auto edit on the section of contig displayed in Fig. 5. All changed characters (for example, position 3369, reading A21A7.S1) are denoted by upper-case letters. Note that apart from position 3375, which has an unresolved consensus, all other changes have been made. These edits were made using a combined consensus for both strands, but the standard version of the program treats each strand separately and will only make a change if the consensus for the two strands agree.

1. Redirect output to disk.
2. Select "Display contig."
3. Identify the contig to edit/display.
4. Close the redirection file.
5. Print the file containing the displayed contig.
6. Check the contig and the original films, and annotate the printout to indicate the required edits.
7. Set the cutoff for the consensus calculation.
8. Select "Auto edit."
9. Identify the contig and the section to edit.
10. The program will display a summary of changes made.
11. Display the contig, and compare it with the annotated printout.
12. Use another editing method to finish the editing.

2.14. Using the Original Editor in SAP

This simple editor can insert, delete, and change gel reading sequences by performing one selected operation at a time. It is used during the interactive entry of new readings and interactive joining of contigs. The commands request the position at which the edit is required and the number of characters to insert, delete, or change.

```
                        3310      3320      3330      3340      3350
-3   L3.SEQ      atggttacgccagactatcaaatatgctgcttgaggcttattcgggcgca
40   A21A7.S1    atggttacgccagactatcaaatatgctgcttgaggcttattcgggcgca
-8   A16A2.S1    atggttacgccagactatcaaatatgctgcttgaggcttattcgggcgca
37   A21A2.S1    atggttacgccagactatcaaatatgctgcttgaggcttattcgggcgca
     CONSENSUS   atggttacgccagactatcaaatatgctgcttgaggcttattcgggcgca

                        3360      3370      3380      3390      3400
-3   L3.SEQ      gatctgaccaagcgacagtttaaa*gtgctgcttgccattctgcgtaaaa
40   A21A7.S1    gatctgaccaagcgacagTttaaagGtgctg
-8   A16A2.S1    gatctgaccaagcgacagtttaaa*gtgctgcttgccattctgcgtaaaa
37   A21A2.S1    gaTctgaccaagcgacagtttaaa*gtgctgcttgccattctgcgtaaaa
-35  A16D12.S1                   gtttaaa-gtgctgcttgccattctgcgtaaaa
-2   L2.SEQ                                       tctgcgtaaaa
     CONSENSUS   gatctgaccaagcgacagtttaaa-gtgctgcttgccattctgcgtaaaa

                        3410      3420      3430      3440      3450
-3   L3.SEQ      cctatgggtggaataaaccaatggacagaatcaccgattctcaacttag
-8   A16A2.S1    cctatgggtggaataaaccaatggacagaatcaccgattctcaacttagc
37   A21A2.S1    cctatgggtggaataaaccaatggacagaatcaccgattctcaacttagc~
-35  A16D12.S1   cctatgggtggaataaaccaatggacagaatcaccgattctcaacttagc
-2   L2.SEQ      cctatgggtggaataaaccaatggacagaatcaccgattctcaacttagc
     CONSENSUS   cctatgggtggaataaaccaatggacagaatcaccgattctcaacttagc
```

Fig. 10. The result of applying the "auto editor" to the section of contig displayed in Fig. 5.

3. Notes

1. As each reading is entered into a project database it is given a unique number. The first is numbered 1, the second 2, and so on. Their original file names (known as "archives" because they are kept outside the database and never edited) are also copied into the database. During assembly, contigs are constantly being changed and reordered, so the program identifies them by the numbers or names of the readings they contain. Whenever the program asks users to identify a contig or reading, they can type its number or its archive name. If they type its archive name they must precede the name by a slash "/" symbol to denote that it is a name rather than a number. For example, if the archive name is fred.gel with number 99, users should type /fred.gel or 99 when asked to identify the contig. Generally, when it asks for the reading to be identified, the program will offer the user a default name, and if the user types only return, that contig will be accessed. When a database is opened, the default contig will be the longest one, but if another is accessed, it will subsequently become the current default.

2. An XDAP database is made from five separate files: the "archive names" file *.ARN, the "relationships" file *.RLN, the "sequences" file *.SQN, the "tag" file *.TGN, and the "comments" file *.CCN. If the database is called FRED, then version 0 of database FRED comprises files FRED.AR0, FRED.RL0, FRED.SQ0, FRED.TG0, and FRED.CC0. The version is the last symbol in the file names. If the "copy database" option is used, it will ask the user to define a new "version." The normal strategy is to use version 0 for all work and to use other versions as backups. Program SAP uses databases formed from only the first three of these files. Normally, the program is used to handle DNA sequences, but many of the functions also work on protein sequences. The choice of sequence type is made when the database is started.

3. The vector sequence should be stored in a simple text file with up to 80-characters of data per line.

4. Almost all readings are assembled automatically in their first pass through the assembly routine. Those that are not can be dealt with in two ways. Either they can be put through assembly again as single named readings (users should type n when asked "Use file of file names"), with the parameters set to allow the reading in, or they can be entered through the assembly routine using the "Put all readings in new contigs" mode, and then joined to the contig they overlap using the Contig Joining Editor. If it is found that readings are not being assembled in their first pass through the assembler, then it is likely that the contigs require some editing to improve the consensus. Also, it may be that poor-quality data is being used, possibly by users overinterpretting films or traces. In the long term, it can be more efficient to stop reading early and save time on editing. For those using fluorescent sequencing machines, the unused data can be incorporated after assembly.

5. Obviously, a script cannot be used to operate a program that expects to be controlled by mouse clicks! The program DAP is an xterm version of XDAP, which can be used from a script.

6. There is a remote possibility of a join being missed by the "Find internal joins" routine. If a small contig is wholly contained within a larger one, such that its ends are further than ("Probe length"—"Minimum initial match length") from the ends of the larger contig, and the consensus for the small contig lies to the left of the consensus for large contig, the overlap will not be discovered. (*See* Section 2.5.)

7. For those using fluorescent sequencing machines and XDAP, the combination of the contig editor and the graphical displays of consensus "quality" will probably be sufficient for checking and editing contigs, since everything can be done at the computer screen. For those using

autoradiographs, the facility to produce printouts of "display" and "high-light disagreements" options for use while checking films, and the auto edit command are most appropriate.

8. In general, the quality of a reading deteriorates along the length of the gel, so it is also possible to use a length cutoff for the quality calculation. Only the data from the first section of each reading will be included in the calculation.

9. There are some limitations on the changes that can be made to the contigs when using the SAP screen editor. Alignments must be maintained during editing. Whole lines of sequence should not be deleted or added unless the order of the gel readings in the contig is preserved. Each line in the contig display consists of gel reading numbers, their names, and 50-character sections of sequence. Insertions are limited in the following way. No line of sequence can be extended rightward more than five characters beyond the end of a full-length line (a full-length line is 50-characters long). Only one character can be added to the left end of full-length lines, but sections of sequence beginning further into a line can be extended leftward up to an equivalent position. Do not delete any nonsequence lines in the file. Before returning the contig to the database, the program checks that the rules have been obeyed. If an error is found, the number of the erroneous line in the file is displayed, and the contig will not be changed.

10. The following is a justification for using the auto edit function. The general strategy employed when collecting shotgun sequence data is to keep sequencing until the redundancy in the contigs is fairly high, and then to get a printout of a contig, check problems against the films, note corrections on the printout, and make the changes using an interactive editor. In general, the consensus is correct except for places where padding characters have been used to accommodate a single gel with an extra character or where the consensus is dash. The important point for the auto editor is that most edits simply make the gel readings conform to the consensus or remove columns of pads. The auto editor calculates a consensus for the contig (or part of a contig) to be edited, and then uses this consensus to direct the editing of the contig in three stages:

1. Stage 1: Find and correct all places where, if the order of two adjacent characters is swapped, they will both agree with the consensus (given that they did not match the consensus before). These corrections are termed "transpositions."

2. Stage 2: Find and correct all places where there is a definite consensus, but the gel reading has a different character. These corrections are termed "changes."

3. Stage 3: Delete all positions in which the consensus is a padding character. These corrections are termed "deletions."

All changed characters are shown in upper-case letters, so it will be obvious which characters have been assigned by the program (except for deletions). The number of each type of correction will be displayed.

11. The "calculate consensus" function, the "display contig" routine, the contig editor, and the "show quality" option use the rules outlined here to calculate a consensus from aligned gel readings. The consensus sequence can contain any of six possible symbols: a, c, g, t, * and −. The last symbol assigned if none of the others makes up a sufficient proportion of the aligned characters at any position in the contig. The following calculation is used to decide which symbol to place in the consensus at each position. Each uncertainty code contributes a score to one of a, c, g, t, * and also to the total at each point. Symbols like r and y, which do not correspond to a single base type, contribute only to the total at each point.

Definite assignments, i.e., A,C,G,T,a,c,g,t,b,d,h,v,k,l,m,n,a, c,g,t,* = 1, probable assignments, i.e., 1,2,3,4 = 0.75, other uncertainty codes including r,y,5,6,7,8,- = 0.1, and a cutoff score between 51 and 100% is set by the user. (When the program starts, this is set to 75%.). At each position in the contig, calculate the total score for each of the five symbols a,c,g,t, and * (denote these by X_i, where i = a,c,g,t, or *), and also the sum of these totals (denote this by S). Then if $100\ X_i/S$ > the cutoff for any i, symbol i is placed in the consensus; otherwise − is assigned. For the "examine quality" algorithm each strand is treated separately, but the calculation is the same.

12. Databases can become corrupted if the machine crashes so the programs contain a function "Check database for logical consistency" which checks to see if all the relational data are internally consistent. Some routines automatically perform this check before they start. Users are advised to make frequent copies of their databases using the "Copy database" option. Simple routines are provided for removing readings from the database or breaking contigs in two.

13. Many of the most important or complicated operations performed by SAP and XDAP have been covered, but several others have not been mentioned. These include those for creation of consensus sequence files for processing by other programs, and complementing contigs, both of which are trivial. There is also a set of routines for fixing corrupted databases.

Notes Added in Proof

14. When operated from the X version of the program, the "Find internal joins" option automatically calls up the contig joining editor with the two contigs aligned in the edit windows.
15. The contig editor now includes an option to select oligonucleotide primers and templates for primer walking experiments. Tags are created for each primer. The algorithm used is described in ref. 5.
16. The program now includes a function that detects segments of contigs that only have data from one strand of the DNA. It then tries to "double strand" such segments by aligning any "unused" data from adjacent readings. If it aligns sufficiently well the data is redefined as "used." Significant amounts of a cosmid can be confirmed in this way, and as the procedure is fully automatic, it saves a great deal of time.
17. The program now includes a function that detects segments of contigs that only have data from one strand of the DNA. It then selects primers and templates that should help to "double strand" the detected segments. The procedure is fully automatic and produces a file containing a list of templates and primers.
18. The VAX version of SAP will only allow one person to access a sequencing database at a time—producing an "unable to open database" error message if a second person tries. On UNIX machines there is no such check in program SAP so users need to make sure that simultaneous use does not occur. Otherwise, the data will be corrupted. Program DAP prevents more than one person from using a database at any time. It does so using the following mechanism. When a user requests to open a particular copy (say 0) of a database (say DB) the program checks for the existence of a file named DB_BUSY0 in the current directory. In normal circumstances, if the file exists, it indicates that somebody else is currently using the database and the program displays the message "Sorry database busy" and does not open the files. If the file does not exist, the program creates it and opens the database. When a user stops using the database (usually by quitting the program) the "busy file" is deleted, hence, allowing others to use the database. If the program terminates abnormally, the busy file will not be deleted and so the database will not be usable until the busy file is explicitly deleted using the rm command. Obviously, it is dangerous to delete the file before checking if another user is using the database.
19. We have recently devised our own file format (called SCF) for storing traces, sequences, and confidence values for data produced by automated sequence readers (6). For ABI data these typically reduce the storage required to 30% of the original. Data from the ABI 373A and

the Pharmacia A.L.F. can be converted to this form using the program makeSCF. Note that A.L.F. files must first be processed by program alfsplit, which splits the original data into one file per reading. Sequences can be extracted for SCF files in a form suitable for assembly by use of the program trace2seq. To locate and mark regions of sequence from an automated sequence reader that are of too low a quality to be used for assembly we use the script clip-seqs. This script takes as input a file of reading file names. For each reading it renames the original file "original-filename~" and writes a new file called "original-filename" in which the poor quality regions are marked.

References

1. Staden, R. (1982) Automation of the computer handling of gel reading data produced by the shotgun method of DNA sequencing. *Nucleic Acids Res.* **10(15)**, 4731–4751.
2. Staden, R. (1990) An improved sequence handling package that runs on the Apple Macintosh. *Comput. Applic. Biosci.* **4**, 387–393.
3. Dear, S. and Staden, R. (1991) A sequence assembly and editing for efficient management of large projects. *Nucleic Acids Res.* **19**, 3907–3911.
4. Gleeson, T. and Hillier, L. (1991) A trace display and editing program for data from fluorescence based sequencing machines. *Nucleic Acids Res.* **19**, 6481–6488.
5. Hillier, L. and Green P. (1991) OSP: an oligonucleotide selection program. *PCR Meth. Appl.* **1**, 124–128.
6. Dear, S. and Staden, R. (1992) A standard file format for data from DNA sequencing instruments. *DNA Sequence* **3**, 107–110.

CHAPTER 5

Staden: Statistical and Structural Analysis of Nucleotide Sequences

Rodger Staden

1. Introduction

This chapter deals with performing simple statistical and structural analysis of nucleotide sequences, and also describes some more unusual tests. Base, dinucleotide and codon compositions, potential amino acid compositions, and the relative frequencies of each base in each position of codons are covered. This chapter describes how to produce plots to show regions of unusual composition and to measure the codon bias for a gene. In addition, it describes a set of functions for finding "structures" in nucleotide sequences, including short-range inverted repeats or stem-loops, long-range inverted repeats, long-range direct repeats, and Z DNA. All the methods are contained in the program NIP.

2. Methods

2.1. Calculating the Base Composition

Select "Calculate base composition." The composition of the active region is shown.

2.2. Calculating the Dinucleotide Composition

Select "Calculate dinucleotide composition." The dinucleotide composition of the active region and an expected dinucleotide composition are shown. The expected composition is calculated from the base composition assuming a random order of bases in the sequence. *See* Fig. 1.

From: *Methods in Molecular Biology, Vol. 25: Computer Analysis of Sequence Data, Part II*
Edited by: A. M. Griffin and H. G. Griffin Copyright ©1994 Humana Press Inc., Totowa, NJ

	T		C		A		G	
	Obs	Expected	Obs	Expected	Obs	Expected	Obs	Expected
T	5.86	5.97	6.18	5.99	4.24	5.91	8.14	6.56
C	6.10	5.99	5.14	6.02	5.91	5.93	7.38	6.59
A	5.57	5.91	5.64	5.93	7.91	5.84	5.05	6.49
G	6.90	6.56	7.56	6.59	6.11	6.49	6.30	7.22

Fig. 1. The dinucleotide composition display.

2.3. Calculating the Codon Composition

This function counts codons, amino acid composition, protein molecular weights, hydrophobicity, and base compositions. Users select the segments of the sequence to be analyzed. The segments can be defined on the keyboard or from an EMBL/GenBank feature table.

1. Select "Calculate codon composition."
2. Accept "Show observed counts." The alternative displays its codon tables so that the total for each amino acid sums to 100. This makes it easier to see any bias present in the codon usage.
3. Accept "Define segments using keyboard." The alternative is to use a feature table.
4. Define "From." The start of the segment to be analyzed.
5. Define "To." This is the end of the segment to be analyzed. The results will be displayed as in Fig. 2, and then the program will again ask "From." The user should define a zero value for "From" when all segments of interest have been analyzed. The program will then display a cumulative total for all the values it calculates. The counts are broken down into several figures. Apart from the codon counts, we see the base composition by position in codon expressed as a percentage of each base's own frequency; base composition by position in codon expressed as a percentage of the overall base composition of the segment; base composition expected for the observed amino acid composition expressed if there were no codon preference; percentage deviations of the observed amino acid composition from an average amino acid composition *(1)*; the molecular weight and hydrophobicity *(2)* of the putative amino acid sequence.

2.4. Creating a Codon Usage File

This method writes a file of codon usage in the form of a codon table (*see* Fig. 2). Such tables can be used by several other methods contained within the programs. If required, the user can start with an existing file and add to it.

```
Calculate base, codon and amino acid compositions
? Show observed counts (y/n) (y) =
? Define segments using keyboard (y/n) (y) =

? From (0-8134) (0) =1
? To (1-8134) (8134) =1000
? + strand (y/n) (y) =
```

```
=============================================
   F TTT   5. S TCT   7. Y TAT   4. C TGT   2.
   F TTC  17. S TCC   3. Y TAC   5. C TGC   3.
   L TTA   3. S TCA   4. * TAA   3. * TGA   1.
   L TTG   4. S TCG   3. * TAG   0. W TGG   7.
=============================================
   L CTT   3. P CCT   6. H CAT   6. R CGT   3.
   L CTC   1. P CCC   1. H CAC   4. R CGC   2.
   L CTA   0. P CCA   4. Q CAA   3. R CGA   1.
   L CTG  36. P CCG   6. Q CAG   5. R CGG   4.
=============================================
   I ATT  12. T ACT   3. N AAT   6. S AGT   0.
   I ATC  13. T ACC   5. N AAC   7. S AGC   7.
   I ATA   1. T ACA   2. K AAA   9. R AGA   0.
   M ATG   9. T ACG   7. K AAG   3. R AGG   1.
=============================================
   V GTT   6. A GCT   5. D GAT   7. G GGT   9.
   V GTC   3. A GCC   6. D GAC   6. G GGC   9.
   V GTA   7. A GCA   2. E GAA   5. G GGA   5.
   V GTG   9. A GCG   7. E GAG   3. G GGG   3.
=============================================
```

```
Total codons=      333.
            T          C          A          G

1         25.00      34.27      40.28      35.94
2         45.42      28.63      36.02      22.27
3         29.58      37.10      23.70      41.80
          -----      -----      -----      -----
=         100%       100%       100%       100%

1         21.32      25.53      25.53      27.63  = 100%
2         38.74      21.32      22.82      17.12  = 100%
3         25.23      27.63      15.02      32.13  = 100%
%         28.43      24.82      21.12      25.63  Observed, overall totals
%         29.65      23.25      23.95      23.15  Expected, even codons per acid
```

```
           A     C     D     E     F     G     H     I     K     L
          20.    5.   13.    8.   22.   26.   10.   26.   12.   47.
O-E %    -27.  -11.  -25.  -61.   71.   10.   38.   52.  -36.   59.

           M     N     P     Q     R     S     T     V     W     Y
           9.   13.   17.    8.   11.   24.   17.   25.    7.    9.
O-E %     14.  -10.    1.  -39.  -41.    6.  -11.   15.   64.  -15.
Total acids=   329. Molecular weight=       36493. Hydrophobicity=  64.7
```

Fig. 2. A worked example of calculating codon, base, and amino acid compositions.

1. Select "Calculate a codon table and write it to disk."
2. Accept "Start with empty table."
3. Accept "Show observed counts." The alternative is to have the counts for each amino acid type sum to 100.
4. Accept "Define segments using keyboard." The alternative is to use an EMBL/GenBank feature table.
5. Define "From." This is the start of the segment to count over.
6. Define "To." This is the end of the segment.
7. Accept "+ strand," or alternatively, the minus strand. The table will appear on the screen, and the program will cycle around to step 5. When all segments have been defined, a zero value for "From" will instruct the program to display on the screen a table file is the sum of all the individual tables.
8. Define "Name for codon table file." Give the name of the file in which to save the final table.

2.5. Plotting the Base Composition

This function plots the base composition for each "window length" of the sequence. The frequency of any combinations of bases can be plotted.

1. Select "Plot base composition."
2. Select which combination of bases to plot. The default is A + T, but any single base or combination of bases can be used.
3. Select "Odd window length." This is the size of window over which each count is made; it is "odd," so that the plotted point exactly corresponds to the center of each window. The count is made over the window, and then the window is moved on by one base and the count repeated.
4. Define "Plot interval." Especially when using long windows, it is unnecessary to plot the results for every point along the sequence. A plot interval of five will mean the value for every fifth point will be plotted. The plot will appear in the form shown in Fig. 3.

2.6. Searching for Anomalous Compositions

This "search" is performed by comparing a standard composition against each segment of the sequence and plotting the difference. The difference between the observed and expected composition at each point is expressed as the chi-square value. Any one of the base, dinucleotide, or trinucleotide compositions can be used as the standard. No expected level of divergence is used, so the program always dis-

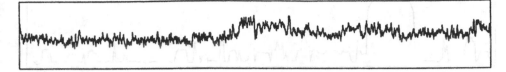

Fig. 3. A typical base composition plot. This is an A + T plot for bacteriophage Lambda and shows that one-half is A + T-rich and the other G + C rich.

plays the results so that the plots fill the allotted space on the screen. At the end, the observed range is displayed.

1. Select "Plot dinucleotide composition differences as chi squared." Alternatively, select base or trinucleotides.
2. Define "Start." Define the position of the first base to be used in the standard.
3. Define "End." Define last base of the standard. The default standard region is the whole sequence.
4. Define "Odd window length."
5. Define "Plot interval." The plot will appear as in Fig. 4.

2.7. Search for Anomalous Word Usage

This function is designed to examine the abundances of short words in a nucleotide sequence to see if particular ones are either under- or overrepresented *(3)*. It compares the observed and expected frequencies, and plots them for each segment of the sequence. There has been some work on the relative abundances of CG dinucleotides in eukaryotic sequences (e.g., ref. *[14]*), and this routine can be used to examine such biases or any others that might be of interest.

1. Select "Plot observed-expected word usage."
2. Define "String." That is the word to search for. The default is CG.
3. Define "Odd window length."
4. Define "Plot interval."
5. Define "Maximum plot value." Define the maximum expected value for the plot.
6. Define "Minimum plot value." The plot will appear as in Fig. 5.

2.8. Calculate Codon Constraint

This method measures the level of constraint imposed on a sequence by coding for a protein. The codon constraint is the difference between the observed codon improbability and the mean improbability

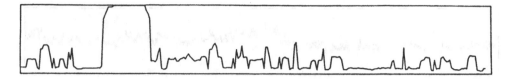

Fig. 4. An anomalous composition plot. This shows an immunoglobulin switch region, and the plateau corresponds to a segment composed entirely of A and G bases.

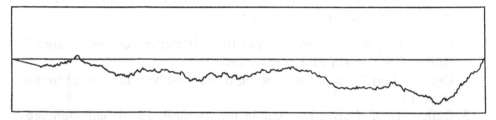

Fig. 5. A plot of anomalous word usage. This shows a plot of CG usage for the human CMV immediate-early region. The frequency of CG is much lower than would be expected from the composition.

for a random sequence of the same composition. That is, it is a measure of the codon bias, and the program performs the calculation over windows of 99-codon length. *See* ref. *5.* The user can select segments to analyze either by defining them on the keyboard or by using an EMBL/GenBank feature table. The result for each selected segment, which is simply a single number, is displayed.

1. Select "Calculate codon constraint."
2. Accept "Define segments using keyboard."
3. Define "From." This is the start of the segment.
4. Define "To." This is the end of the segment.
5. Accept "+ strand."

The result will be displayed, and the program will ask for the next segment to be defined.

2.9. Searching for Stem-Loop Structures

This routine finds simple putative stem-loop structures having a minimum number of base pairs in their stems. Results can be listed or plotted.

```
                   g
                 g.t
                 t.g
                 c-g
                 a-t
                 t.g
                 t.g
                 g-c
                 t.g
                 g.t
                 g.t
                 t.g
                 t.g
                 g-c
                 t.g
           tggcga gttttaa
                 843
```

Fig. 6. A typical textual display from the routine for finding simple hairpin loops.

1. Select "Search for hairpin loops."
2. Define "Minimum loop size."
3. Define "Maximum loop size."
4. Define "Minimum number of base pairs."
5. Reject "Plot results." The alternative writes out the stem-loops as shown in Fig. 6. The plotted output marks the position of each stem, the height of the mark showing the length of the stem;,

2.10. Searching for Long-Range Inverted Repeats

This method finds inverted repeats. It allows for no mismatches, insertions, or deletions within the matching segments.

1. Select "Find long range inverted repeats."
2. Accept "Plot results." The alternative lists out all the matching segments.
3. Define "Start." This is the beginning of the region to analyze. In general the whole sequence will be analyzed.
4. Define "End."
5. Define "Minimum inverted repeat." This is the length of the minimum match.

Fig. 7. A plot of direct or inverted repeats. Each matching segment is joined by a rectangular line. Here we show, the direct repeats of at least 25 bases in a mouse immunoglobulin switch region.

The results will now be plotted in an unusual way as shown in Fig. 7, in which the positions of matching segments are joined by rectangular lines.

2.11. Searching for Long-Range Repeats

This method finds direct repeats. It allows for no mismatches, insertions, or deletions within the matching segments.

1. Select "Find long-range repeats.
2. Accept "Plot results." The alternative lists out all the matching segments.
3. Define "Start." This is the beginning of the region to analyze. In general the whole sequence will be analyzed.
4. Define "End."
5. Define "Minimum repeat." This is the length of the minimum match. The results will now be plotted in an unusual way as shown in Fig. 7, in which the positions of matching segments are joined by rectangular lines.

2.12. Searching for Possible Z DNA

The program contains three algorithms for searching for sequences with the potential for forming Z DNA. In varying ways they look for segments of alternating purines and pyrimidines, and they all plot their results. A typical result is shown in Fig. 8.

3. Notes

1. Whenever the program reads a sequence file, it always displays the base composition to provide the user with a check on the correctness of the file.
2. The search for anomalous words function operates in the following way. Users select a "word"—say CG—and a window length. The program examines each successive window length along the sequence, with each window overlapping the previous one by window-length-1 bases. For each window position, the program calculates the base composition and the number of occurrences of the chosen word. From the base composition, it calculates an expected number of occurrences of the chosen word by simply multiplying the relevant frequencies and assuming random

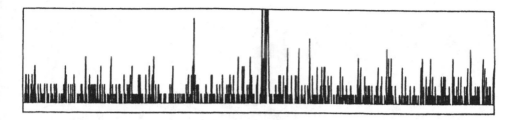

Fig. 8. A plot of predictions for potential Z DNA containing some high peaks produced by regions of alternating purines and pyrimidines.

ordering. It plots observed-expected, hence showing regions that are enriched or depleted in the chosen word.

3. The codon constraint calculation offers a measure of the codon bias that is independent of any set tables of expected codons. Although some users may find the underlying mathematics difficult to understand, the values obtained provide an interesting measure. It was shown *(5)* for a set of *E. coli* genes that their values of codon constraint correlated with their levels of expression.

4. The algorithm for finding possible stem loops counts A-T, G-C, and G-T pairs as matching, but will only find stems with no mismatches or loopouts.

5. The long-range inverted and direct repeat routines are fast, but only find exact matches. More flexible and exhaustive methods are described in Chapter 14 on sequence comparisons.

6. It is also possible to use the pattern-searching routines to define and search for inverted and direct repeats. They are particularly useful for finding specific structures—for example, tRNA folds.

References

1. McCaldon, P. and Argos, P. (1988) Oligopeptide biases in protein sequences and their use in predicting protein coding regions in nucleotide sequences. *Proteins* **4,** 99–122.

2. Sweet, R. M. and Eisenberg, D. (1983) Correlation of sequence hydrophobicity measures similarity in three-dimensional protein structure. *J. Mol. Biol.* **171,** 479–488.

3. Honess, R. W., Gompels, U. A., Barrell, B. G., Craxton, M., Cameron, K. R., Staden, R., Chang, Y.-N., and Hayward, G. S. (1989) Deviations from expected frequencies of CpG dinucleotides in herpesvirus DNAs may be diagnostic of differences in the states of their latent genomes. *J. Gen. Virol.* **70,** 837–855.

4. Bird, A. P. (1980) DNA methylation and the frequency of CpG in animal DNA. *Nucleic Acids Res.* **8,** 1499–1504.

5. McLachlan, A. D., Staden, R., and Boswell, D. R. (1984) A method for measuring the non-random bias of a codon usage table. *Nucleic Acids Res.* **12,** 9567–9575.

CHAPTER 6

Staden: Searching for Restriction Sites

Rodger Staden

1. Introduction

The program NIP contains a routine for finding and displaying the positions of the cut sites of restriction enzyme recognition sequences. Linear or circular sequences can be searched, and the results can be listed in various forms or displayed graphically. The recognition sequences to be searched for can be typed on the keyboard or read from files. The format of these files is given in Note 1. The end of the chapter also describes how to produce back-translations of protein sequences so that these routines can be used to search them for restriction sites.

2. Methods

2.1. Search for Restriction Enzyme Sites and List Them Enzyme by Enzyme

1. Select "Search."
2. Select "Input source" as "All enzymes file." A number of standard files are available, and users may also have their own.
3. Accept "Search for all names."
4. Select "Order results enzyme by enzyme."
5. Accept "List matches."
6. Accept "The sequence is linear." The alternative is circular.
7. Accept "Search for definite matches." The alternative is to search for possible matches in a sequence containing IUB redundancy codes.

The results will then appear in the form shown in Fig. 1. Each match is numbered, and its enzyme name given, followed by the

From: *Methods in Molecular Biology, Vol. 25: Computer Analysis of Sequence Data, Part II*
Edited by: A. M. Griffin and H. G. Griffin Copyright ©1994 Humana Press Inc., Totowa, NJ

```
Matches found=     3
        Name              Sequence                           Position Fragment length
     1 AccII              cg'cg                                    313      312      51
     2 AccII              cg'cg                                    364       51     188
     3 AccII              cg'cg                                    552      188     312
                                                                            449     449
Matches found=     6
        Name              Sequence                           Position Fragment length
     1 AciI               cc'gc                                    503      502      12
     2 AciI               gc'gg                                    553       50      12
     3 AciI               gc'gg                                    714      161      50
     4 AciI               gc'gg                                    872      158     105
     5 AciI               gc'gg                                    884       12     158
     6 AciI               cc'gc                                    896       12     161
                                                                            105     502
Matches found=     3
        Name              Sequence                           Position Fragment length
     1 AcyI               gg'cgtc                                  698      697       5
     2 AcyI               gg'cgtc                                  765       67      67
     3 AcyI               ga'cgcc                                  996      231     231
```

Fig. 1. Typical output from "List enzyme by enzyme."

matching sequence with the cut site indicated by a ' symbol. The position of the cut site is given followed by the length of the potential fragment ending at that site, followed by a list of fragment sizes sorted on length.

2.2. Search for Restriction Enzyme Sites and List Them by Position

1. Select "Search."
2. Select "Input source" as "All enzymes file."
3. Accept "Search for all names."
4. Select "Order results by position."
5. Accept "List matches."
6. Accept "The sequence is linear."
7. Accept "Search for definite matches."

The results will then appear in the form shown in Fig. 2. Each match is numbered, and its enzyme name given, followed by the matching sequence with the cut site indicated by a ' symbol. The position of the cut site is given followed by the length of the potential fragment ending at that site.

	Name	Sequence	Position	Fragment length
1	HapII	c'cgg	2	1
2	HpaII	c'cgg	2	0
3	MspI	c'cgg	2	0
4	MseI	t'taa	14	12
5	HincII	gtt'aac	15	1
6	HindII	gtt'aac	15	0
7	HpaI	gtt'aac	15	0
8	DsaV	'ccagg	23	8
9	EcoRII	'ccagg	23	0
10	TspAI	'ccagg	23	0
11	ApyI	cc'agg	25	2
12	BstNI	cc'agg	25	0
13	MvaI	cc'agg	25	0
14	ScrFI	cc'agg	25	0
15	MaeIII	'gttac	47	22
16	BsrI	actggt'	49	2
17	MseI	t'taa	55	6
18	MaeII	a'cgt	63	8
19	SfaNI	gcatcaacaa'gata	86	23
20	MaeII	a'cgt	91	5

Fig. 2. Typical output from "List by position."

2.3. Search for Restriction Enzyme Sites and List Their Names Above the Sequence

1. Select "Search."
2. Select "Input source" as "All enzymes file."
3. Accept "Search for all names."
4. Select "Show names above the sequence."
5. Reject "Hide translation."
6. Accept "Use 1 letter codes."
7. Define "Line length." This is the number of bases that will appear on each line of output. It must be a multiple of 30.
8. Accept "The sequence is linear."
9. Accept "Search for definite matches."

The results will then appear in the form shown in Fig. 3. The sequence is listed with a three-phase translation underneath and every tenth base numbered. Above the sequence the positions of the cut sites of restriction enzymes are marked.

```
Search for restriction enzyme sites
 Select operation
X  1 Search
   2 List enzyme file
   3 Clear text
   4 Clear graphics
 ? Selection  (1-4) (1) =
Select input source
   1 All enzymes file
X  2 Six cutter file
   3 Four cutter file
   4 Personal file
   5 Keyboard
 ? Selection  (1-5) (2) =1
 ? Search for all names (y/n) (y) =
 Select results display mode
X  1 Order results enzyme by enzyme
   2 Order results by position
   3 Show only infrequent cutters
   4 Show names above the sequence
 ? Selection  (1-4) (1) =4
 ? Hide translation (y/n) (y) =n
  ? Use 1 letter codes (y/n) (y) =
  ? Line length (30-90) (60) =
 ? The sequence is linear (y/n) (y) =
  ? Search for definite matches (y/n) (y) =


HapII
HpaII
MspI         MseI
  .          .HincII
  .          .HindII
  .          .HpaI     DsaV
  .          ..        EcoRII
  .          ..        TspAI
  .          ..        . ApyI
  .          ..        . BstNI
  .          ..        . MvaI
  .          ..        . ScrFI                MaeIII
  .          ..        . .                    BsrI  MseI
ccggttagactgttaacaacaaccaggttttctactgatataactggttacatttaacgc
       10        20        30        40        50        60
   P  V  R  L  L  T  T  T  R  F  S  T  D  I  T  G  Y  I  *  R
    R  L  D  C  *  Q  Q  P  G  F  L  L  I  *  L  V  T  F  N  A
     G  *  T  V  N  N  N  Q  V  F  Y  *  Y  N  W  L  H  L  T  P
```

Fig. 3. Typical dialog and output for a "Names above the sequence" search.

2.4. Search for Restriction Enzyme Sites and Plot Their Positions

1. Select "Search."
2. Select "Input source" as "All enzymes file."
3. Accept "Search for all names."
4. Select "Order results by position."
5. Reject "List matches."
6. Accept "The sequence is linear."
7. Accept "Search for definite matches."

The results will then appear in the form shown in Fig. 4. Each enzyme that has a match is named at the left edge of the display, and its cut sites are marked by short vertical lines. If the display window fills up, the bell will ring. Users may then take a screen dump before typing return. The program then displays the message "? Restart plotting from bottom of frame." To do so, type return. To quit, type !

2.5. Finding Restriction Enzymes That Cut Infrequently

1. Select "Search."
2. Select "Input source" as "All enzymes file."
3. Accept "Search for all names."
4. Select "Show only infrequent cutters."
5. Define "Maximum number of cuts."
6. Accept "The sequence is linear."
7. Accept "Search for definite matches."

The names and number of cut sites of all enzymes that cut less than or equal to the "Maximum number of cuts" will then be displayed.

2.6. Producing a Back-Translation From a Protein Sequence

The routine for producing back-translations is contained in the program PIP. It back-translates protein sequences into DNA using the standard genetic code. The translation can use either the IUB symbols or a set of codon preferences. If a set of codon preferences is used, they must conform to the format of codon tables produced by the nucleotide interpretation program, and the back-translation will

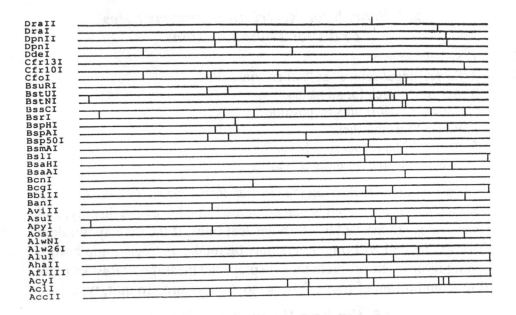

Fig. 4. Typical output from "Plot positions."

contain the favored codons. If for any amino acid there is no favored codon, the IUB symbols will be employed. The program will plot the redundancy along the sequence and, hence, can be used to find the best sequences to use as primers. The DNA sequence can be saved to a file and analyzed using the nucleotide analysis program.

1. Select "Back-translate."
2. Accept "No codon preference." The alternative will cause the program to ask for "File name of codon table," which should be in the same format as those created by the nucleotide interpretation program.
3. Reject "Plot redundancy." The alternative will ask for a window length to use for the plot. The window length is in codons. A plot will appear in which the best primers are sited at the peaks and the worst at the troughs.
4. Accept "Save DNA to disk."
5. Define "File name for DNA sequence." This file can later be read into program NIP, and all the searches described above employed.

3. Notes

1. The file containing the definitions of the restriction enzymes names and their recognition sequences uses the standard IUB redundancy symbols and has the following format. Each name is followed by a /, and then each of its recognition sequences is followed by a /. The last recognition sequence for each enzyme is followed by //. The cut sites should be indicated by a '. If the cut site is not contained in the recognition sequence, the recognition sequence should be extended by sufficent N symbols. For example, the two lines from the standard file shown below define the enzymes Alw21I and Alw26I. These files are kindly updated each month by Rich Roberts.

Alw21I/GWGCW'C//

Alw26I/GTCTCN'NNNN/'NNNNNGATCC//

2. To search for a subset of the restriction enzymes in a file, the user should reject "Search for all names," and the program will ask for the names of the enzymes wanted and extract their recognition sequences from the file. Alternatively, if a user was always using the same subset, then a file containing only those enzymes could be created by editing the standard file. This file would then be selected as "Personal file" for "Input source."

3. The routine also allows names and recognition sequence to be entered on the keyboard. This is selected as "Keyboard" for "Input source," and the program will prompt for names and their recognition sequences. In this way, the routine can be used to search for exact matches to any short sequence. Again, IUB redundancy codes can be used.

4. When back-translating from proteins, it is often useful to produce a back-translation using both a table of codon preferences and one using the IUB symbols. This is because the restriction enzyme search program can distinguish between definite and possible cuts in the sequence. Those matches that the program terms "definite matches" are ones in which the specification of the recognition sequence corresponds exactly to that of the back-translation. The program will also find what it terms "possible matches," which are ones that depend on the particular codons chosen for each amino acid. These are sites at which recognition sequences could be engineered to produce a cut in the DNA without changing the amino acid, but which are not necessarily found in the original sequence.

CHAPTER 7

Staden: Translating and Listing Nucleic Acid Sequences

Rodger Staden

1. Introduction

This chapter deals with producing simple listings from nucleotide sequences. All functions are contained in the program NIP. Sequences can be listed alone, in single- or double-stranded format, or with translations to protein. The translations can be of all six phases, all open reading frames, or of specified segments. The positions of these segments can be defined on the keyboard or read from a EMBL/GenBank feature table. Translations can use the one-letter or three-letter codes. In addition, files can be produced containing only the protein translations and which are suitable for processing by other programs. Again, the positions of the translated segments can be defined on the keyboard, read from a feature table, or be all open reading frames. For the user, producing all these results is very simple, so only examples of "methods" are given and what the results look like is shown. All outputs that list the sequence can be produced from the menu option named "Translate and list."

2. Methods

2.1. Listing the Sequence with All Six Reading Frames Translated

1. Select "Translate and list."
2. Accept "Show translation."
3. Select "The segments to translate will be All six frames."
4. Accept "Use 1 letter codes."

From: *Methods in Molecular Biology, Vol. 25: Computer Analysis of Sequence Data, Part II*
Edited by: A. M. Griffin and H. G. Griffin Copyright ©1994 Humana Press Inc., Totowa, NJ

5. Define "Start." This is where to list from.
6. Define "End." This is where to list to.
7. Define "Line length." This is the number of characters in each line of output.
8. Reject "Number ends of lines." This alternative writes the positions underneath each line.

The listing will then appear. Given the choices taken, it will look the same as Fig. 1.

2.2. Listing the Sequence with Its Open Reading Frames Translated

1. Select "Translate and list."
2. Accept "Show translation."
3. Select "The segments to translate will be Open reading frames."
4. Define "Minimum open frame in amino acids."
5. Accept "Use 1 letter codes."
6. Define "Start." This is where to list from.
7. Define "End." This is where to list to.
8. Define "Line length." This is the number of characters in each line of output.
9. Select "Both strands."
10. Accept "Number ends of lines."

A typical result is shown in Fig. 2.

2.3. Listing the Sequence with Defined Segments Translated

1. Select "Translate and list."
2. Accept "Show translation."
3. Select "The segments to translate will be Typed on the keyboard."
4. Accept "Use 1 letter codes."
5. Define "Start." This is where to list from.
6. Define "End." This is where to list to.
7. Define "Line length." This is the number of characters in each line of output.
8. Select "Both strands."
9. Accept "Number ends of lines."
10. Define "Translate from." Define the start of the next segment to translate—say, the next exon.
11. Define "Translate to." Define the end of the next segment to translate.

```
Q  D Y  I  G  H H  H  L N  N  L Q  L  D  L  R  T  F  S  L
   R  I  T  *  D  T  T  *  I  T  F  S  W  T  C  V  H  S  R  W
   G  L  H  R  T  P  P  E  *  P  S  A  G  P  A  Y  I  L  A
caggattacataggacaccacctgaataaccttcagctggacctgcgtacattctcgctg
    1010      1020      1030      1040      1050      1060
gtcctaatgtatcctgtggtggacttattggaagtcgacctggacgcatgtaagagcgac
   L  I  V  Y  S  V  V  Q  I  V  K  L  Q  V  Q  T  C  E  R  Q
   P  N  C  L  V  G  G  S  Y  G  E  A  P  G  A  Y  M  R  A  P
   S  *  M  P  C  W  R  F  L  R  *  S  S  R  R  V  N  E  S

V  D  P  Q  N  P  P  A  T  F  W  T  I  N  I  D  S  M  F  F
   W  I  H  K  T  P  Q  P  P  S  G  Q  S  I  L  T  P  C  S  S
   G  G  S  T  K  P  P  S  H  L  L  D  N  Q  Y  *  L  H  V  L
gtggatccacaaaaccccccagccaccttctggacaatcaatattgactccatgttcttc
    1070      1080      1090      1100      1110      1120
cacctaggtgttttgggggggtcggtggaagacctgttagttataactgaggtacaagaag
   H  I  W  L  V  G  W  G  G  E  P  C  D  I  N  V  G  H  E  E
   P  D  V  F  G  G  L  W  R  R  S  L  *  Y  Q  S  W  T  R  R
   T  S  G  C  F  G  G  A  V  K  Q  V  I  L  I  S  E  M  N  K

S  V  V  L  G  L  L  F  L  V  L  F  R  S  V  A  K  K  A  T
   R  W  C  W  V  C  C  S  W  F  Y  S  V  A  *  P  K  R  R  P
   L  G  G  A  G  S  V  V  P  G  F  I  P  *  R  S  Q  K  G  D
tcggtggtgctgggtctgttgttcctggtttattccgtagcgtagccaaaaaggcgacc
    1130      1140      1150      1160      1170      1180
agccaccacgacccagacaacaaggaccaaaataaggcatcgcatcggttttttccgctgg
   R  H  H  Q  T  Q  Q  E  Q  N  *  E  T  A  Y  G  F  L  R  G
   P  P  A  P  D  T  T  G  P  K  I  G  Y  R  L  W  F  P  S  W
   E  T  T  S  P  R  N  N  R  T  K  N  R  L  T  A  L  F  A  V

S  G  V  P  G  K  F  Q  T  A  I  E  L  V  I  G  F  V  N  G
   A  V  C  Q  V  S  F  R  P  R  L  S  W  *  S  A  L  L  M  V
   Q  R  C  A  R  *  V  S  D  R  D  *  A  G  D  R  L  C  *  W
agcggtgtgccaggtaagtttcagaccgcgattgagctggtgatcggctttgttaatggt
    1190      1200      1210      1220      1230      1240
tcgccacacggtccattcaaagtctggcgctaactcgaccactagccgaaacaattacca
   A  T  H  W  T  L  K  L  G  R  N  L  Q  H  D  A  K  N  I  T
   R  H  A  L  Y  T  E  S  R  S  Q  A  P  S  R  S  Q  *  H  Y
   L  P  T  G  P  L  N  *  V  A  I  S  S  T  I  P  K  T  L  P
```

Fig. 1. A six phase translation using the one-letter codes.

12. Select "Strand." Since both strands have been selected above, the program will allow either to be translated for each defined segment.

The program will now cycle around through steps 10, 11, and 12 until a zero value is defined for "Translate from," at which point the listing will appear. Given the choices made, it will look the same as Fig. 2.

```
       Q  D  Y  I  G  H  H  L  N  N  L  Q  L  D  L  R  T  F  S  L
      caggattacataggacaccacccgaataaccttcagctggacctgcgtacattctcgctg      1060
         .    :    .    :    .    :    .    :    .    :    .    :
      gtcctaatgtatcctgtggtggacttattggaagtcgacctggacgcatgtaagagcgac
       L  I  V  Y  S  V  V  Q  I  V  K  L  Q  V  Q  T  C  E  R  Q
                                  *  S  S  R  R  V  N  E  S


       V  D  P  Q  N  P  P  A  T  F  W  T  I  N  I  D  S  M  F  F
      gtggatccacaaaacccccagccaccttctggacaatcaatattgactccatgttcttc      1120
         .    :    .    :    .    :    .    :    .    :    .    :
      cacctaggtgttttggggggtcggtggaagacctgttagttataactgaggtacaagaag
       H  I  W  L  V  G  W  G  G  E  P  C  D  I  N  V  G  H  E  E
       T  S  G  C  F  G  G  A  V  K  Q  V  I  L  I  S  E  M  N  K


       S  V  V  L  G  L  L  F  L  V  L  F  R  S  V  A  K  K  A  T
      tcggtggtgctgggtctgttgttcctggttttattccgtagcgtagccaaaaaggcgacc      1180
         .    :    .    :    .    :    .    :    .    :    .    :
      agccaccacgacccagacaacaaggaccaaaataaggcatcgcatcggttttttccgctgg
       R  H  H  Q  T  Q  Q  E  Q  N  *  E  T  A  Y  G  F  L  R  G
       E  T  T  S  P  R  N  N  R  T  K  N  R  L  T  A  L  F  A  V


       S  G  V  P  G  K  F  Q  T  A  I  E  L  V  I  G  F  V  N  G
      agcggtgtgccaggtaagtttcagaccgcgattgagctggtgatcggctttgttaatggt      1240
         .    :    .    :    .    :    .    :    .    :    .    :
      tcgccacacggtccattcaaagtctggcgctaactcgaccactagccgaaacaattacca
       A  T  H  W  T  L  K  L  G  R  N  L  Q  H  D  A  K  N  I  T
       L  P  T  G  P  L  N  *  V  A  I  S  S  T  I  P  K  T  L  P


       S  V  K  D  M  Y  H  G  K  S  K  L  I  A  P  L  A  L  T  I
      agcgtgaaagacatgtaccatggcaaaagcaagctgattgctccgctggccctgacgatc      1300
         .    :    .    :    .    :    .    :    .    :    .    :
      tcgcacttтctgtacatggtaccgtttтcgttcgactaacgaggcgaccgggactgctag
       A  H  F  V  H  V  M  A  F  A  L  Q  N  S  R  Q  G  G  Q  R  D
       L  T  F  S  M  Y  W  P  L  L  L  S  I  A  G  S  A  R  V  I
```

Fig. 2. A listing showing the translation of open reading frames from both strands of a sequence from position 1001 to 1300.

2.4. Listing the Sequence with Translated Segments Defined From a Feature Table

1. Select "Translate and list."
2. Accept "Show translation."
3. Select "The segments to translate will be Read from a feature table."
4. Define "Feature table file name." Type the name of the file containing the appropriate feature table in EMBL/GenBank format.
5. Define "Operator." This defines which feature table operators should be employed when selecting the segments to translate.
6. Accept "Use 1 letter codes."
7. Define "Start." This is where to list from.

8. Define "End." This is where to list to.
9. Define "Line length." This is the number of characters in each line of output.
10. Select "Both strands."
11. Accept "Number ends of lines."

The program will now read the feature table file and translate the segments defined using the selected operator(s). The listing will appear as in Fig. 2.

2.5. Producing a File of Protein Sequences for All Open Reading Frames

1. Select "Translate and write protein sequences to disk."
2. Reject "Translate selected regions." The alternative is "Open reading frames."
3. Define "Minimum open frame in amino acids."
4. Select "Both strands."
5. Define "File name for translation."

A typical results file is shown in Fig. 3. It shows that the file is written in PIR format, i.e., an entry name line starting with a > symbol (here the first entry name is 1052, the start of the DNA segment), followed by a one-line title (here in EMBL feature table format giving the start and end of the DNA that produced the protein), followed by the sequence terminated by an *. *See* Note 6.

2.6. Producing a File of Protein Sequences for Segments Defined From a Feature Table

1. Select "Translate and write protein sequences to disk."
2. Accept "Translate selected regions."
3. Reject "Define segments using keyboard." The alternative is to use a feature table.
4. Define "Feature table file name." Type the name of the file containing the appropriate feature table in EMBL/GenBank format.
5. Define "Operator." This defines which feature table operators should be employed when selecting the segments to translate.
6. Define "File name for translation."

The program will now read the feature table file and translate the segments defined using the selected operator(s).

3. Notes

1. To produce a listing without translation, the "Translate and list" function can be used with the "Show translation" option rejected. Alternatively, the function "List the sequence" can be used.

```
>P1;    956
FT    CDS              956..1789
QQRVKGIMASENMTPQDYIGHHLNNLQLDLRTFSLVDPQNPPATFWTINIDSMFFSVVLG
LLFLVLFRSVAKKATSGVPGKFQTAIELVIGFVNGSVKDMYHGKSKLIAPLALTIFVWVF
LMNLMDLLPIDLLPYIAEHVLGLPALRVVPSADVNVTLSMALGVFILILFYSIKMKGIGG
FTKELTLQPFNHWAFIPVNLILEGVSLLSKPVSLGLRLFGNMYAGELIFILIAGLLPWWS
QWILNVPWAIFHILIITLQAFIFMVLTIVYLSMASEEH*
>P1;    3159
FT    CDS              3159..4709
GDWSMQLNSTEISELIKQRIAQFNVVSEAHNEGTIVSVSDGVIRIHGLADCMQGEMISLP
GNRYAIALNLERDSVGAVVMGPYADLAEGMKVKCTGRILEVPVGRGLLGRVVNTLGAPID
GKGPLDHDGFSAVEAIAPGVIERQSVDQPVQTGYKAVDSMIPIGRGQRELIIGDRQTGKT
ALAIDAIINQRDSGIKCIYVAIGQKASTISNVVRKLEEHGALANTIVVVATASESAALQY
LARMPVALMGEYFRDRGEDALIIYDDLSKQAVAYRQISLLLRRPPGREAFPGDVFYLHSR
LLERAARVNAEYVEAFTKGEVKGKTGSLTALPIIETQAGDVSAFVPTNVISITDGQIFLE
TNLFNAGIRPAVNPGISVSRVGGAAQTKIMKKLSGGIRTALAQYRELAAFSQFASDLDDA
TRKQLDHGQKVTELLKQKQYAPMSVAQQSLVLFAAERGYLADVELSKIGSFEAALLAYVD
RDHAPLMQEINQTGGYNDEIEGKLKGILDSFKATQSW*
>P1;    2250
FT    CDS              complement(2250..3023)
QTFFHRSRNFRELLFTQCGSGNDVYLSGSLTHGTQVNKLLQNIRERVKTTIFSHNPNQVL
TVFVQLLTTNCDKRLGERFWRKRAREKLCHLFVFGYLGGKRQHVLPAFYTLVFDGKVKSC
FGVGASYRNKFRHQPLPPYSSATSLSTMSLLAASSTERSMIFSAPATARIATCLRSSSRA
RLRSASISACAWATILVRSCSASAFASSRICERRLFACSMITWASAFAFFSWSVALAFAR
SRSLCARSAEARPSAISF*
```

Fig. 3. The contents of a file containing the protein sequences of the open reading frames found by the program (*see* Note 6).

2. Some users may be confused by the fact that the program asks "Where to list from, and to" and also "Define segments to translate." This allows for 5' and 3' untranslated regions to be included in the listing.

3. The feature table file employed by the programs is a simple text file containing the data for the current sequence. Because of the multiplicity of different sequence library formats, the facility of reading such data directly from libraries has not been provided. The feature tables for individual library entries must be extracted (*see* Chapter 2), or files can be created for new sequences.

4. The current feature tables use "operators," such as "join" or "order," to specify which segments should be translated together to make a complete protein sequence. The program allows users to select which ones to employ, the default being "Use all operators."

5. The program contains a function "Set genetic code," which allows users to choose from a menu of codes, or to define their own by specifying amino acid and codon pairs. This sets the code for all functions.

6. Note added in proof: The file of translated protein sequences is now written in FASTA format.

CHAPTER 8

Staden: Searching for Motifs in Nucleic Acid Sequences

Rodger Staden

1. Introduction

The program NIP contains several ways of defining and searching for motifs *(1–4)*, and also contains a number of "hardwired" motifs that are already defined and can be selected as separate searches. Searches for percentage matches to consensus sequences, the use of score matrices, and the creation and use of nucleotide and dinucleotide weight matrices (*see* Note 7) are described here. In addition, details are given of the "hardwired" motifs available from the program. Chapter 6 has covered searches for exact matches to consensus sequences by describing how to find restriction enzyme recognition sequences. When searching for exact matches, percentage matches, or using a score matrix the search string or consensus sequence may include IUB redundancy codes. All of the searches produce both listed and graphical output. The listed output displays the matching sequence and its position, and the graphical output draws a box to represent the length of the sequence and plots vertical lines within the box at the positions of matches. The heights of the lines are proportional to the match score (*see* Fig. 1).

2. Methods

2.1. Searching for Percentage Matches to Consensus Sequences

1. Select "Find percentage matches."
2. Accept "Type in strings." The alternative allows the string to be extracted from a named file.

From: *Methods in Molecular Biology, Vol. 25: Computer Analysis of Sequence Data, Part II*
Edited by: A. M. Griffin and H. G. Griffin Copyright ©1994 Humana Press Inc., Totowa, NJ

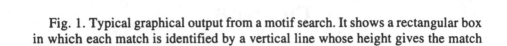

Fig. 1. Typical graphical output from a motif search. It shows a rectangular box in which each match is identified by a vertical line whose height gives the match score and whose *x* coordinate indicates the position in the sequence.

3. Reject "Keep picture." This will cause the graphics window to be cleared. The alternative leaves it unchanged.
4. Define "String." Type in the search string. When the program cycles around to this point again, the previous string will be offered as a default.
5. Accept "This sense." The alternative directs the program to search for the complement of the string.
6. Define "Percent match." The search is performed, the results are presented graphically (*see* Fig. 1), the number of matches displayed, and the scores and positions of the top 10 matches displayed.
7. Define the number of matches to "Display." For the number of matches chosen, the program will display the search string and matching sequence written one above the other with matching characters indicated by asterisk symbols. The program now cycles around to step 3. *See* Fig. 2.

2.2. Searching for Consensus Sequences Using a Score Matrix

A score matrix gives a score for the alignment of each possible pair of sequence symbols. The matrix used by this program includes all the IUB redundancy codes and gives scores that represent the level of redundancy. The matrix is shown in Fig. 3.

1. Select "Find matches using a score matrix."
2. Accept "Type in strings." The alternative allows the string to be extracted from a named file.
3. Reject "Keep picture." This will cause the graphics window to be cleared. The alternative leaves it unchanged.
4. Define "String." Type in the search string. When the program cycles around to this point again, the previous string will be offered as a default.
5. Accept "This sense." The alternative directs the program to search for the complement of the string. The program displays the maximum possible score for the string.
6. Define "Score." The search is performed, the results are presented graphically (*see* Fig. 1), the number of matches is displayed, and the scores and positions of the top 10 matches are displayed.
7. Define the number of matches to "Display." For the number of matches chosen, the program will display the search string and matching sequence

```
Find percentage matches
? Type in string (y/n) (y) =
 ? Keep picture (y/n) (y) =
 ? String=AAAATTTT
STRING=AAAATTTT
? This sense (y/n) (y) =
 ? Percent match (1.00-100.00) (70.00) =

Total scoring positions above 70.000 percent =  41
Scores          7      7      7       7      6      6      6      6      6      6
Positions     428    534   2994    7026    130    191    192    372    427    429
? Display (0-41) (0) =4

    428
      aaaatatt
      ***** **
      AAAATTTT
      1

    534
      aaagtttt
      *** ****
      AAAATTTT
      1

   2994
      aaaatttc
      *******
      AAAATTTT
      1

   7026
      aaaactttt
      **** ***
      AAAATTTT
      1
```

Fig. 2. Worked example for the percentage match search.

written one above the other with matching characters indicated by asterisk symbols. The program now cycles around to step 3. The dialog shown in Fig. 2 is almost exactly the same as that for "Searching for consensus sequences using a score matrix."

2.3. Using Weight Matrices for Searching Nucleotide Sequences

A weight matrix is the most sensitive way of defining a motif. It is a table of values that gives scores for each base type in each position along a motif. For a motif of length eight bases the weight matrix would be a table eight positions long and four deep. The simplest way of choosing the values for the table is to take an alignment of all

	T	C	A	G	-	R	Y	W	S	M	K	H	B	V	D	N	?
T	36	0	0	0	9	0	18	18	0	0	18	12	12	0	12	9	0
C	0	36	0	0	9	0	18	0	18	18	0	12	12	12	0	9	0
A	0	0	36	0	9	18	0	18	0	18	0	12	0	12	12	9	0
G	0	0	0	36	9	18	0	0	18	0	18	0	12	12	12	9	0
-	9	9	9	9	36	18	18	18	18	18	18	27	27	27	27	36	0
R	0	0	18	18	18	36	0	9	9	9	9	6	6	12	12	18	0
Y	18	18	0	0	18	0	36	9	9	9	9	12	12	6	6	18	0
W	18	0	18	0	18	9	9	36	0	9	9	12	6	6	12	18	0
S	0	18	0	18	18	9	9	0	36	9	9	6	12	12	6	18	0
M	0	18	18	0	18	9	9	9	9	36	0	12	6	12	6	18	0
K	18	0	0	18	18	9	9	9	9	0	36	6	12	6	12	18	0
H	12	12	12	0	27	6	12	12	6	12	6	36	8	8	8	27	0
B	12	12	0	12	27	6	12	6	12	6	12	8	36	8	8	27	0
V	0	12	12	12	27	12	6	6	12	12	6	8	8	36	8	27	0
D	12	0	12	12	27	12	6	12	6	6	12	8	8	8	36	27	0
N	9	9	9	9	36	18	18	18	18	18	18	27	27	27	27	36	0
?	0	0	0	0	0	0	0	0	0	0	0	0	0	0	0	0	0

Fig. 3. The DNA score matrix using IUB symbols.

known examples of the motif and to count the frequency of occurrence of each base type at each position. These frequencies can be used as the table of weights. When the table is used to search a new sequence, the program calculates a score for each position along the sequence by adding, or multiplying (*see* Note 6) the relevant values in the table. All positions that exceed some cutoff score are reported as matching the original set of motifs.

How can we select a suitable cutoff score? The simplest way is to apply the weight matrix to all the known occurrences of the motif, i.e., the set of sequence segments used to create the table, and to see what scores they achieve. The cutoff can be selected accordingly. For convenience the weight matrix is stored as a file along with its cutoff score, a title that is displayed when the file is read, and a few other values needed by the program. A routine for creating weight matrix files from sets of aligned sequences is included in the program. When a search using the weight matrix is performed, the program will either list the matching sequence segments or plot their positions as for the other motif search methods.

2.3.1. Creating a Weight Matrix File
from a Set of Aligned Sequences

1. Select "Motif search using weight matrix."
2. Select "Make weight matrix."
3. Define "Name of aligned sequences file." Assume the file of aligned sequences has already been created (*see* Note 3). The program reads and displays the contents of the file, numbering each sequence as it goes. Then it displays the length of the longest sequence.
4. Accept "Sum logs of weights." The alternative is to sum the weights when calculating scores (*see* Note 4).
5. Accept "Use all motif positions." The alternative allows the user to define a "mask," which identifies positions within the motif that should be ignored when the matrix is created (*see* Note 5). The program now calculates the weights and applies them in turn to each of the sequences in the file. The number and score for each sequence are displayed, followed by the top, bottom, and mean scores and the standard deviation. In addition, the mean ± 3 SD is displayed.
6. Define "Cutoff score." The default is the mean − 3 SD, but users may, for example, decide to use the lowest score obtained by the sequences in the file.
7. Define "Top score for scaling plots." This parameter is used by the graphics output routine when scaling the plots. Its value will influence the height of lines plotted to represent matches.
8. Define "Position to identify." When a search is performed, it is not always appropriate to report the position of a match relative to the leftmost base in the motif. For example, when performing a splice junction search, we may want to know the position of the G in the conserved GT, rather than the position of the first base in the matrix. The "Position to identify" allows the user to define which base is marked. The bases in the table are number 1,2,3, and so on.
9. Define a "Title." This is a title that will be displayed when the matrix file is read prior to performing a search. It is limited to 60 characters.
10. Define "Name for new weight matrix file." Give a name for the weight matrix file. Typical dialog is shown in Fig. 4.

2.3.2. Searching Using a Weight Matrix

Once a weight matrix has been stored in a file, it can be used to search any sequence. Results can be displayed graphically, or the matching sequence segments can be listed out with their scores.

```
Motif search using weight matrix
Select operation
X  1 Use weight matrix
   2 Make weight matrix
   3 Rescale weight matrix
? Selection  (1-3) (1) =2
? Name of aligned sequences file=heatshock.seq
        1 ATAAAGAATATTCTAGAA
        2 CTCGAGAAATTTCTCTGG 144
        3 TTCTCGTTGCTTCGAGAG  36
        4 GCCTCGAATGTTCGCGAA  15
        5 GACTGGAATGTTCTGACC  45 DROSOPHILA HSP68
        6 ATCTCGAATTTTCCCCTC  12
        7 ATCCAGAAGCCTCYAGAA  35 DROSOPHILA HSP83
        8 CTCTAGAAGTTTCTAGAG  25
        9 TTCTAGAGACTTCCAGTT  15
       10 CCCCAGAAACTTCCACGG 147 DROSOPHILA HSP22
       11 GCGAAGAAAATTCGAGAG  46
       12 TGCCGGTATTTTCTAGAT  26
       13 CCCGAGAAGTTTCGTGTC  97 DROSOPHILA HSP23
       14 TTCCGGACTCTTCTAGAA  13 DROSOPHILA HSP26
       15 CTCGAGAAAGCTCGCGAA 204 XENOPUS HSP70
       16 CTCGCGAATCTTCCGCGA 194
       17 CTCGCGAAAGTTCTTCGG 139
       18 CTCGGGAAACTTCGGGTC  72
       19 TGCCAGAAGTTGCTAGCA 124 XENOPUS HSP30
       20 CTCGGGAACGTCCCAGAA  14
       21 ATCCCGAAACTTCTAGTT 129 SOYBEAN HSP17
       22 GTCCAGAATGTTTCTGAA  98
       23 TTTCAGAAAATTCTAGTT  78
       24 CCCAAGGACTTTCTCGAA  28
       25 TTTTAGAATGTTCTAGAA 179 DICTYOSTELIUM DIRS-1
       26 TTCTAGAACATTCGAAGA 169
Length of motif    18
? Sum logs of weights (y/n) (y) =
? Use all motif positions (y/n) (y) =
Applying matrix to input sequences
    1        -15.609 ATAAAGAATATTCTAGAA
    2        -15.965 CTCGAGAAATTTCTCTGG
```

Fig. 4. An example run of creating a weight matrix.

1. Select "Motif search using weight matrix."
2. Select "Use weight matrix."
3. Define "Motif weight matrix file." This is the name of the file containing the weight matrix. The program reads the file and displays its title.
4. Define "Cutoff score." The default will be the value set when the weight matrix file was created. If the score is negative, the program will calculate sums of logs of frequencies; otherwise, it will add frequencies.
5. Accept "Plot results." Alternatively, they will be listed. The results will appear as in Fig. 5.

```
 3     -18.186  TTCTCGTTGCTTCGAGAG
 4     -15.331  GCCTCGAATGTTCGCGAA
 5     -20.897  GACTGGAATGTTCTGACC
 6     -17.347  ATCTCGAATTTTCCCCTC
 7     -16.271  ATCCAGAAGCCTCYAGAA
 8     -12.227  CTCTAGAAGTTTCTAGAG
 9     -15.933  TTCTAGAGACTTCCAGTT
10     -15.604  CCCCAGAAACTTCCACGG
11     -17.866  GCGAAGAAAATTCGAGAG
12     -17.159  TGCCGGTATTTTCTAGAT
13     -16.399  CCCGAGAAGTTTCGTGTC
14     -14.646  TTCCGGACTCTTCTAGAA
15     -14.801  CTCGAGAAAGCTCGCGAA
16     -16.163  CTCGCGAATCTTCCGCGA
17     -16.280  CTCGCGAAAGTTCTTCGG
18     -15.598  CTCGGGAAACTTCGGGTC
19     -17.721  TGCCAGAAGTTGCTAGCA
20     -16.257  CTCGGGAACGTCCCAGAA
21     -14.243  ATCCCGAAACTTCTAGTT
22     -16.456  GTCCAGAATGTTTCTGAA
23     -15.453  TTTCAGAAAATTCTAGTT
24     -17.443  CCCAAGGACTTTCTCGAA
25     -13.335  TTTTAGAATGTTCTAGAA
26     -15.914  TTCTAGAACATTCGAAGA
Top score    -12.227  Bottom score     -20.897
Mean      -16.119  Standard deviation      1.636
Mean minus 3.sd     -21.028  Mean plus 3.sd     -11.210
? Cutoff score (-999.00-9999.00) (-21.03) =
? Top score for scaling plots (-21.03-999.00) (-11.21) =
? Position to identify (0-18) (1) =
? Title=Heatshock weights 24-10-91
? Name for new weight matrix file=heatshock.wts
```

Fig. 4 *(continued)*.

2.4. Using "Hardwired" Motif Searches

The program contains predefined motif definitions for the following:

E. coli promoters;
Prokaryotic ribosome binding sites;
mRNA splice junctions
Eukaryotic ribosome binding sites; and
Polyadenylation sites.

All except the polyadenylation site, which is simply defined as an exact match to the string AATAAA, are represented as weight matrices. Each search is performed simply by the user selecting the appropriate option from the menu, and each plots its results in its own

```
Motif search using weight matrix
Select operation
X  1 Use weight matrix
   2 Make weight matrix
   3 Rescale weight matrix
? Selection  (1-3) (1) =
? Motif weight matrix file=heatshock.wts
 Heatshock weights 24-10-91
? Cutoff score (-9999.00-9999.00) (-21.03) =
? Plot results (y/n) (y) =


     619    -20.84 gctcggaagcttctgctc
     818    -20.74 ttggcgaagctttcaaag
    1190    -21.02 gccaggtaagtttcagac
    1601    -20.91 tttgcgactgttcggtaa
    2387    -20.24 cgctcgcagattctggac
    2534    -20.87 gccgagaagatcatcgaa
    2890    -16.38 ctcccggatgttctggag
    2989    -19.54 ctcgcgaaaatttctgct
    3451    -20.76 atcctggaagttccggtt
    6020    -20.73 tctcaggaactgctggaa
    6335    -20.51 gctgagaaattccgtgac
    7107    -20.31 ctctggtctggtcgagaa
    7117    -19.61 gtcgagaaaatccaggta
    7892    -20.18 cttccgaaagtgctgcat
```

Fig. 5. An example run of a search using a weight matrix to produce text output.

graphics window. The ribosome binding site searches are reading-frame-specific, so they normally plot their results to fit nicely with the output from the "gene search by content" methods described in Chapter 10 on finding genes. Likewise, the splice junction searches produce separate output for each of the three reading frames. Below, as an example of using the hardwired motifs, is shown how to perform such a search.

2.4.1. Searching for Splice Junctions

1. Select "Splice search using weight matrix." The program automatically reads in weight matrices that define the donor and acceptor sites, and displays their titles.
2. Define "Donor cutoff score." The default is stored in the file.
3. Define "Acceptor cutoff score." The default is stored in the file.
4. Accept "Plot results." The alternative lists the results giving the position, score, matching sequence, and reading frame. A typical plotted result appears in Fig. 6.

3. Notes

1. For this program, a motif is a short segment of sequence of fixed length. More complex structures, termed "patterns," which are defined as sets of motifs separated by varying gaps, are covered in Chapter 9. The current chapter should be read before the chapter on patterns.

2. It is debatable whether the gain in sensitivity that is afforded by the use of a score matrix is of value for searching nucleotide sequences; however it is very important for protein sequences.

3. The files of aligned sequences used to make weight matrices have the following format. Each sequence should be on a separate line. The sequence should start in column 2, and is terminated by a new line or a space. Anything after the space is treated as a comment. The files can be created by previous searches or using an editor.

4. The frequencies in the weight matrix can be used in two ways to calculate scores for sequences. Some users prefer to add the frequencies to give a total score and others to multiply them by summing their logs. If we regard the frequencies as probabilities, then multiplication seems the correct procedure. The user chooses which method will be employed when the weight matrix is created; however, the choice can be overridden when the matrix is used. If multiplication is selected, then all results will be presented as sums of logs.

5. Masking the weight matrix is particularly useful in cases where a limited number of examples of a motif are available or when the motif may have several components. In the first case, the limited number of examples may make the matrix unrepresentative of the motif, because the bases in the unconserved positions may bias the results of searches. When a large number of examples is available to create the matrix, the unconserved positions should tend toward equal base composition and, hence, have no influence on the overall score. It was stated that a motif might have several components: For example, a motif might have both structural and specificity components. The user may want to separate out the two parts, and masking provides such a facility.

6. The weight matrix handling routine contains a further option, "Rescale weight matrix." If the user has edited a weight matrix to change the frequency values, this provides a way of selecting a new cutoff score. It allows users to read in a set of aligned sequences and a weight matrix, and to apply the matrix to the set of sequences to see the range of scores achieved. A new weight matrix file containing the selected cutoff score is written to disk.

7. The program also contains a set of routines identical to those used to create and search for nucleotide weight matrices, but that deal instead with dinucleotide weight matrices.

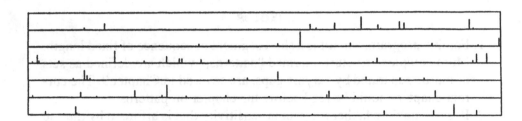

Fig. 6. Typical graphical output from using the hardwired splice junction search. The results are presented in a reading-frame-specific way, so it shows three sets of results for donor sites and three for acceptor sites.

8. The reader is reminded that most options in the program, if selected when in "execute without dialog" mode, will automatically use a set of defaults, and produce a result with little, or no user input. Most motif searches require far less user input than that shown above, where it was attempted to show the scope of the methods.

9. Although the program contains hardwired motifs, it is expected that most sites that use the programs will accumulate their own libraries of motifs and patterns, which users can employ by simply knowing the names of the corresponding files.

References

1. Staden, R. (1984) Computer methods to locate signals in nucleic acid sequences. *Nucleic Acids Res.* **12,** 521–538.
2. Staden, R. (1985) Computer methods to locate genes and signals in nucleic acid sequences, in *Genetic Engineering, Principle and Methods* vol. 7 (Setlow, J. K. and Hollaender, A., eds.), Plenum, New York, pp. 67–114.
3. Staden, R. (1988) Methods to define and locate patterns of motifs in sequences. *CABIOS* **4(1),** 53–60.
4. Staden, R. (1990) Searching for patterns in protein and nucleic acid sequences, in *Methods in Enzymology* vol. 183 (Doolittle, R. F., ed.), Academic, New York, pp. 193–211.

CHAPTER 9

Staden: Using Patterns to Analyze Nucleic Acid Sequences

Rodger Staden

1. Introduction

Here is described one of the most powerful facilities provided by the program NIP: the ability to define and search for complex patterns of motifs *(1–3)*. Chapter 8 gives details of searching for individual motifs, but this chapter shows how to create patterns and libraries of patterns, and to use them to search single sequences and sequence libraries. Once a pattern has been defined and stored in a file, it can be used to search any sequence. In addition, if users want to screen sequences routinely against libraries of patterns, this can be achieved by use of files of file names. The program can produce several alternative forms of output. It will display the segment of sequence matching each individual motif in the pattern, display all the sequence between and including the two outermost motifs, produce a description of the match in the form of an EMBL feature table, or draw a simple graphical plot.

The end of the chapter describes how a related program, NIPL, is used to search libraries of sequences to find patterns. NIPL is capable of producing alignments of sequence families.

Patterns are defined as sets of motifs with variable spacing. Each motif in a pattern can be defined using any of several methods, and their positions relative to one another are defined in terms of minimum and maximum separations. In addition, by the use of logical

From: *Methods in Molecular Biology, Vol. 25: Computer Analysis of Sequence Data, Part II*
Edited by: A. M. Griffin and H. G. Griffin Copyright ©1994 Humana Press Inc., Totowa, NJ

operators, each motif can be declared to be essential (the AND operator), optional (the OR operator), or forbidden (the NOT operator). The following methods (termed "classes" by the program) for defining motifs are provided:

1. Exact match to a short sequence;
2. Percentage match to a short sequence;
3. Match to a short sequence using a score matrix and cutoff score;
4. Match to a weight matrix;
5. Match to the complement of a weight matrix;
6. Inverted repeat or stem-loop;
7. Exact match to a short sequence with a defined step; and
8. Direct repeat.

Classes 1, 2, 3, and 7 permit the use of IUB redundancy codes.

The motifs in a pattern are numbered sequentially, and motif spacing is defined in the following way. When a new motif is added to a pattern, the user specifies the "Reference motif" by its number and then a "Relative start position." The "Relative start position" is defined by taking the first base of the "Reference motif" as position 1, the next as 2, and so on. Then the user defines the allowed variation in the spacing by specifying the "Number of extra positions." Notice that the position of a motif can be defined relative to any other motif and that a negative "Relative start position" declares the motif to be to the left of its "Reference motif."

The probability of finding each individual motif in the current sequence, the product of the probabilities for all the motifs in a pattern "Probability of finding pattern," and the "Expected number of matches" are calculated and displayed by the program. In addition, to the cutoffs used for the individual motifs, users can apply two pattern cutoffs: "Maximum pattern probability" and "Minimum pattern score."

Below is described: how to create a pattern; how to use a pattern file to search a sequence; and how to use a "File of pattern file names" to search a sequence for a whole library of patterns. In order to describe how to create a pattern file, first all the steps to make one containing two motifs are shown, and then, to save space, the parts specific to the individual motif types are sketched in the notes section.

2. Methods

2.1. Creating a Pattern File Containing an Exact Match Motif and Weight Matrix Motif

1. Select "Pattern searcher."
2. Select "Pattern definition mode" as "Use keyboard."
3. Select "Results display mode" as "Motif by motif." The alternatives are listed in Section 1.
4. Select "Motif definition mode" as "Exact match."
5. Define "Motif name." Each motif can be given an eight character name.
6. Define "String." Type in the sequence of the motif. The program will display the probability of finding the motif.
7. Select "Motif definition mode" as "Weight matrix."
8. Define "Motif name."
9. Select "Logical operator" as "AND." The alternatives are "OR" and "NOT."
10. Select "Number of reference motif." At this stage, the only choice is 1, and this is the default.
11. Define "Relative start position." This is the base position relative to the "Reference motif." *See* Section 1.
12. Define "Number of extra positions."
13. Define "Weight matrix file name." Type the name of the file containing the weight matrix. The program now cycles around to step 7, and all subsequent passes around the loop to add further motifs to the pattern would differ only in the details for the different motif "classes."
14. Select "Pattern complete."
15. Accept "Save pattern in a file." The alternative does not save the pattern, so it can only be used once on the current sequence.
16. Define "Pattern definition file." Give a name for the new file.
17. Define "Pattern title." All patterns can have a 60-character title that can be displayed when the pattern file is read and the sequence searched. The program will now display a detailed textual description of the pattern, the "Probability of finding the pattern," and the "Expected number of matches."
18. Define "Maximum pattern probability." Yes maximum: any match with a greater probability of being found will be rejected. If no value is specified, the search will be quicker (*see* Section 3).
19. Define "Minimum pattern score." A minimum pattern score only makes sense if all the motifs in the pattern are defined with compatible scoring

methods. For example, percentage matches and weight matrices using sums of logs are incompatible. Searching will now commence and any matches displayed using the chosen method. A worked example of creating such a pattern and performing a search is shown in Fig. 1, and the actual pattern file is shown in Fig. 2.

2.2. Searching a Sequence Using a Pattern File

1. Select "Pattern searcher."
2. Select "Pattern definition mode" as "Use pattern file."
3. Select "Results display mode" as "Inclusive."
4. Define "Pattern definition file." Type the name of the file containing the pattern. The program will read the file, then display its title, a detailed textual description of the pattern, the "Probability of finding the pattern," and the "Expected number of matches."
5. Define "Maximum pattern probability."
6. Define "Minimum pattern score." Searching will now commence, and any matches will be displayed using the chosen method. A worked example, using the pattern file created in Fig. 1, is shown in Fig. 3.

2.3. Comparing a Sequence Against a Library of Patterns

This mode of operation allows a sequence to be searched, in turn, for any number of patterns, each stored in a separate pattern file. The names of the files containing the individual patterns must be stored in a simple text file. This file is called "a file of pattern file names," and its name is the only user input required to define the search.

1. Select "Pattern searcher."
2. Select "Pattern definition mode" as "Use file of pattern file names."
3. Select "Results display mode" as "Inclusive."
4. Define "File of pattern file names." Type the name of the file containing the list of pattern file names. The program will read the file and then, in turn, all the pattern files it names. Each of these patterns will be compared against the current sequence, but only those that give matches will produce any output. The pattern title and each match will be displayed.

2.4. Searching Sequence Libraries for Patterns

The program NIPL can be used to search sequence libraries for patterns. Its use is similar to the pattern search routine described above, except that it does not have the facility for creating pattern files, so they must be created beforehand using NIP. In addition to its obvious application of finding new occurrences of patterns, or checking on their frequency, it is a useful way of obtaining sequence alignments. It can restrict its search to a list of named entries, or can search all but those on a list of entries. It can restrict its output to showing the highest scoring match in each sequence, but by default, it will show all matches.

Of its modes of output, two require further description. The first, "Padded sections," creates a new file for each match. The file will contain the sequence between and including the two outermost motifs in the pattern. It will be gapped to the furthest extent defined by the pattern, which means that if all the files were subsequently written one above the other, all the motifs in the pattern would be exactly aligned, with the sections between them containing the requisite numbers of padding characters. The second such mode of output is called "Complete padded sequences." Here the user must know the maximum distance between the leftmost motif and the start of all the sequences that match. A trial run in which only the positions of matches are reported is usually required. The user gives this maximum distance to the program. The program then writes a new file containing the full length of all matching sequences, again maximally gapped (including their left ends), so that they would all align if written above one another. For both of these modes of output, the files created are named "entryname" where "entryname" is the name given to the sequence in the sequence library. These modes are best used with the option "Report all matches" rejected, so that only the best match for each sequence is reported. The sequences can be lined up using the sequence assembly program SAP.

1. Select NIPL.
2. Define "Name for results file."
3. Select a library.
4. Select "Search whole library." The alternatives are "Search only a list of entries" and "Search all, but a list of entries." The files containing the list of entries should contain one entry name per line, left justified.
5. Select "Results display mode" as "Inclusive." The alternatives include "Motif by motif," "Scores only," "Complete padded sequences," and "Padded sections."
6. Accept "Report all matches." The alternative only shows the best match for each sequence.

```
    Pattern searcher
Select pattern definition mode
X   1 Use keyboard
    2 Use pattern file
    3 Use file of pattern file names
? Selection   (1-3) (1) =
Select results display mode
X   1 Motif by motif
    2 Inclusive
    3 Graphical
    4 EMBL feature table
? Selection   (1-4) (1) =
Select motif definition mode
X   1 Exact match
    2 Percentage match
    3 Cut-off score and score matrix
    4 Cut-off score and weight matrix
    5 Complement of weight matrix
    6 Inverted repeat or stem-loop
    7 Exact match, defined step
    8 Direct repeat
    9 Pattern complete
? Selection   (1-9) (1) =
? Motif name=T run
? String=TTTTT
Probability of score     5.0000 = 0.870E-03
Select motif definition mode
X   1 Exact match
    2 Percentage match
    3 Cut-off score and score matrix
    4 Cut-off score and weight matrix
    5 Complement of weight matrix
    6 Inverted repeat or stem-loop
    7 Exact match, defined step
    8 Direct repeat
    9 Pattern complete
? Selection   (1-9) (1) =4
? Motif name=heat
Select logical operator
X   1 And
    2 Or
    3 Not
? Selection   (1-3) (1) =
? Number of reference motif (1-1) (1) =
? Relative start position (-1000-1000) (6) =10
? Number of extra positions (0-1000) (0) =20
? Weight matrix file name=heatshock.wts
 Heatshock weights 18-12-90
Probability of score    -21.0280 = 0.117E-02
Select motif definition mode
    1 Exact match
    2 Percentage match
    3 Cut-off score and score matrix
X   4 Cut-off score and weight matrix
    5 Complement of weight matrix
    6 Inverted repeat or stem-loop
    7 Exact match, defined step
```

Fig. 1. Worked example of creating a simple pattern and performing a search.

```
        8 Direct repeat
        9 Pattern complete
? Selection   (1-9) (4) =9
? Save pattern in a file (y/n) (y) =
? Pattern definition file=_paper.pat
? Pattern title=demo pattern
Pattern description

demo pattern
Motif  1 named T run     is of class    1
Which is an exact match to the string
TTTTT
Motif  2 named heat       is of class    4
Which is a match to a weight matrix with score -21.028
and the 5 prime base can take positions      10 to      30
relative to the 5 prime end of motif   1
It is anded with the previous motif.
Probability of finding pattern = 0.1015E-05
Expected number of matches  = 0.1734E+00
? Maximum pattern probability (0.00-1.00) (1.00) =
? Minimum pattern score (-9999.00-9999.00) (-9999.00) =
Working
Match
      505 T run
        ttttt
      528 heat
        ttaaagaaagttttatac
Total matches found        1
Minimum and maximum observed scores        -15.34      -15.34
```

Fig. 1 *(continued)*

```
      demo pattern
      A1               T run     Class
      TTTTT
      @ End of string
      A4               heat      Class
         1             Relative motif
        10             Relative start position
        20             Number of extra positions
      heatshock.wts
```

Fig. 2. The pattern file created by the work shown in Fig. 1.

7. Define "Pattern definition file." This is the name of the file containing the pattern created using NIP. The program displays a textual description of the pattern and the expected number of matches per 1000 residues assuming an average amino acid composition.
8. Define "Maximum pattern probability." The program will run much more quickly if none is given.
9. Define "Minimum pattern score." The search will start.

```
   Pattern searcher
Select pattern definition mode
X  1 Use keyboard
   2 Use pattern file
   3 Use file of pattern file names
? Selection  (1-3) (1) =2
? Pattern definition file=_paper.pat
Select results display mode
X  1 Motif by motif
   2 Inclusive
   3 Graphical
   4 EMBL feature table
? Selection  (1-4) (1) =2
Probability of score      5.0000 = 0.870E-03
 Heatshock weights 18-12-90
Probability of score     -21.0280 = 0.117E-02

Pattern description

 demo pattern
Motif  1 named T run     is of class     1
Which is an exact match to the string
TTTTT
Motif  2 named heat      is of class     4
Which is a match to a weight matrix with score -21.028
and the 5 prime base can take positions      10 to        30
relative to the 5 prime end of motif    1
It is anded with the previous motif.
Probability of finding pattern = 0.1015E-05
Expected number of matches   = 0.1734E+00
? Maximum pattern probability (0.00-1.00) (1.00) =
? Minimum pattern score (-9999.00-9999.00) (-9999.00) =
Working
     505 T run
       tttttgatgcttgactctaagccttaaagaaagttttatac
Total matches found        1
Minimum and maximum observed scores      -15.34       -15.34
```

Fig. 3. Worked example of using a pattern file as input.

3. Notes

1. The "exact match" motif class requires a consensus sequence. The "percentage match" motif class requires a consensus sequence and a cutoff score. The "score matrix" motif class requires a consensus sequence and a cutoff score. The "weight matrix" search and the "complement of a weight matrix" only require the name of the file containing the matrix. The "inverted repeat," or "stem-loop" requires a stem length, minimum and maximum loop sizes, and a cutoff score using scores A – T =

G – C = 2, G – T = 1. Note that if the user defines an inverted repeat as a "Reference motif," the "Relative position" can be defined from either its 5', or 3' ends. The "direct repeat" motif class requires a repeat length, the minimum and maximum gap between the two occurrences of the repeat, and a minimum score.

2. The motif class "Exact match, defined step" is rarely used. A typical use might be to find a start codon followed, for some minimum distance, by no stop codons in the same reading frame. The step would have the value three to keep the reading frame the same as that of the start codon, and the stop codon searches would be included using the NOT operator.

3. The details of the probability calculations are outside the scope of this chapter. They are quite rapid, and are essential both for assessing the statistical significance of any matches found and for allowing meaningful cutoffs to be applied to patterns. Obviously, in general, cutoff scores are inappropriate for patterns containing a mixture of motif classes.

4. The program calculates the "Probability of finding the pattern" and the "Expected number of matches." The first figure is actually the product of the individual motif probabilities, but the latter figure is more useful because it takes into account the allowed variation in spacing between motifs and the length of the current sequence. In both cases, the composition of the current sequence is also used so that different probabilities would be calculated for other sequences.

5. The pattern definition system is very flexible. Assume that a laboratory has a large library of patterns stored in its computer. Different groups or users may want to screen their sequences against different subsets of a pattern library. Each group therefore uses its own "File of pattern file names," which contains only the names of the pattern files that are relevant to their sequences. Of course, a pattern may contain only one motif. Hence, a library of patterns can include both simple and complex patterns. In the same way, a laboratory may have a large library of weight matrices defining different motifs, and different users may want to combine them in different ways to produce their own patterns.

References

1. Staden, R. (1988) Methods to define and locate patterns of motifs in sequences. *CABIOS* **4(1)**, 53–60.
2. Staden, R. (1989) Methods for calculating the probabilities of finding patterns in sequences. *CABIOS* **5(2)**, 89–96.
3. Staden, R. (1990) Searching for patterns in protein and nucleic acid sequences, in *Methods in Enzymology*, vol. 183 (Doolittle, R. F., ed.), Academic, New York, pp. 193–211.

CHAPTER 10

Staden: Analyzing Sequences to Find Genes

Rodger Staden

1. Introduction

Three methods are outlined for finding protein genes and one for locating tRNA genes, as are routines for finding open reading frames and displaying the positions of stop codons. All the methods are contained in the program NIP. The correct interpretation of the analyses presented requires a good understanding of the underlying ideas used by the methods. Despite this, this chapter concentrates here on the use of the techniques, the reader is referred to earlier publications (1–5) for more background information.

The assumption made by the methods for finding protein genes is that protein coding regions, when analyzed in terms of three-letter nonoverlapping "words," will look different to noncoding regions analyzed in the same way. Suppose a sequence is analyzed in one reading frame and its codons counted. Then the "positional base composition" will be defined as the frequency at which each of the four base types occupies each of the three positions in codons. In coding regions, the positional base frequencies will be less random than they are in noncoding regions. This is the basis of method one in Section 2.1.: the "Uneven positional base frequencies method." If this reading frame is coding for a protein, the positional base composition will tend toward a particular bias that is common to the majority of genes. This is the basis of method two in Section 2.2.: the "Positional base preferences method." If the sequence has a very biased base composition, then in protein genes, this may effect the choice of amino acids and will effect the use of bases in the third positions of codons.

From: *Methods in Molecular Biology, Vol. 25: Computer Analysis of Sequence Data, Part II*
Edited by: A. M. Griffin and H. G. Griffin Copyright ©1994 Humana Press Inc., Totowa, NJ

This bias is also utilized by the positional base preferences method. Finally, if the reading frame is coding for a protein, its use of codons is also likely to be nonrandom, and this is the basis of method three in Section 2.3.: the "Codon usage method."

All the methods perform their analyses over segments of the sequence of size "window," and then move the window on by three bases and repeat the calculation. The "Uneven positional base frequencies" method only produces a single value for each segment, and hence, cannot distinguish between frames or strand—it only measures the probability that a region is coding and nothing more. The other two methods produce different values for each of the three potential reading frames and, hence, can help to decide which is coding. Their results are plotted in three separate boxes arranged one above the other. For these, which of the three reading frames is the highest scoring at each position along the sequence is also indicated. This is done by plotting a single dot at the midheight of the box that contains the highest score, so that if one frame is the highest scoring for many consecutive positions, the dots will produce a solid line at the midheight of its box. The positions of stop codons are also marked. These are represented by short vertical lines and are positioned so that they bisect the midheight of each box. Start codons are marked at the base of the box for each reading frame. The search for tRNA genes involves looking for segments that could fold into the cloverleaf structure and that have the expected conserved bases in the appropriate positions.

Notice that searches for relevant "signals," like promoters, or splice junctions, which are also useful for finding genes, have not been mentioned. These searches are described in the chapter on searching for motifs. In Chapter 8, the only "signal" included is the stop codon. However, since all results are presented graphically, it is easy for users to overlay the displays of signal searches with those presented here and so effectively combine them.

2. Methods

2.1. The Uneven
Positional Base Frequencies Method

This method produces a single value for each segment of the sequence, and would give the same result if applied to each reading frame or to the complementary strand. The results are plotted in a box that is cut by a horizontal line. This line is labeled 76%, and 76% of noncoding sequences are expected to score below this line and 76% of coding sequences to score above it. Of the methods described, this one makes the fewest assumptions and so is a good unbiased indicator of the probability that a sequence is coding.

1. Select "Uneven positional base frequencies."
2. Define "Odd window length."
3. Define "Plot interval."

The plot will appear as in Fig. 1. In the example shown, the 5' end of the sequence codes for several proteins, and the 3' end codes for ribosomal RNAs.

2.2. The Positional Base Preferences Method

As a result of the genetic code and the relative frequencies with which amino acids are used in proteins, DNA sequences coding for proteins have a particular bias in their positional base frequencies. This method scans DNA sequences and measures the closeness of each reading frame to this bias in their positional base frequencies. The closeness to the expected bias is expressed as a "score." By default, the program will use a "global" set of expected values for the positional base frequencies that are derived from average amino acid compositions in known proteins. Alternatively, users may create their own set of expected values by analyzing known genes from the same genome. In addition, users can combine the "global" values for the first two positions in codons with third position values derived from other genes of the same genome.

In order to use a nonglobal standard, a codon table in the format described in Chapter 5 on statistical analysis of nucleic acid sequences can be created using the method "Creating a codon usage file." Alternatively, a section of the sequence being analyzed can be scanned to produce an internal standard. The method is particularly useful for selecting which reading frame is coding.

2.2.1. Using the Global Standard

1. Select "Positional base preferences method."
2. Select "Standard source" as "Global."
3. Define "Window length." The default length of 67 should be used for most cases. Shorter windows give noisier plots, and the longer the window, the more chance there is of missing a short exon.
4. Define "Plot interval."

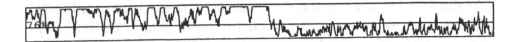

Fig. 1. Example output from the "Uneven positional base frequencies method."
The 5' end codes for proteins, and the 3' end contains ribosomal RNA genes.

The plot will appear as in Fig. 2. This shows a 10,000-base section
of sequence that codes for several proteins in each of the three reading
frames. *See* Section 1. for an explanation of the plotting scheme used.

2.2.2. Using a Nonglobal Standard

1. Make an appropriate codon usage file as described in the chapter on
   statistical analysis of nucleotide sequences.
2. Select "Positional base preferences method."
3. Select "Standard source" as "Codon usage table."
4. Define "File name of standard." The file will be read and displayed on
   the screen.
5. Select "Normalization" as "Combine with global standard." This alternative means the values for the first two positions of codons combined
   with the third-position values from the codon table will be used. Otherwise ("Use observed frequencies") all three positions from the codon
   table will be used. The positional base frequencies to be used will be
   displayed.
6. Accept "Use 1.0 for positional weights." The alternative allows users
   to give greater or lesser emphasis to any of the three positions by defining weights for each. The program displays the "Expected scores per
   codon in each frame."
7. Define "Window length." Windows shorter than the default of 67 may
   be useful if the bias is sufficiently strong. Look at the "Expected scores
   in each frame" to help decide.
8. Define "Plot interval."
9. Accept "Plot relative scores." This means that for each frame, the value
   plotted is frame score divided by the scores for all three frames. It
   produces smoother plots than the alternative "Plot absolute scores,"
   which simply plots the scores for each frame. The minimum and maximum expected scores for the given standard and window length are
   displayed.
10. Accept "Leave scaling values unchanged." The expected scores just
    displayed will be used to scale the plots. If required, the user can change
    the scaling values at this point. The plot will now appear as in Fig. 2.
    Typical dialog is shown in Fig. 3.

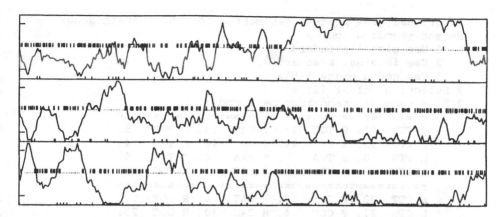

Fig. 2. Example output from the "Positional base preferences method." Most of the sequence is coding for proteins.

2.3. The Codon Usage Method

The codon usage method scans along a sequence and measures the closeness of each reading frames codon composition to an expected set of codons. Of the methods described, it is the most sensitive, but consequently has to make the strongest assumption, namely, that we know the approximate codon usage for the genes being searched for. The codon usage will depend on the codon preferences and the amino acid composition of the protein product. For this reason, the program contains three methods of "normalization." The table of codon usage may be used as read "Observed frequencies;" the table may be transformed to reflect an average amino acid composition "Normalize to average amino acid composition;" the table may be transformed to have no amino acid bias "Normalize to no amino acid bias." The table can be read from a file produced by "Creating a codon usage file" as described in the chapter on statistical analysis of nucleic acid sequences, or an "internal standard" can be used by the user defining a region of the current sequence. In the latter case, the program will calculate the codon usage for the defined region.

1. Select "Codon usage method."
2. Reject "Define internal standard." If an internal standard is used, the program will ask for the end points of the segments over which to calculate the codon usage.
3. Define "File name of standard." The file will be read and displayed on the screen.

```
Positional.base preferences method to find protein genes
Select standard source
X   1 Use global standard
    2 Use internal standard
    3 Use codon usage table
? Selection   (1-3) (1) =3
? File name of standard=atpase.cods
        ===============================================
      F TTT   21.  S TCT   33.  Y TAT   15.  C TGT    5.
      F TTC   55.  S TCC   40.  Y TAC   40.  C TGC    4.
      L TTA    8.  S TCA    7.  * TAA    8.  * TGA    0.
      L TTG   19.  S TCG   12.  * TAG    1.  W TGG   17.
        ===============================================
      L CTT   22.  P CCT   17.  H CAT    6.  R CGT   73.
      L CTC   21.  P CCC    4.  H CAC   30.  R CGC   23.
      L CTA    1.  P CCA   10.  Q CAA   19.  R CGA    5.
      L CTG  168.  P CCG   48.  Q CAG   80.  R CGG    3.
        ===============================================
      I ATT   47.  T ACT   14.  N AAT   17.  S AGT    8.
      I ATC   98.  T ACC   54.  N AAC   52.  S AGC   26.
      I ATA    6.  T ACA    7.  K AAA   85.  R AGA    0.
      M ATG   75.  T ACG   13.  K AAG   28.  R AGG    0.
        ===============================================
      V GTT   67.  A GCT   56.  D GAT   41.  G GGT   90.
      V GTC   29.  A GCC   53.  D GAC   66.  G GGC   66.
      V GTA   49.  A GCA   59.  E GAA  101.  G GGA    5.
      V GTG   57.  A GCG   64.  E GAG   41.  G GGG    8.
        ===============================================
Select normalisation
X   1 Use observed frequencies
    2 Combine with global standard
? Selection   (1-2) (1) =2
              T       C       A       G      Range
        1   0.177   0.211   0.277   0.336   0.159
        2   0.271   0.238   0.310   0.182   0.128
        3   0.242   0.301   0.168   0.289   0.132
? Use 1.0 for positional weights (y/n) (y) =
  Expected scores per codon in each frame
          0.785      0.736      0.736
? odd span length (31-101) (67) =
? plot interval (1-11) (5) =
? Plot relative scores (y/n) (y) =

   Minimum   maximum    range
     0.3219    0.3519    0.0214
? Leave scaling values unchanged (y/n) (y) =
```

Fig. 3. Typical dialog from the "Positional base preferences method" using a nonglobal standard in the form of a codon table to specify the values for the third positions in codons.

4. Select "Normalization" as "Average amino acid composition." The program will display the expected values for each reading frame for the window lengths 21, 31, and 41 codons.
5. Select "Window length."
6. Select "Plot interval."

The plot will appear as in Fig. 4. This shows a 10,000-base section of sequence that codes for several proteins in each of the three reading frames. *See* Section 1. for an explanation of the plotting scheme used.

2.4. Searching for Open Reading Frames

This routine finds all open reading frames of some minimum length and writes its results in the form of an EMBL feature table.

1. Select "Find open reading frames."
2. Define "Minimum open frame in amino acids."
3. Select "Strands." The alternatives are: + strand only, – strand only, or both strands. Typical output is shown in Fig. 5.

2.5. Searching for tRNA Genes

tRNA genes have two classes of features that can be used to locate them in genomic sequences: their ability to fold into the cloverleaf secondary structure and the presence of specific "conserved" bases at particular positions relative to this structure. The level of congruence with the canonical structure is quite variable: Some tRNA genes contain intervening sequences, and others, particularly those from organelles, have few of the conserved bases. The program searches for potential cloverleaf-forming structures and optionally the presence of conserved bases. The user can define the range of loop sizes, the minimum numbers of potential base pairs, a range of intron sizes, and which, if any, of the conserved bases should be present. The results are presented either textually or graphically.

1. Select "tRNA search."
2. Define "Maximum tRNA length."
3. Define "Aminoacyl stem score." *See* Note 8.
4. Define "Tu stem score."
5. Define "Anticodon stem score."
6. Define "D stem score."
7. Define "Minimum base pairing total."

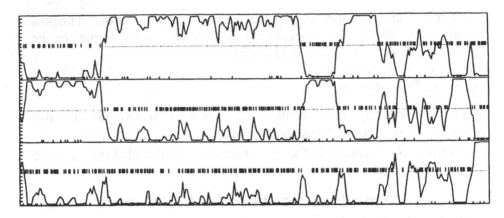

Fig. 4. Example output from the "Codon usage method." Most of the sequence is coding for proteins.

8. Define "Minimum intron length."
9. Define "Maximum intron length."
10. Define "Minimum length for TU loop."
11. Define "Maximum length for TU loop."
12. Accept "Skip search for conserved bases." *See* Section 3.
13. Reject "Plot results." This gives listed output in which the potential cloverleafs are displayed. The alternative plotted output simply draws a vertical line to represent the score for the potential gene, at the position it has been found. Typical dialog and the beginning of some listed output are shown in Fig. 6.

3. Notes

1. In general, for finding protein genes, the use of all the methods is recommended. The "Uneven positonal base frequencies method" can show which regions are likely to be coding, but not which strand or frame. The "Positional base preferences method" can show the correct frame and also help to find which regions are coding. The "Codon usage method" has the greatest resolution, having been used successfully with windows of 11 codons, and can help find small exons and pinpoint exon/intron boundaries.

2. When the "Uneven positional base frequencies" calculation was applied to all the sequences in the 1984 version of the EMBL library, 14% of noncoding segments failed to reach the value represented by the base of the box, whereas all coding segments did. The top value of the box was not reached by any noncoding segments, but was exceeded by 16% of coding sequences. Seventy-six percent of noncoding segments failed

```
FT    CDS              525..965
FT    CDS              956..1789
FT    CDS              2128..2607
FT    CDS              2604..3155
FT    CDS              3159..4709
FT    CDS              4733..5623
FT    CDS              5539..7032
FT    CDS              7044..7454
FT    CDS              7797..8134
FT    CDS              complement(2227..2634)
FT    CDS              complement(2250..3023)
FT    CDS              complement(3027..3899)
FT    CDS              complement(3903..4760)
FT    CDS              complement(4327..4626)
FT    CDS              complement(4646..5332)
FT    CDS              complement(5345..5647)
FT    CDS              complement(5635..6012)
FT    CDS              complement(6016..6441)
FT    CDS              complement(6445..7083)
FT    CDS              complement(7035..7445)
FT    CDS              complement(7406..7777)
```

Fig. 5. Typical output from "Find open reading frames."

to reach the line labeled 76%, but 76% of coding segments fell above it. It would not be expected that this result would change significantly if it were to be recalculated on the current libraries.

3. When the "Positional base preferences method," using "global" values, was applied to all the *E. coli* genes in the 1984 version of the EMBL library, it chose the correct reading frame for 91% of coding segments. *E. coli* sequences were used for technical rather than scientific reasons, and there is no reason to believe that other organisms should give significantly different results. This result used only the values for the first two positions in codons, so for genes with a strongly biased base composition even better discrimination would be expected.

4. If the codon table used by the "Codon usage method" is normalized to have average amino acid composition, it retains its codon preference bias for each amino acid type, but now the amino acid composition is the average of all proteins. In general, this is optimal: We have the expected codon preference bias plus an expected amino acid bias. If the user normalizes to no amino acid bias, the user is safeguarding against missing a protein of anomalous composition, but at the expense of not employing all of the useful information for distinguishing coding from noncoding.

```
tRNA search
 ? Maximum trna length (70-130) (92) =
 ? Aminoacyl stem score (0-14) (11) =
 ? Tu stem score (0-10) (8) =
 ? Anticodon stem score (0-10) (8) =
 ? D stem score (0-8) (3) =
 ? Minimum base pairing total (30-44) (30) =
 ? Minimum intron length (0-30) (0) =
 ? Maximum intron length (0-30) (0) =
 ? Minimum length for TU loop (4-12) (6) =
 ? Maximum length for TU loop (6-12) (9) =
 ? Skip search for conserved bases (y/n) (y) =n
Give a score for each base, then a minimum total at the end
 ? Base  8, T is 100% conserved. Score (0-100) (0) =
 ? Base 10, G is  95% conserved. Score (0-100) (0) =
 ? Base 11, Y is  96% conserved. Score (0-100) (0) =
 ? Base 14, A is 100% conserved. Score (0-100) (0) =
 ? Base 15, R is 100% conserved. Score (0-100) (0) =
 ? Base 21, A is  97% conserved. Score (0-100) (0) =
 ? Base 32, Y is 100% conserved. Score (0-100) (0) =
 ? Base 33, T is  98% conserved. Score (0-100) (0) =
 ? Base 37, A is  91% conserved. Score (0-100) (0) =
 ? Base 48, Y is 100% conserved. Score (0-100) (0) =
 ? Base 53, G is 100% conserved. Score (0-100) (0) =
 ? Base 54, T is  95% conserved. Score (0-100) (0) =
 ? Base 55, T is  97% conserved. Score (0-100) (0) =
 ? Base 56, C is 100% conserved. Score (0-100) (0) =
 ? Base 57, R is 100% conserved. Score (0-100) (0) =
 ? Base 58, A is 100% conserved. Score (0-100) (0) =
 ? Base 60, Y is  92% conserved. Score (0-100) (0) =
 ? Base 61, C is 100% conserved. Score (0-100) (0) =
 ? Minimum total conserved base score (0-0) (0) =
 ? Plot results (y/n) (y) =n

            264
                    t
                  t-a
                  c-g
                  a-t
                  t+g
                  a-t
                  a a
                  a-t      gta
                c   aacgc
        a     t     !!!!   c
          cgt     gtgcg   a
          !!!       t   cga
        a gca        c
          g     t       g
                  c aa   t
                  a-t a
                  t-a  t a
                  t-a
                  t-a
                  g   t
                  c   g
                  caa
```

Fig. 6. Typical dialog and textual output from "Find tRNA genes."

5. The program also contains a graphical version of Fickett's method *(6)*, except here a window is used to analyze each segment of the sequence rather than giving a single value for each open reading frame. The tables used are those from the original publication.

6. If the results from the "Find open reading frames" option are directed to disk *(see* Chapter 2), the file can be used by the routines that use feature tables as input.

7. The program also contains several routines for plotting the positions of stop and start codons for either strand of the sequence. One form of the output is included in Figs. 2 and 4.

8. The tRNA gene search uses a simple scoring system for base pairing: A–T and G–C base pairs each score 2 and G–T scores 1. The use of a "Minimum base pairing total" allows low cutoffs to be set for each individual stem, but that overall some reasonable level of stability is possible. In this way, a low score for one stem can be compensated by a high score in another.

9. The conserved bases are selected by the user defining a score for each one and then giving an overall minimum total. If, in the genome being scanned, certain of the conserved base positions are known not to be canonical, their scores can be set to zero and hence ignored. When prompting the user for scores for each of the conserved bases, the program displays the level of conservation found in the compilations of tRNA sequences published in *Nucleic Acids Research.*

10. The cloverleaf is composed of four base-paired stems and four loops. Three of the stems are of fixed length, but the fourth, the dhu stem, which usually has 4 base pairs, sometimes has only 3. All of the loops can vary in size. The following relationships between the stems in the cloverleaf are assumed in the program:

    a. There are no bases between one end of the aminoacyl stem and the adjoining tuc stem;

    b. There are two bases between the aminoacyl stem and the dhu stem;

    c. There is one base between the dhu stem and the anticodon stem; and

    d. There are at least three bases between the anticodon stem and the tuc stem.

References

1. Staden, R. and McLachlan, A. D. (1982) Codon preference and its use in identifying protein coding regions in long DNA sequences. *Nucleic Acids Res.* **10,** 151–156.

2. Staden, R. (1984) Measurements of the effects that coding for a protein has on a DNA sequence and their use for finding genes. *Nucleic Acids Res.* **12,** 551–567.

3. Staden, R. (1985) Computer methods to locate genes and signals in nucleic acid sequences, in *Genetic Engineering, Principle and Methods,* vol. 7 (Setlow, J. K. and Hollaender, A., eds.), Plenum, New York, pp. 67–114.
4. Staden, R. (1990) Finding Protein Coding Regions in Genomic Sequences, in *Methods in Enzymology,* vol. 183 (Doolittle, R. F., ed.), Academic, New York, pp. 163–180.
5. Staden, R. (1980) A computer program to search for tRNA genes. *Nucleic Acids Res.* **8,** 817–825.
6. Fickett, J. W. 1982. Recognition of protein coding regions in DNA sequences. *Nucleic Acids Res.* **10,** 5303–5318.

CHAPTER 11

Staden: Statistical and Structural Analysis of Protein Sequences

Rodger Staden

1. Introduction

This chapter describes the use of routines for plotting hydrophobicity, charge, and hydrophobic moments, drawing helix wheels and predicting secondary structure. Use of all these routines is very straightforward, and they are contained in the program PIP.

2. Methods

2.1. Plotting Hydrophobicity

This method uses the values of Kyte and Doolittle *(1)*.

1. Select "Plot hydrophobicity."
2. Define "Window length."
3. Define "Plot interval." The plot will appear as in Fig. 1.

2.2. Plotting Charge

1. Select "Plot charge."
2. Define "Window length."
3. Define "Plot interval." The plot will appear and will be similar to that shown in Fig. 1.

2.3. Plotting Hydrophobic Moment and Hydrophobicity

This method plots the hydrophobic moment and the hydrophobicity as defined by Eisenberg et al. *(2)*.

1. Select "Plot hydrophobic moment."

From: *Methods in Molecular Biology, Vol. 25: Computer Analysis of Sequence Data, Part II*
Edited by: A. M. Griffin and H. G. Griffin Copyright ©1994 Humana Press Inc., Totowa, NJ

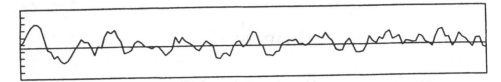

Fig. 1. A hydrophobicity plot using the values of Kyte and Doolittle *(1)*.

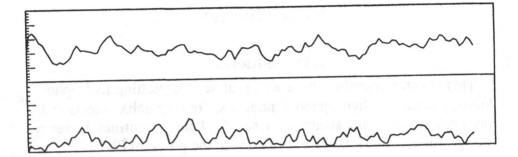

Fig. 2. A hydrophobic moment (below) and hydrophobicity plot. The hydrophobicity plot displays the mean values on a scale of –1.5–1.0 and the hydrophobic moment on a scale of 0.0 –1.5.

2. Define "Angle." This is the angle between the residues when the helix is viewed end on. The default value of 100° is that found in α helices.
3. Define "Window length." The default of 18, if used in conjunction with the default "Angle," is equivalent to five turns of the helix.
4. Define "Plot interval."

The plot will appear as in Fig. 2 with the hydrophobicity shown above the hydrophobic moment. The scale for the hydrophobicity runs from –1.0 to 1.5 and for the hydrophobic moment from 0.0 to 1.5. The program plots the mean values for each window position with the value at position x representing the segment from x-window length + 1 to x.

2.4. Drawing Helical Wheels

This method draws helical wheels for any segment of the sequence *(3)*. In addition, it displays the hydrophobic moment for the segment *(2)*.

1. Select "Draw helix wheel."
2. Define "Angle." The default angle of 100° is that found in α helices.
3. Define "Window length." The default of 18, if used in conjunction with the default "Angle," is equivalent to five turns of the helix. The display for the current position in the sequence will appear as in Fig. 3, and the

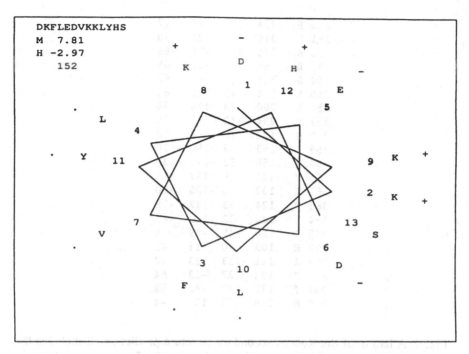

Fig. 3. A typical helix wheel display using a window of only 13 residues. The display includes a schematic of the helix showing the links between residues, with each vertex numbered according to position; the residue type at each vertex; and a symbol denoting a classification as hydrophobic (.), positively charged (+), negatively charged (−), or otherwise ( ). The residue number of the first sequence element in the current window is displayed at the top left corner along with the sequence. Below this is the total hydrophobicity and hydrophobic moment according to Eisenberg et al. *(2)*.

bell will ring. The program now allows the user to "step' through the sequence displaying the helix wheel for each position.

4. Define "Step." To produce a display for a sequence position *N* bases from the current one, type N, and the display will appear in place of the previous one. The default value of *N* is 1, so by repeatedly hitting carriage return, the user can step, residue by residue, through the sequence.

2.5. Producing
a Robson Secondary Structure Prediction

This method uses the method of Garnier et al. *(4)* to predict the positions of α helices, β sheets, turns, and random coil. The results can be either plotted or listed.

1. Select "Robson secondary structure prediction."

```
350 P    274 -178  -84  -77
351 L    316 -192  -21  -38
352 K    371 -223  -75  -68
353 L    365 -152 -101  -65
354 S    331  -82  -84  -63
355 K    311  -43 -110  -88
356 A    280  -23 -110  -80
357 V    234  -12 -135  -75
358 H    177  -10 -143  -92
359 K    153    2 -180 -138
360 A    158   52 -175 -130
361 V    144   78 -187 -115
362 L    132   58 -186  -80
363 T    124   63 -142  -78
364 I    144   32 -111  -43
365 D    120  -49  -29    5
366 E    103  -80   13   43
367 K    111 -113   23   42
368 G    132 -127  -13   64
369 T    172 -132  -42   52
370 E    216 -170 -122   -4
```

Fig. 4. A listing of the Robson secondary structure prediction. It includes the sequence position, the residue type, and the values for the four structure classes.

2. Accept "Plot results." The alternative produces a listing like that shown in Fig. 4.

The plot will appear as in Fig. 5, and the program also prints a count of the number of positions at which each of the four structure types is the highest scoring.

2.6. Calculating the Composition and Molecular Weight of a Sequence

Select "Count amino acid composition." The composition and molecular weight are displayed as in Fig. 6. Each column contains the one-letter code for the amino acid, the number of occurrences of that amino acid in the sequence, the number expressed as a percentage, and its molecular weight.

3. Notes

1. The methods described Chapters 12 and 13 on motif and pattern searching can also be used to search for specific structures. For example, a sequence can be searched for all the structures contained in the PROSITE motif library.

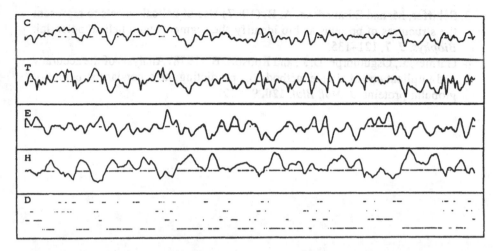

Fig. 5. A secondary structure plot using the method of Robson. The likelihood that each 17-residue segment of the sequence forms one of the four structure classes—helix (H), extended (E) normally termed sheet, turn (T), and coil (C)—is plotted out across the screen in four strips. Below this is a "decision" strip (D), in which a single dot is plotted for the higest scoring structure class at each point. Here we see a sequence that is predicted to be predominantly helical.

Sequence composition

A	C	S	T	P	A	G	N	D	E	Q	B	Z	H
N	0.	14.	19.	12.	30.	26.	3.	10.	11.	4.	0.	0.	0.
%	0.0	5.3	7.3	4.6	11.5	9.9	1.1	3.8	4.2	1.5	0.0	0.0	0.0
W	0.	1219.	1921.	1165.	2132.	1483.	342.	1151.	1420.	513.	0.	0.	0.

A	R	K	M	I	L	V	F	Y	W	-	X	?	
N	7.	7.	10.	15.	39.	23.	13.	11.	8.	0.	0.	0.	0.
%	2.7	2.7	3.8	5.7	14.9	8.8	5.0	4.2	3.1	0.0	0.0	0.0	0.0
W	1093.	897.	1312.	1697.	4413.	2280.	1913.	1795.	1490.	0.	0.	0.	0.

Total molecular weight= 28256.254

Fig. 6. A typical molecular-weight and composition display. It includes the residue type, their number, their percentage, and their contribution to the molecular weight.

2. It is often convenient to produce displays in which several of the plots described above appear together on the screen.

References

1. Kyte, J. and Doolittle, R. F. (1982) A simple method for displaying the hydropathic character of a protein. *J. Mol. Biol.* **157,** 105–132.
2. Eisenberg, D., Schwarz, E., Komaromy, M., and Wall, R. (1984) Analysis of membrane and surface protein sequences with the hydrophobic moment plot. *J. Mol. Biol.* **179,** 125–142.

3. Schiffer, M. and Edmundson, A. B. (1967) Use of helical wheels to represent the structures of proteins and to identify the segments with helical potential. *Biophys. J.* **7,** 121–135.

4. Garnier, J., Osguthorpe, D. J., and Robson, B. (1978) Analysis of the accuracy and implications of simple methods for predicting the secondary structure of globular proteins. *J. Mol. Biol.* **120,** 97–120.

CHAPTER 12

Staden: Searching for Motifs in Protein Sequences

Rodger Staden

1. Introduction

The program PIP contains several ways of defining and searching for motifs *(1,2)*. Searches for exact matches and percentage matches, the use of score matrices, and the creation and use of weight matrices are described here. All of the searches produce both listed and graphical output.

2. Methods

2.1. Searching for Exact Matches

The routine for finding and displaying the positions of exact matches to sequences can display its results in various forms. It is equivalent to the restriction enzyme search routine in the nucleotide analysis programs. The sequences to be searched for can be typed on the keyboard or read from files. The format of these files is given in Section 3. Here, only a single example is given of the use of the routine, which shows how to produce a plot of the positions of all amino acid types in a sequence.

1. Select "Search."
2. Select "Input source" as "All acids file." A number of standard files are available, and users may also have their own. The one selected simply contains the one-letter codes for all the standard amino acids.
3. Accept "Search for all names." The alternative allows users to select a subset of the entries in the file by name.
4. Select "Order results name by name."

From: *Methods in Molecular Biology, Vol. 25: Computer Analysis of Sequence Data, Part II*
Edited by: A. M. Griffin and H. G. Griffin Copyright ©1994 Humana Press Inc., Totowa, NJ

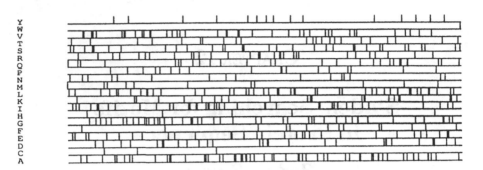

Fig. 1. Typical graphical output from "Search for exact matches" in which the position of each matching string (here individual amino acid types) is marked.

5. Reject "List matches." If results are listed, the output gives the name and position of each match, and also the separations between matches. The results will then appear in the form shown in Fig. 1.

2.2. Searching for Percentage Matches to Sequences

1. Select "Find percentage matches."
2. Accept "Type in strings." The alternative allows the string to be extracted from a named file.
3. Reject "Keep picture." This will cause the graphics window to be cleared. The alternative leaves it unchanged.
4. Define "String." Type in the search string. When the program cycles around to this point again, the previous string will be offered as a default.
5. Define "Percent match." The search is performed, the results are presented graphically, the number of matches displayed, and the scores and positions of the top-10 matches displayed.
6. Define the number of matches to "Display." For the number of matches chosen, the program will display the search string and matching sequence written one above the other with matching characters indicated by asterisk symbols. The program now cycles around to step 3.

2.3. Searching for Sequences Using a Score Matrix

A score matrix gives a score for the alignment of each possible pair of sequence symbols. This method is more sensitive than the simple percentage match search. The default matrix MDM78 used by this program is shown in Fig. 2.

	C	S	T	P	A	G	N	D	E	Q	B	Z	H	R	K	M	I	L	V	F	Y	W	-	X	?	
C	22	10	8	7	8	7	6	5	5	5	5	5	7	6	5	5	8	4	8	6	10	2	10	10	10	10
S	10	12	11	11	11	11	11	10	10	9	10	10	9	10	10	8	9	7	9	7	7	8	10	10	10	10
T	8	11	13	10	11	10	10	10	10	9	10	10	9	9	10	9	10	8	10	7	7	5	10	10	10	10
P	7	11	10	16	11	9	9	9	9	10	9	10	10	10	9	8	8	7	9	5	5	4	10	10	10	10
A	8	11	11	11	12	11	10	10	10	10	10	9	8	9	9	9	9	8	10	6	7	4	10	10	10	10
G	7	11	10	9	11	15	10	11	10	9	10	10	8	7	8	7	7	6	9	5	5	3	10	10	10	10
N	6	11	10	9	10	10	12	12	11	11	12	11	12	10	11	8	8	7	8	6	8	6	10	10	10	10
D	5	10	10	9	10	11	12	14	13	12	13	12	11	9	10	7	8	6	8	4	6	3	10	10	10	10
E	5	10	10	9	10	10	11	13	14	12	12	13	11	9	10	8	8	7	8	5	6	3	10	10	10	10
Q	5	9	9	10	10	9	11	12	12	14	11	13	13	11	11	9	8	8	8	5	6	5	10	10	10	10
B	5	10	10	9	10	10	12	13	12	11	13	11	11	10	10	8	8	6	8	5	7	4	10	10	10	10
Z	5	10	10	10	10	10	11	12	13	13	11	14	12	10	10	8	8	8	8	5	6	4	10	10	10	10
H	7	9	9	10	9	8	12	11	11	13	11	12	16	12	10	8	8	8	8	8	10	7	10	10	10	10
R	6	10	9	10	8	7	10	9	9	11	10	10	12	16	13	10	8	7	8	6	6	12	10	10	10	10
K	5	10	10	9	9	8	11	10	10	10	10	10	10	13	15	10	8	7	8	5	6	7	10	10	10	10
M	5	8	9	8	9	7	8	7	8	9	8	8	8	10	10	16	12	14	12	10	8	6	10	10	10	10
I	8	9	10	8	9	7	8	8	8	8	8	8	8	8	8	12	15	12	14	11	9	5	10	10	10	10
L	4	7	8	7	8	6	7	6	7	8	6	8	8	7	7	14	12	16	12	12	9	8	10	10	10	10
V	8	9	10	9	10	9	8	8	8	8	8	8	8	8	8	12	14	12	14	9	8	4	10	10	10	10
F	6	7	7	5	6	5	6	4	5	5	5	5	8	6	5	10	11	12	9	19	17	10	10	10	10	10
Y	10	7	7	5	7	5	8	6	6	6	7	6	10	6	6	8	9	9	8	17	20	10	10	10	10	10
W	2	8	5	4	4	3	6	3	3	5	4	4	7	12	7	6	5	8	4	10	10	27	10	10	10	10
-	10	10	10	10	10	10	10	10	10	10	10	10	10	10	10	10	10	10	10	10	10	10	10	10	10	10
X	10	10	10	10	10	10	10	10	10	10	10	10	10	10	10	10	10	10	10	10	10	10	10	10	10	10
?	10	10	10	10	10	10	10	10	10	10	10	10	10	10	10	10	10	10	10	10	10	10	10	10	10	10
	10	10	10	10	10	10	10	10	10	10	10	10	10	10	10	10	10	10	10	10	10	10	10	10	10	10

Fig. 2. The amino acid score matrix MDM78.

1. Select "Find matches using a score matrix."
2. Accept "Type in strings." The alternative allows the string to be extracted from a named file.
3. Reject "Keep picture." This will cause the graphics window to be cleared. The alternative leaves it unchanged.
4. Define "String." Type in the search string. When the program cycles around to this point again, the previous string will be offered as a default. The program displays the minimum and maximum possible scores for the string.
5. Define "Score." The search is performed, the results are presented graphically, the number of matches displayed, and the scores and positions of the top-10 matches displayed.
6. Define the number of matches to "Display." For the number of matches chosen, the program will display the search string and matching sequence written one above the other with matching characters indicated by asterisk symbols. The program now cycles around to step 3. An example run is shown in Fig. 3.

```
Find matches using a score matrix
? Keep picture (y/n) (y) =
  ? String=ALPHA
Minimum score=     23 Maximum score=    72
? Score (23-72) (72) =60

For score    60 the number of matches=      5
Scores        62      62      62      61      61
Positions    120     217     420      54     326
? Display (0-5) (0) =

        120
        PLDHD
         * *

        ALPHA
        1

        217
        ALANT
        **

        ALPHA
        1

        420
        QLDHG
         * *

        ALPHA
        1

        54
        SLPGN
         **

        ALPHA
        1

        326
        ALPII
        ***

        ALPHA
        1
? Keep picture (y/n) (y) =
 Default String=ALPHA
? String=!
```

Fig. 3. An example of the listed output from "Search using a score matrix."

2.4. Using Weight Matrices
for Searching Protein Sequences

A weight matrix is the most sensitive way of defining a motif. It is a table of values that gives scores for each amino acid type in each position along a motif. For a motif of length eight amino acids, the weight matrix would be a table eight positions long and, allowing for 26 amino acid symbols, 26 deep. The simplest way of choosing the values for the table is to take an alignment of all known examples of the motif and to count the frequency of occurrence of each amino acid type at each position. These frequencies can be used as the table of weights. When the table is used to search a new sequence, the program calculates a score for each position along the sequence by adding or multiplying (*see* Section 3) the relevant values in the table. All positions that exceed some cutoff score are reported as matching the original set of motifs.

How can we select a suitable cutoff score? The simplest way is to apply the weight matrix to all the known occurrences of the motif, i.e., the set of sequence segments used to create the table, and to see what scores they achieve. The cutoff can be selected accordingly. For convenience, the weight matrix is stored as a file along with its cutoff score, a title that is displayed when the file is read, and a few other values needed by the program. A routine for creating weight matrix files from sets of aligned sequences is included in the program. When a search using the weight matrix is performed, the program will either list the matching sequence segments or plot their positions as for the other motif search methods.

2.4.1. Creating a Weight Matrix File
from a Set of Aligned Sequences

1. Select "Motif search using weight matrix."
2. Select "Make weight matrix."
3. Define "Name of aligned sequences file." It is assumed that the file of aligned sequences has already been created (*see* Note 5). The program reads and displays the contents of the file, numbering each sequence as it goes. Then it displays the length of the longest sequence.
4. Accept "Sum logs of weights." The alternative is to sum the weights when calculating scores (*see* Note 6).

5. Accept "Use all motif positions." The alternative allows the user to define a "mask," which identifies positions within the motif that should be ignored when the matrix is created (*see* Note 7). The program now calculates the weights and applies them in turn to each of the sequences in the file. The number and score for each sequence are displayed, followed by the top, bottom, and mean scores, and the standard deviation. In addition, the mean ± 3 SD is displayed.

6. Define "Cutoff score." The default is the mean – 3 SD, but users may, for example, decide to use the lowest score obtained by the sequences in the file.

7. Define "Top score for scaling plots." This parameter is used by the graphics output routine when scaling the plots. Its value will influence the height of lines plotted to represent matches.

8. Define "Position to identify." When a search is performed, it is not always appropriate to report the position of a match relative to the leftmost amino acid in the motif. For example, when performing a helix-turn-helix motif search, the user may want to know the position of the well-conserved glycine rather than the position of the first amino acid in the matrix. The "Position to identify" allows the user to define which amino acid is marked. The amino acids in the table are numbers 1, 2, 3, and so on.

9. Define a "Title." This is a title that will be displayed when the matrix file is read prior to performing a search. It is limited to 60 characters.

10. Define "Name for new weight matrix file." Give a name for the weight matrix file.

See the example run in Fig. 4.

2.4.2. Searching Using a Weight Matrix

Once a weight matrix has been stored in a file, it can be used to search any sequence. Results can be displayed graphically, or the matching sequence segments can be listed out with their scores.

1. Select "Motif search using weight matrix."
2. Select "Use weight matrix."
3. Define "Motif weight matrix file." This is the name of the file containing the weight matrix. The program reads the file and displays its title.
4. Accept "Use frequencies as weights." The alternative will use the weight matrix file as a definition of a "Membership of set" motif (*see* Note 10).
5. Define "Cutoff score." The default will be the value set when the weight matrix file was created. If the score is negative, the program will calculate sums of logs of frequencies; otherwise, it will add frequencies.

6. Accept "Plot results." Alternatively, they will be listed. The results will appear.

3. Notes

1. The files containing the definitions of peptides that can be searched for by the exact match search routine have the following format. Each name is followed by a /, and then each of its peptide sequences is followed by a /. The last peptide sequence for each name is followed by //. For example, a file might contain the following.

   Acidic/D/E//
   Basic/R/K/H//
   Glyco/N-S/N-T//

   Users could then search for these named sets of sequences. Note that the symbol "–" matches any amino acid.
2. To search for a subset of the names in a file employed by exact match routine, the user should reject "Search for all names," and the program will ask for the names wanted and extract their sequences from the file. Alternatively, if a user was always using the same subset, then a file containing only those names could be created. This file would then be selected as "Personal file" for "Input source."
3. The exact match routine also allows names and their sequences to be entered on the keyboard. This is selected as "Keyboard" for "Input source," and the program will prompt for names and their sequences. In this way, the routine can be used to search for exact matches to any short sequence.
4. For this program, a motif is a short segment of sequence of fixed length. More complex structures termed "patterns," defined as sets of motifs separated by varying gaps, are covered in Chapter 13. The current chapter should be read before the chapter on patterns.
5. The files of aligned sequences used to make weight matrices have the following format. Each sequence should be on a separate line. The sequence should start in column 2, and is terminated by a new line or a space. Anything after the space is treated as a comment. The files can be created by previous searches or using an editor.
6. The frequencies in the weight matrix can be used in two ways to calculate scores for sequences. Some users prefer to add the frequencies to give a total score, and others to multiply them by summing their logs. If the frequencies are regarded as probabilities, then multiplication seems the correct procedure. The user chooses which method will be used when the weight matrix is created; however, the choice can be overridden when the matrix is used. If multiplication is selected, then all results will be presented as sums of logs.

```
    Motif search using weight matrix
    Select operation
X   1 Use weight matrix
    2 Make weight matrix
    3 Rescale weight matrix
?   Selection  (1-3) (1) =2
?   Name of aligned sequences file=atpbinding.seq
        1 GETLGIVGESGSGKSQSLR
        2 GESLGVVGESGGGKSTFAR OppF
        3 GDVISIDGSSGSGKSTFLR HisP
        4 GEFVVFVGPSGGGKSTLLR MalK E. coli
        5 NQVTAFIGPSGGGKSTLLR PstB
        6 GRVMALVGENGAGKSTMMK RbsA(N)
        7 GEVIGIVGRSGSGKSTLTK HlyB
        8 GECFGLLGPNGAGKSTITR NodI R. leguminosarum
        9 GEMAFLTGHSGAGKSTLLK FtsE E. coli
       10 GQRELIIGDRQTGKTALAI ATPase
       11 GGKVGLFGGAGVGKTVNMM ATPase
       12 GRIVEIYGPESSGKTTLTL RecA
       13 RSNLLVLAGAGSGKTRVLV UvrD
       14 GGKIGLFGGAGVGKTVGIM ATPase Bovine
       15 SKIIFVVGGPGSGKGTQCE Adenylate Kinase Rabbit
       16 NQSILITGESGAGKTVNTK Myosin Rabbit
       17 HVNVGTIGHVDHGKTTLTA EF-Tu E. coli
       18 YRNIGISAHIDAGKTTERI EF-G E. coli
       19 EYKLVVVGARGVGKSALTI v-ras (HARVEY)
       20 EYKLVVVGASGVGKSALTI v-ras (KIRSTEN)
       21 EYKLVVVGAVGVGKSALTI pEJ BLADDER CARCINOMA TRANSFORMING
       22 EYKLVVVGAGGVGKSALTI pEJ BLADDER CARCINOMA CELLULAR
    Length of motif    19
?   Sum logs of weights (y/n) (y) =
?   Use all motif positions (y/n) (y) =
```

Fig. 4. An example run of the creation of a weight matrix from a set of aligned sequences.

7. Masking the weight matrix is particularly useful in cases where a limited number of examples of a motif are available or when the motif may have several components. In the first case, the limited number of examples may make the matrix unrepresentative of the motif because the amino acids in the unconserved positions may bias the results of searches. It was stated that a motif might have several components: for example, it might have both structural and specificity components. The user may want to separate out the two parts, and again masking provides such a facility.

8. The weight matrix handling routine contains a further option, "Rescale weight matrix." If the user has edited a weight matrix to change the frequency values, this provides a way of selecting a new cutoff score. It allows users to read in a set of aligned sequences and a weight matrix, and to apply the matrix to the set of sequences to see the range of scores achieved. A new weight matrix file containing the selected cutoff score is written to disk.

```
Applying weights to input sequences
    1       -36.651 GETLGIVGESGSGKSQSLR
    2       -35.780 GESLGVVGESGGGKSTFAR
    3       -38.180 GDVISIDGSSGSGKSTFLR
    4       -35.403 GEFVVFVGPSGGGKSTLLR
    5       -39.039 NQVTAFIGPSGGGKSTLLR
    6       -40.653 GRVMALVGENGAGKSTMMK
    7       -34.017 GEVIGIVGRSGSGKSTLTK
    8       -37.454 GECFGLLGPNGAGKSTITR
    9       -36.474 GEMAFLTGHSGAGKSTLLK
   10       -43.431 GQRELIIGDRQTGKTALAI
   11       -40.210 GGKVGLFGGAGVGKTVNMM
   12       -40.720 GRIVEIYGPESSGKTTLTL
   13       -45.143 RSNLLVLAGAGSGKTRVLV
   14       -40.684 GGKIGLFGGAGVGKTVGIM
   15       -45.197 SKIIFVVGGPGSGKGTQCE
   16       -39.098 NQSILITGESGAGKTVNTK
   17       -43.832 HVNVGTIGHVDHGKTTLTA
   18       -44.817 YRNIGISAHIDAGKTTERI
   19       -36.305 EYKLVVVGARGVGKSALTI
   20       -35.101 EYKLVVVGASGVGKSALTI
   21       -36.305 EYKLVVVGAVGVGKSALTI
   22       -36.711 EYKLVVVGAGGVGKSALTI
Top score      -34.017   Bottom score       -45.197
Mean      -39.146   Standard deviation        3.441
Mean minus 3.sd     -49.470  Mean plus 3.sd      -28.822
? Cutoff score (-999.00-9999.00) (-49.47) =
? Top score for scaling plots (-49.47-999.00) (-28.82) =
? Position to identify (0-19) (1) =13
? Title=ATP binding motif
? Name for new weight matrix file=atpbinding.wts
```

Fig. 4 *(continued)*.

9. The program contains no hardwired motifs, since it was expected that most sites that would use the programs would accumulate their own libraries of motifs and patterns, and would use the PROSITE library, both of which users can employ by simply knowing the names of the corresponding files.

10. The weight matrix search can also be used as a "Membership of a set" search. This means that at each position in the motif, any amino acid type that is nonzero in the weight matrix is counted as a match and scores a value 1. *See* Chapter 13 on searching protein sequences for patterns.

References

1. Staden, R. (1988) Methods to define and locate patterns of motifs in sequences. *CABIOS* **4(1)**, 53–60.
2. Staden, R. (1990) Searching for patterns in protein and nucleic acid sequences, in *Methods in Enzymology*, vol. 183 (Doolittle, R. F., ed.), Academic, New York, pp. 193–211.

CHAPTER 13

Staden: Using Patterns
to Analyze Protein Sequences

Rodger Staden

1. Introduction

Described here is one of the most powerful facilities provided by
the program PIP: the ability to define and search sequences or libraries of sequences for complex patterns of motifs. Another chapter gives
details of searching for individual motifs, but this chapter shows how
to create individual patterns and libraries of patterns, and how to use
them to search sequences. Once a pattern has been defined and stored
in a file, it can be used to search any sequence. In addition, if users
want to screen sequences against libraries of patterns routinely, this
can be achieved by use of files of file names. For example, the program can use the PROSITE protein motif library. The program can
produce several alternative forms of output. It will display the segment of sequence matching each individual motif in the pattern, display all the sequence between and including the two outermost motifs,
produce a description of the match in the form of a SWISSPROT
feature table, or draw a simple graphical plot.

The end of the chapter will describe how a related program, PIPL,
is used to search libraries of sequences to find patterns. This program
can produce alignments of sequence families.

Patterns are defined as sets of motifs with variable spacing. Each
motif in a pattern can be defined using any of several methods, and
their positions relative to one other are defined in terms of minimum

From: *Methods in Molecular Biology, Vol. 25: Computer Analysis of Sequence Data, Part II*
Edited by: A. M. Griffin and H. G. Griffin Copyright ©1994 Humana Press Inc., Totowa, NJ

and maximum separations. In addition, by the use of logical operators, each motif can be declared to be essential (the AND operator),
optional (the OR operator), or forbidden (the NOT operator). The
following methods (termed "classes" by the program) for defining
motifs are provided:

1. Exact match to a short sequence;
2. Percentage match to a short sequence;
3. Match to a short sequence using a score matrix and cutoff score;
4. Match to a weight matrix;
5. Direct repeat; and
6. Membership of a set.

The motifs in a pattern are numbered sequentially, and motif spacing is defined in the following way. When a new motif is added to a
pattern, the user specifies the "Reference motif" by its number and
then a "Relative start position." The "Relative start position" is defined
by taking the first amino acid of the "Reference motif" as position
one, the next as two, and so on. Then the user defines the allowed
variation in the spacing by specifying the "Number of extra positions." Notice that the position of a motif can be defined relative to
any other motif and that a negative "Relative start position" declares
the motif to be to the left of its "Reference motif."

The probability of finding each individual motif in the current
sequence, the product of the probabilities for all the motifs in a pattern "Probability of finding pattern," and the "Expected number of
matches" are calculated and displayed by the program. In addition to
the cutoffs used for the individual motifs, users can apply two pattern
cutoffs: "Maximum pattern probability" and "Minimum pattern
score."

Described below are: how to create a pattern; how to use a pattern
file to search a sequence; how to use a "File of pattern file names" to
search a sequence for a whole library of patterns; how to use a pattern file to search a whole library of sequences; and how to reformat
the PROSITE motif library into a form compatible with these search
programs. To describe how to create a pattern file, first all the steps
are shown to make one containing two motifs, and then, to save space,
the parts specific to the individual motif types are sketched in Section 3.

```
ID   2FE2S_FERREDOXIN; PATTERN.
AC   PS00197;
DT   APR-1990 (CREATED); APR-1990 (DATA UPDATE); APR-1990 (INFO UPDATE).
DE   2Fe-2S ferredoxins, iron-sulfur binding region signature.
PA   C-x(1,2)-[STA]-x(2)-C-[STA]-{P}-C.
NR   /RELEASE=14,15409;
NR   /TOTAL=69(69); /POSITIVE=63(63); /UNKNOWN=0(0); /FALSE_POS=6(6);
NR   /FALSE_NEG=5(5);
CC   /TAXO-RANGE=A?EP?; /MAX-REPEAT=1;
CC   /SITE=1,iron_sulfur; /SITE=5,iron_sulfur; /SITE=8,iron_sulfur;
DR   P15788, FER$APHHA , T; P00250, FER$APHSA , T; P00223, FER$ARCLA , T;
DR   P00227, FER$BRANA , T; P07838, FER$BRYMA , T; P13106, FER$BUMFI , T;
DR   P00247, FER$CHLFR , T; P07839, FER$CHLRE , T; P00222, FER$COLES , T;
DO   PDOC00175;
//
```

Fig. 1. A typical entry from the PROSITE library.

1.1. Introduction to the PROSITE Motif Library

A library of protein motifs (in the author's terminology, because they include variable gaps, many would be called patterns) has recently become available from Amos Bairoch, Departement de Biochimie Medicale, University of Geneva. Currently, it contains over 500 patterns/motifs, and arrives on tape or cdrom in two files: a.DAT file and a.DOC file. There is also a user documentation file PROSITE.USR. Here the library structure and what is required to prepare the PROSITE library for use by the author's programs are outlined. A typical entry in the.DAT file is shown in Fig. 1.

Each entry has an accession number (in Fig. 1, PS00197), a pattern definition (in Fig. 1, C-x(1,2)-[STA]-x(2)-C-[STA]-{P}-C), and a documentation file crossreference (in Fig. 1, PDOC00175). This pattern means: C, gap of one or two, any of STA, gap of two, C, any of STA, not P, C.

It is necessary to convert all of these patterns into the author's pattern definitions (as membership of a set, with the appropriate gap ranges) and write each into a separate pattern file with corresponding "membership of a set" weight matrices. After the conversion, each pattern file is named accession_number.PAT (here PS00197.PAT). The corresponding matrix files are accession_number.WTSA, accession_number.WTSB, and so on, for however many are needed (here PS00197.WTSA and PS00197.WTSB): two are needed because of the variable gap.

```
N-glycosylation site.                                              00001,00001
Glycosaminoglycan attachment site.                                 00002,00002
Tyrosine sulfatation site.                                         00003,00003
cAMP- and cGMP-dependent protein kinase phosphorylation site.      00004,00004
```

Fig. 2. The start of the index created by the conversion program.

In addition, the.DAT and.DOC files can be optionally split into separate files, one for each entry, with names accession_number.DAT and accession_number.DOC. Also, an index is created for the library, which gives a one-line description of each pattern, and ends with the pattern file and documentation file numbers. The start of the file is shown in Fig. 2. So, referring to Fig. 2, the name of the pattern file for Glycosaminoglycan attachment site is PS00002.PAT, and for the documentation file, PDOC00002.DOC.

Finally, a file of file names for all the patterns in the library is created. If this file of file names is PROSITE.NAM, then to use the complete PROSITE library from program PIP, users select "pattern searcher," choose the option "use file of pattern file names," and give the file name PROSITE.NAM. For any matches found, the accession_number and pattern title will be displayed.

In order to make the PROSITE library usable by the search programs, it is only necessary to run a program named SPLITP3. Two other programs, SPLITP1 and SPLITP2, only make the original files marginally easier to manage and produce an index. SPLITP1 splits the PROSITE.DAT file to create a separate file for each entry. Each file is automatically named PSentry_number.DAT. In addition, it creates an index for the library.

SPLITP2 performs the same operation for the PROSITE.DOC file, except that no index is created. Files are named PSentry_number.DOC.

SPLITP3 creates a separate pattern file and weight matrix files for each PROSITE entry from the file PROSITE.DAT. Pattern files are named PSentry_number.PAT, weight matrix files PSentry_number.WTSA, PSentry_number.WTSB, and so forth. The pattern title is the one-line description of the motif. SPLITP3 also creates a file of file names. Notice that it will ask for a path name, so that the path can be included in the file of file names. This is the path to the directory in which the pattern files are stored.

2. Methods

2.1. Creating a Pattern File Containing a Weight Matrix Motif and a Membership of a Set Motif

1. Select "Pattern searcher."
2. Select "Pattern definition mode" as "Use keyboard."
3. Select "Results display mode" as "Inclusive." The alternatives are listed in Section 1.
4. Select "Motif definition mode" as "Weight matrix."
5. Define "Motif name." Each motif can be given an eight-character name.
6. Define "Weight matrix file name." Type in the name of the file containing the weight matrix. The program will display the probability of finding the motif.
7. Select "Motif definition mode" as "Membership of a set."
8. Define "Motif name."
9. Select "Logical operator" as "AND." The alternatives are "OR" and "NOT."
10. Select "Number of reference motif." At this stage, the only choice is 1, and this is the default.
11. Define "Relative start position." This is the position relative to the "Reference motif." *See* Section 1.
12. Define "Number of extra positions."
13. Select input mode as "Keyboard." The alternative is an existing file in the form of a weight matrix.
14. Define "String." Type in the sets of allowed residue types using the one-letter code. *See* Note 1.
15. Define the "Minimum matches." This is the number of positions within the motif that must match. The default is that all positions must match, but users may want to allow some flexibility by giving a lower score. The program now cycles around to step 7, and all subsequent passes around the loop to add further motifs to the pattern would differ only in the details for the different motif "classes."
16. Select "Pattern complete."
17. Accept "Save pattern in a file." The alternative does not save the pattern, so it can only be used once on the current sequence.
18. Define "Pattern definition file." Give a name for the new file.
19. "Define "Pattern title." All patterns can have a 60-character title that can be displayed when the pattern file is read and the sequence searched.

```
Pattern searcher
Select pattern definition mode
X  1 Use keyboard
   2 Use pattern file
   3 Use file of pattern file names
? Selection  (1-3) (1) =1
Select results display mode
X  1 Motif by motif
   2 Inclusive
   3 Graphical
   4 SWISSPROT feature table
? Selection  (1-4) (1) =2
Select motif definition mode
X  1 Exact match
   2 Percentage match
   3 Cut-off score and score matrix
   4 Cut-off score and weight matrix
   5 Direct repeat
   6 Membership of set
   7 Pattern complete
? Selection  (1-7) (1) =4
? Motif name=atp
? Weight matrix file name=atpbinding.wts
 ATP binding
Probability of score    -47.8010 = 0.302E-04
Select motif definition mode
   1 Exact match
   2 Percentage match
   3 Cut-off score and score matrix
X  4 Cut-off score and weight matrix
   5 Direct repeat
   6 Membership of set
   7 Pattern complete
? Selection  (1-7) (4) =6
? Motif name=hydro
Select logical operator
X  1 And
   2 Or
   3 Not
? Selection  (1-3) (1) =
? Number of reference motif (1-1) (1) =
? Relative start position (-1000-1000) (20) =22
? Number of extra positions (0-1000) (0) =5
Select input mode
X  1 Keyboard
   2 File
? Selection  (1-2) (1) =
Separate sets with commas
? String=ivl,ivl,,,rkhde
```

Fig. 3. The creation and use of a pattern containing a weight matrix motif and a membership of a set motif.

```
? Minimum matches (1.00-5.00) (3.00) =
Probability of score      3.000 = 0.145E-01
Select motif definition mode
    1 Exact match
    2 Percentage match
    3 Cut-off score and score matrix
    4 Cut-off score and weight matrix
    5 Direct repeat
X   6 Membership of set
    7 Pattern complete
? Selection  (1-7) (6) =7
? Save pattern in a file (y/n) (y) =
? Pattern definition file=_paper.pat
? Pattern title=atpbinding plus

? Weight matrix file name=_hydro.wts
Weight matrix needs a title
? Title=hydrophobic and + spot

Pattern description

atpbinding plus
Motif  1 named atp       is of class    4
Which is a match to a weight matrix with score -47.801
Motif  2 named hydro     is of class    6
Which is membership of a set with score   3.000
It is anded with the previous motif.
Probability of finding pattern = 0.4368E-06
Expected number of matches  = 0.1350E-02
? Maximum pattern probability (0.00-1.00) (1.00) =
? Minimum pattern score (-9999.00-9999.00) (-9999.00) =

    162
        GQRELIIGDRQTGKTALAIDAIINQR
Total matches found       1
Minimum and maximum observed scores       -38.35       -38.35
```

Fig. 3 *(continued)*.

20. Define "Weight matrix file name." The membership of a set motifs are stored in the form of weight matrices, so the program needs the user to define a file name.
21. Define "Title." Type in a title for the weight-matrix-like file. The title will be displayed when the file is read. The program will now display a detailed textual description of the pattern, the "Probability of finding the pattern," and the "Expected number of matches" (*see* Fig. 3).
22. Define "Maximum pattern probability." Yes maximum: any match with a greater probability of being found will be rejected. If no value is specified, the search will be quicker (*see* Section 3).

```
atpbinding plus
A4              atp        Class
atpbinding.wts
A6              hydro      Class
        1       Relative motif
       22       Relative start position
        5       Number of extra positions
_hydro.wts
```

Fig. 4. The pattern file created in the worked example shown in Fig. 3.

23. Define "Minimum pattern score." A minimum pattern score only makes sense if all the motifs in the pattern are defined with compatible scoring methods. For example, memberships of a set of motifs and weight matrices using sums of logs are incompatible. Searching will now commence and any matches will be displayed using the chosen method. Figure 3 shows a typical run in which a pattern containing a weight matrix and a membership of a set motif is created and stored on disk. Figure 4 shows the contents of the pattern file.

2.2. Searching a Sequence Using a Pattern File

1. Select "Pattern searcher."
2. Select "Pattern definition mode" as "Use pattern file."
3. Select "Results display mode" as "Inclusive."
4. Define "Pattern definition file." Type the name of the file containing the pattern. The program will read the file, then display its title, a detailed textual description of the pattern, the "Probability of finding the pattern," and the "Expected number of matches."
5. Define "Maximum pattern probability."
6. Define "Minimum pattern score." Searching will now commence, and any matches will be displayed using the chosen method. Figure 5 shows a typical run using a pattern file and output in the form of a SWISSPROT feature table.

2.3. Comparing a Sequence Against a Library of Patterns Including PROSITE

This mode of operation allows a sequence to be searched, in turn, for any number of patterns, each stored in a separate pattern file. The names of the files containing the individual patterns must be stored in a simple text file. This file is called "a file of pattern file names," and its name is the only user input required to define the search. The

```
   Pattern searcher
Select pattern definition mode
X  1 Use keyboard
   2 Use pattern file
   3 Use file of pattern file names
? Selection  (1-3) (1) =2
? Pattern definition file=_paper.pat
Select results display mode
X  1 Motif by motif
   2 Inclusive
   3 Graphical
   4 SWISSPROT feature table
? Selection  (1-4) (1) =4
 ATP binding sequences
Probability of score   -47.8010 = 0.302E-04
 hydrophobic and + spot
Probability of score    3.0000 = 0.145E-01

Pattern description

 atpbinding plus
Motif  1 named atp      is of class    4
Which is a match to a weight matrix with score -47.801
Motif  2 named hydro    is of class    6
Which is membership of a set with score    3.000
It is anded with the previous motif.
Probability of finding pattern = 0.4368E-06
Expected number of matches   = 0.1350E-02
? Maximum pattern probability (0.00-1.00) (1.00) =
? Minimum pattern score (-9999.00-9999.00) (-9999.00) =

FT    atp        162    187      Program

 Total matches found         1
 Minimum and maximum observed scores      -38.35      -38.35
```

Fig. 5. Worked example of using a pattern file to search a sequence, and writing the results in the form of a SWISSPROT feature table.

file of file names could contain references to entries in the PROSITE motif library and also include the names of other patterns.

1. Select "Pattern searcher."
2. Select "Pattern definition mode" as "Use file of pattern file names."
3. Select "Results display mode" as "Inclusive."
4. Define "File of pattern file names." Type the name of the file containing the list of pattern file names. The program will read the file and then, in turn, all the pattern files it names. Each of these patterns will be

compared against the current sequence, but only those that give matches will produce any output. The pattern title and each match will be displayed.

2.4. Searching Libraries for Patterns

The program PIPL can be used to search whole sequence libraries for patterns. Its use is similar to the pattern search routine described above, except that it does not have the facility for creating pattern files, so they must be created beforehand using PIP. In addition to its obvious application of finding new occurrences of patterns or checking on their frequency, it is a useful way of obtaining sequence alignments. It can restrict its search to a list of named entries, or can search all but those on a list of entries. It can restrict its output to showing the highest scoring match in each sequence, but by default, it will show all matches.

Of its modes of output, two require further description. The first, which is called "Padded sections," creates a new file for each match. The file will contain the sequence between and including the two outermost motifs in the pattern. It will be gapped to the furthest extent defined by the pattern, which means that if all the files were subsequently written one above the other, all the motifs in the pattern would be exactly aligned, with the sections between them containing the requisite numbers of padding characters. The second such mode of output is called "Complete padded sequences." Here the user must know the maximum distance between the leftmost motif and the start of all the sequences that match. A trial run in which only the positions of matches are reported is usually required. The user gives this maximum distance to the program. The program then writes a new file containing the full length of all matching sequences, again maximally gapped (including their left ends) so that they would all align if written above one another. For both of these modes of output, the files created are named "entryname" where "entryname" is the name given to the sequence in the sequence library. These modes are best used with the option "Report all matches" rejected, so that only the best match for each sequence is reported. The sequences can be lined up using the sequence assembly program SAP.

The searches, which have recently been recoded, are very rapid. For example, a search of the current SWISSPROT library for a pattern defining the globin family as six weight matrices with widely varying gaps finds only globins and takes <4 min using a single processor on an Alliant FX2800. This time includes reading in the whole library as stored in EMBL CDROM format. A worked example of this search is shown in Fig. 6.

1. Select PIPL.
2. Define "Name for results file."
3. Select a library.
4. Select "Search whole library." The alternatives are "Search only a list of entries" and "Search all but a list of entries." The files containing the list of entries should contain one entry name per line, left justified.
5. Select "Results display mode" as "Inclusive." The alternatives include "Motif by motif," "Scores only," "Complete padded sequences," and "Padded sections."
6. Accept "Report all matches." The alternative only shows the best match for each sequence.
7. Define "Pattern definition file." This is the name of the file containing the pattern created using PIP. The program displays a textual description of the pattern and the expected number of matches per 1000 residues assuming an average amino acid composition.
8. Define "Maximum pattern probability." The program will run much more quickly if none is given.
9. Define "Minimum pattern score." The search will start.

2.5. Preparing the PROSITE Motif Library for Use by the Programs

Only the program SPLITP3 is essential for preparing the PROSITE library for use by our programs.

1. Select SPLITP3.
2. Define "Prosite library file." Type the name of the file containing the prosite library (usually PROSITE.DAT).
3. Define "Name for file of pattern file names." This is the file of file names that users will employ to search the whole library. It will be convenient for them if an environment variable is defined for this file name.
4. Define "Path name of motif directory." This is the full path name, including the final /, to the directory in which the converted library will be stored.

3. Notes

1. The "exact match" motif class requires a consensus sequence. The "percentage match" motif class requires a consensus sequence and a cutoff score. The "score matrix" motif class uses the MDM78 matrix, and requires a consensus sequence and a cutoff score. The "weight matrix"

```
PIPL (Protein interpretation program (library)) V4.1 Jul 1991
Author: Rodger Staden
Searches protein libraries for patterns of motifs

? Name for results file=globin.res
Select a library
    1 EMBL nucleotide library
X   2 SWISSPROT protein library
    3 Personal file in PIR format
? Selection  (1-3) (2) =
Library is in EMBL format with indexes
Select a task
X   1 Search whole library
    2 Search only a list of entries
    3 Search all but a list of entries
? Selection  (1-3) (1) =
Select results display mode
X   1 Motif by motif
    2 Inclusive
    3 Scores only
    4 Complete padded sequences
    5 Padded sections
? Selection  (1-5) (1) =5
? (y/n) (y) Report all matches n
? Pattern definition file=globin.pat
 globin 1
Probability of score    -34.5300 = 0.197E-02
 globin 2
Probability of score    -44.6000 = 0.409E-02
 globin 3
Probability of score    -75.1000 = 0.293E-01
 globin 4
Probability of score    -36.1000 = 0.147E-01
 globin 5
Probability of score    -73.7000 = 0.375E-01
 globin 6
Probability of score    -55.9000 = 0.483E-01
```

Fig. 6. A typical run of PIPL using a pattern of six weights.

search only requires the name of the file containing the matrix. The "direct repeat" motif class requires a repeat length, the minimum and maximum gap between the two occurrences of the repeat, and a minimum score. The "membership of a set" motif class defines sets of residue types that are allowed at each position in the motif. When they are first entered into the pattern, they are normally typed on the keyboard, but when they are stored in a file, they are written in the same format as a weight matrix. To enter them on the keyboard, use the following format. Type the one-letter codes for the set of residue types allowed at

```
Pattern description

Globin pattern file
Motif  1 named g1        is of class    4
Which is a match to a weight matrix with score -34.530
Motif  2 named g2        is of class    4
Which is a match to a weight matrix with score -44.600
and the N-terminal residue can take positions     17 to       22
relative to the N-terminal end of motif    1
It is anded with the previous motif.
Motif  3 named g3        is of class    4
Which is a match to a weight matrix with score -75.100
and the N-terminal residue can take positions     27 to       35
relative to the N-terminal end of motif    2
It is anded with the previous motif.
Motif  4 named g4        is of class    4
Which is a match to a weight matrix with score -36.100
and the N-terminal residue can take positions     29 to       53
relative to the N-terminal end of motif    3
It is anded with the previous motif.
Motif  5 named g5        is of class    4
Which is a match to a weight matrix with score -73.700
and the N-terminal residue can take positions     12 to       16
relative to the N-terminal end of motif    4
It is anded with the previous motif.
Motif  6 named g6        is of class    4
Which is a match to a weight matrix with score -55.900
and the N-terminal residue can take positions     29 to       33
relative to the N-terminal end of motif    5
It is anded with the previous motif.
Probability of finding pattern = 0.6273E-11
Expected number of matches per 1000 residues = 0.2119E-03
? Maximum pattern probability (0.00-1.00) (1.00) =
? Minimum pattern score (-9999.00-9999.00) (-9999.00) =
```

Fig. 6 *(continued)*.

each position terminated by a comma (,). For positions where any residue type is allowed, simply type an extra comma. For example, VLI,FY,,,DE means any of Valine, Leucine, or Isoleucine in the first position, either Phenylalanine or Tyrosine in the next position, anything in the next two positions, and Aspartic acid or Glutamic acid in the next. When the pattern is stored on the disk, the program will request a name for the file and a title for the motif.

2. The details of the probability calculations are outside the scope of this chapter. They are quite rapid, and are essential both for assessing the

statistical significance of any matches found and for allowing meaningful cutoffs to be applied to patterns. Obviously, in general, cutoff scores are inappropriate for patterns containing a mixture of motif classes.

3. The program calculates the "Probability of finding the pattern" and the "Expected number of matches." The first figure is actually the product of the individual motif probabilities, but the latter figure is more useful because it takes into account the allowed variation in spacing between motifs and the length of the current sequence. In both cases, the composition of the current sequence is also used so that different probabilities would be calculated for other sequences.

4. The pattern definition system is very flexible. Assume that a laboratory has a large library of patterns stored in its computer. Different groups or users may want to screen their sequences against different subsets of a pattern library. Each group therefore uses its own "File of pattern file names," which contains only the names of the pattern files that are relevant to their sequences. Of course a pattern may contain only one motif. Hence, a library of patterns can include both simple and complex patterns. In the same way, a laboratory may have a large library of weight matrices defining different motifs, and different users may want to combine them in different ways to produce their own patterns. Also, of course, a library does not have to be used solely for performing mass screenings: Each individual entry can be used as a single pattern by giving the name of its pattern file, e.g., **pathname/PS00002.PAT**.

5. Note that five of the PROSITE motifs contain the symbols > or <, which means that the motifs must appear exactly at the N or C termini of the sequences. Currently, the author's methods have no mechanism for such definitions and, for example, KDEL motifs will be permitted to occur anywhere throughout a sequence.

6. Further information about the methods is given in refs. *1–3*.

References

1. Staden, R. (1988) Methods to define and locate patterns of , ! motifs in sequences. *CABIOS* **4(1)**, 53–60.
2. Staden, R. (1989) Methods for calculating the probabilities of finding patterns in sequences. *CABIOS* **5(2)**, 89–96.
3. Staden, R. (1990) Searching for patterns in protein and nucleic acid sequences, in *Methods in Enzymology* vol. 183 (Doolittle, R. F., ed.), Academic, New York, pp. 193–211.

CHAPTER 14

Staden: Comparing Sequences

Rodger Staden

1. Introduction

This chapter describes methods for comparing and aligning pairs of nucleic acid or protein sequences. The program described (SIP), the original version of which was first described in 1982 *(1)*, is based around several methods for producing "dot matrix" plots and includes routines for assessing the statistical significance of the plots, plus a dynamic programming algorithm for finding optimal alignments. The end of the chapter describes a program, SIPL, that is used for comparing a single sequence against a whole library of sequences.

It is assumed the reader is familiar with the general principle of dot matrix diagrams. The program uses a number of different algorithms to calculate the score for each point in a dot matrix, and the user defines a minimum score so that only those points in the diagram for which the score is at least this value will be marked with a dot. The first scoring method finds uninterrupted sections of perfect identity, i.e., those that contain no mismatches, insertions or deletions. Generally, this method, termed "the identities algorithm," is of limited value, but runs very quickly.

The second method looks for sections where a proportion of the characters in the sequence are similar, again allowing no insertions or deletions. For a thorough analysis, this method, termed "the proportional algorithm," is the best. The original method of this type was first described by McLachlan *(2)*, and involves calculating a score for each position in the matrix by summing points found when looking forward and backward along a diagonal line of a given length

From: *Methods in Molecular Biology, Vol. 25: Computer Analysis of Sequence Data, Part II*
Edited by: A. M. Griffin and H. G. Griffin Copyright ©1994 Humana Press Inc., Totowa, NJ

```
     C  S  T  P  A  G  N  D  E  Q  B  Z  H  R  K  M  I  L  V  F  Y  W  -  X  ?
C   22 10  8  7  8  7  6  5  5  5  5  5  7  6  5  5  8  4  8  6 10  2 10 10 10 10
S   10 12 11 11 11 11 11 10 10 10  9 10 10  9 10 10  8  9  7  9  7  7  8 10 10 10 10
T    8 11 13 10 11 10 10 10 10 10  9 10 10  9  9 10  9 10  8 10  7  7  5 10 10 10 10
P    7 11 10 16 11  9  9  9  9  9 10  9 10 10 10  9  8  8  7  9  5  5  4 10 10 10 10
A    8 11 11 11 12 11 10 10 10 10 10 10  9  8  9  9  9  8 10  6  7  4 10 10 10 10
G    7 11 10  9 11 15 10 11 10  9 10 10  8  7  8  7  7  6  9  5  5  3 10 10 10 10
N    6 11 10  9 10 10 12 12 11 11 12 11 12 10 11  8  8  7  8  6  8  6 10 10 10 10
D    5 10 10  9 10 11 12 14 13 12 13 12 11  9 10  7  8  6  8  4  6  3 10 10 10 10
E    5 10 10  9 10 10 11 13 14 12 12 13 11  9 10  8  8  7  8  5  6  3 10 10 10 10
Q    5  9  9 10 10  9 11 12 12 14 11 13 13 11 11  9  8  8  8  5  6  5 10 10 10 10
B    5 10 10  9 10 10 12 13 12 11 13 11 11 10 10  8  8  6  8  5  7  4 10 10 10 10
Z    5 10 10 10 10 10 11 12 13 13 11 14 12 10 10  8  8  8  8  5  6  4 10 10 10 10
H    7  9  9 10  9  8 12 11 11 13 11 12 16 12 10  8  8  8  8  8 10  7 10 10 10 10
R    6 10  9 10  8  7 10  9  9 11 10 10 12 16 13 10  8  7  8  6  6 12 10 10 10 10
K    5 10 10  9  9  8 11 10 10 11 10 10 10 13 15 10  8  7  8  5  6  7 10 10 10 10
M    5  8  9  8  9  7  8  7  8  9  8  8  8 10 10 16 12 14 12 10  8  6 10 10 10 10
I    8  9 10  8  9  7  8  8  8  8  8  8  8  8  8 12 15 12 14 11  9  5 10 10 10 10
L    4  7  8  7  8  6  7  6  7  8  6  8  8  7  7 14 12 16 12 12  9  8 10 10 10 10
V    8  9 10  9 10  9  8  8  8  8  8  8  8  8  8 12 14 12 14  9  8  4 10 10 10 10
F    6  7  7  5  6  5  6  4  5  5  5  5  8  6  5 10 11 12  9 19 17 10 10 10 10 10
Y   10  7  7  5  7  5  8  6  6  7  6 10  6  6  8  9  9  8 17 20 10 10 10 10 10
W    2  8  5  4  4  3  6  3  3  5  4  4  7 12  7  6  5  8  4 10 10 27 10 10 10 10
-   10 10 10 10 10 10 10 10 10 10 10 10 10 10 10 10 10 10 10 10 10 10 10 10 10 10
X   10 10 10 10 10 10 10 10 10 10 10 10 10 10 10 10 10 10 10 10 10 10 10 10 10 10
?   10 10 10 10 10 10 10 10 10 10 10 10 10 10 10 10 10 10 10 10 10 10 10 10 10 10
    10 10 10 10 10 10 10 10 10 10 10 10 10 10 10 10 10 10 10 10 10 10 10 10 10 10
```

Fig. 1. The amino acid score matrix MDM78.

(the window). The algorithm does not simply look for identity, but uses a score matrix that contains scores for every possible pair of characters. For comparing amino acid sequences, the score matrix MDM78 *(3),* which is shown in Fig. 1, is usually used. It is also possible to use other matrices, including an identity matrix for proteins. For nucleic acids, an identity matrix is usually used.

For the proportional method, plotting dots at the centers of windows that reach the cutoff leads to a persistence effect that, to some extent, can be mitigated by a variation on the method. If, for example, all the high scoring amino acids are clustered at the left end of a particular diagonal segment, dots will continue to be plotted to their right until the window score drops below the cutoff. Instead of plotting a single point for each window that reaches the cutoff score, the variant method plots points for all the identities that lie in windows that reach the cutoff. Obviously, the persistence effect can be more pronounced for long windows and low cutoff scores, but note that the variant method will plot nothing if there are no identities present, and so similar regions could be missed! A further variant, useful for comparing a sequence against itself, ignores the main diagonal.

The third comparison method, called "quick scan," is really a combination of the first two, and is similar to the FASTP program of Lipman and Pearson *(4)*, but produces a dot matrix diagram. The algorithm is as follows. The dot matrix positions are found for all words of some minimum length (obviously length one is most sensitive) that are common to both sequences. Imagine a diagonal line running from corner to corner of the diagram, at right angles to the diagonals in the dot matrix. The scores for the common words (according to the current score matrix, e.g., MDM78) are accumulated at the appropriate positions on that imaginary line, hence, producing a histogram. The histogram is analyzed to find its mean and standard deviation. The diagonals that lie above some cutoff score (defined in standard deviation units) are rescanned using the proportional algorithm, and a diagram produced. The method is very fast and is also employed by the library comparison program (*see* Section 2.8.).

The dynamic programming alignment algorithm contained in the program is based on that of Myers and Miller *(5)*. It guarantees to produce alignments with the optimum score given a score matrix, a gap start penalty, and a gap extension penalty. It is very useful to have the dot matrix methods and the alignment routine together in the same program, because it allows users to produce a dot matrix diagram to help select which regions of the sequence they wish to align. Selection is made by use of the crosshair. The crosshair is positioned first at the bottom left-hand end of the segment to be aligned and then at the top right of the segment. When the alignment routine is selected, the segment will be aligned. The alignment can replace the original segment of the sequence. By repeated plotting of dot matrices, followed by alignment, very long sequences can easily be aligned.

2. Methods

2.1. Producing a Dot Matrix Plot of Exact Matches

This method is relatively fast and can be useful for very similar sequences. It marks the position of every exact match of some minimum length with a dot.

Fig. 2. A dot matrix for two related protein sequences using the "Identities algorithm" and a score of two. Notice that the similarity is not apparent.

1. Select "Apply identities algorithm."
2. Define "Identity score." The plot will appear as in Fig. 2, which shows a comparison of two protein sequences using a score of two.

2.2. Producing a Dot Matrix Plot Using the Proportional Algorithm

This method gives the most thorough analysis.

1. Select "Apply proportional algorithm."
2. Define "Odd window length." The size of window over which the scores for each point are summed.
3. Define "Proportional score." All points achieving at least this score will be marked with a dot in the diagram. The plot will appear as in Fig. 3.

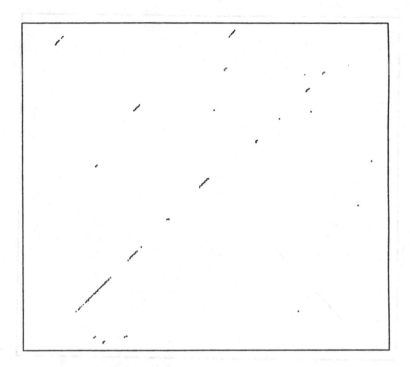

Fig. 3. A dot matrix for the two related protein sequences shown in Fig. 2, but here using the "Proportional algorithm" with a window of 21 and a score of 240. Notice that the similarity is now apparent.

2.3. Producing a Dot Matrix Plot
Using the Quick Scan Algorithm

This method is very fast. Using the current score matrix, it accumulates the scores for all the exact matches that lie on each diagonal. The mean diagonal score and its standard deviation are calculated, and those diagonals that have scores more than a chosen number of s.d. above the mean are rescanned using the proportional algorithm, and the points above the proportional algorithms cutoff are plotted.

1. Select "Apply quick scan algorithm."
2. Define "Identity score." This is the minimum number of consecutive identical sequence symbols that count as a match.
3. Define "Odd window length." This is the size of window over which the scores for each point are summed when the proportional algorithm is applied to the best diagonals.

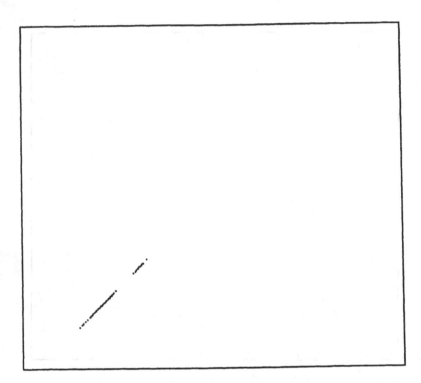

Fig. 4. A dot matrix for the two related protein sequences shown in Figs. 2 and 3, but here using the "Quick scan algorithm" with an identity score of one and a window of 21, and a score of 240 for the proportional algorithm. Notice that the similarity is now apparent, but the absence of background "noise" is misleading.

4. Define "Proportional score." For the best diagonals, all points achieving at least this score will be marked with a dot in the diagram.
5. Define "Number of s.d. above mean." Diagonals with scores above the minimum number of s.d. are rescanned using the proportional algorithm. The plot will appear as in Fig. 4.

2.4. Producing a List of All Matching Segments Using the Proportional Algorithm

1. Select "List matching segments."
2. Define "Odd window length." This is the size of window over which the scores for each point are summed.
3. Define "Proportional score." All segments achieving at least this score will be listed out with the two sequences written one above the other. *See* Fig. 5.

```
List matching segments
? Odd window length (1-401) (11) =
? Proportional score (1-567) (252) =
Working
       62
GLRRGLDVKDLEHPIEVPVGK
DLAEGMKVKCTGRILEVPVGR
       81
       63
LRRGLDVKDLEHPIEVPVGKA
LAEGMKVKCTGRILEVPVGRG
       82
       65
RGLDVKDLEHPIEVPVGKATL
EGMKVKCTGRILEVPVGRGLL
       84
       66
GLDVKDLEHPIEVPVGKATLG
GMKVKCTGRILEVPVGRGLLG
       85
       67
LDVKDLEHPIEVPVGKATLGR
MKVKCTGRILEVPVGRGLLGR
       86
```

Fig. 5. A typical run of "List matching segments."

2.5. Calculating the Expected Scores for the Proportional Algorithm

This function calculates the probability of achieving each possible score using the proportional algorithm. Hence, it provides a method of setting cutoff scores and assessing the statistical significance of the scores found. The algorithm calculates the "Double matching probability" described by McLachlan *(2)*, which is defined as the probability of finding the scores in two infinitely long sequences of the same composition as the pair being compared. It is very much faster than the alternative of repeatedly scrambling and recomparing the sequences. The program offers three ways for the user to see the results of the calculation: The user can type a score, and the program will display its probability; the user can type a probability, and the program will display the corresponding score; alternatively, the program will list the full range of scores and probabilities.

```
Calculate expected proportional scores
? Odd window length (1-401) (21) =
Working
Average score=   196.99062
Select probability display mode
    1 Show probability for a score
X   2 Show score for a probability
    3 List scores and probabilities
? Selection   (1-3) (2) =3
? Number of steps between scores (1-10) (5) =
```

5	0.10000E+01	200	0.40004E+00	395	0.00000E+00
10	0.10000E+01	205	0.24037E+00	400	0.00000E+00
15	0.10000E+01	210	0.12555E+00	405	0.00000E+00
20	0.10000E+01	215	0.56905E-01	410	0.00000E+00
25	0.10000E+01	220	0.22402E-01	415	0.00000E+00
30	0.10000E+01	225	0.76821E-02	420	0.00000E+00
35	0.10000E+01	230	0.23031E-02	425	0.00000E+00
40	0.10000E+01	235	0.60614E-03	430	0.00000E+00
45	0.10000E+01	240	0.14064E-03	435	0.00000E+00
50	0.10000E+01	245	0.28888E-04	440	0.00000E+00
55	0.10000E+01	250	0.52741E-05	445	0.00000E+00
60	0.10000E+01	255	0.85917E-06	450	0.00000E+00
65	0.10000E+01	260	0.12534E-06	455	0.00000E+00
70	0.10000E+01	265	0.16433E-07	460	0.00000E+00
75	0.10000E+01	270	0.19425E-08	465	0.00000E+00
80	0.10000E+01	275	0.20772E-09	470	0.00000E+00
85	0.10000E+01	280	0.20155E-10	475	0.00000E+00
90	0.10000E+01	285	0.17801E-11	480	0.00000E+00
95	0.10000E+01	290	0.14353E-12	485	0.00000E+00
100	0.10000E+01	295	0.10599E-13	490	0.00000E+00
105	0.10000E+01	300	0.71886E-15	495	0.00000E+00
110	0.10000E+01	305	0.44920E-16	500	0.00000E+00
115	0.10000E+01	310	0.25938E-17	505	0.00000E+00
120	0.10000E+01	315	0.13881E-18	510	0.00000E+00

Fig. 6. A typical run of "Calculate expected proportional scores." The scores are listed in three columns alongside their probabilities, e.g., score 250 has a probability 0.527×10^{-5}.

1. Select "Calculate expected scores."
2. Define "Odd window length." The calculation takes a noticeable time.
3. Select "List scores and probabilities."
4. Define "Number of steps between scores." This allows, say, every fifth score to be listed if the user defines the number of steps to be five. The list will appear as in Fig. 6.

2.6. Calculating the Observed Scores for the Proportional Algorithm

This function applies the proportional algorithm, but instead of producing a dot matrix, it accumulates the scores and their frequencies of occurrence. It provides a method of setting cutoff scores and assessing the statistical significance of the scores found. The program offers three ways for the user to see the results of the calculation: The user can type a score, and the program will display its frequency; the user can type a frequency, and the program will display the corresponding score; alternatively, the program will list the full range of scores and frequencies. The frequencies are expressed as percentages.

1. Select "Calculate observed scores."
2. Define "Odd window length." The calculation takes a noticeable time.
3. Select "List scores and percentages."
4. Define "Number of steps between scores." This allows, say, every fifth score to be listed if the user defines the number of steps to be five. The list will appear as in Fig. 7.

2.7. Producing an Optimal Alignment

This function produces an optimal alignment for any segments of the two sequences using the algorithm of Myers and Miller *(5)*. It guarantees to produce alignments with the optimum score, given a score matrix, a "gap start penalty," and a "gap extension penalty." That is, starting a gap costs a fixed penalty, F, and each residue added to the gap costs a further penalty, E, so for gap of length K residues, the penalty is F + KE. Gaps at the ends of sequences incur no penalty. The size of the segments of sequence that can be aligned at once is limited to 5000 characters. The user can select the start and end of the segments by use of the crosshair simply by clicking on any dot matrix plot. After the alignment has been produced, the user can elect to have it replace the original sequence segments. By alternate use of dot matrix plotting and alignment, very long sequences can be aligned.

1. Select "Align sequences." The crosshair will appear in the graphics window.
2. Position the crosshair on the bottom left of the segment to be aligned, and hit the space bar on the keyboard. The bell will ring.

```
      Calculate observed proportional scores
      ? Odd window length (1-401) (21) =
      Working
      Maximum observed score is     285
      Select score display mode
      X   1 Show percentage reaching a score
          2 Show score for a percentage
          3 List scores and percentages
      ? Selection   (1-3) (1) =3
       ? Number of steps between scores (1-10) (5) =
          156    236949  0.99998E+02
          161    236938  0.99993E+02
          166    236792  0.99932E+02
          171    235882  0.99548E+02
          176    232582  0.98155E+02
          181    222875  0.94058E+02
          186    203232  0.85769E+02
          191    171507  0.72380E+02
          196    131216  0.55376E+02
          201     89194  0.37642E+02
          206     52791  0.22279E+02
          211     27315  0.11528E+02
          216     12117  0.51137E+01
          221      4890  0.20637E+01
          226      1774  0.74867E+00
          231       656  0.27685E+00
          236       263  0.11099E+00
          241       111  0.46845E-01
          246        66  0.27854E-01
          251        36  0.15193E-01
          256        23  0.97065E-02
          261        16  0.67524E-02
          266        15  0.63303E-02
          271        10  0.42202E-02
          276         6  0.25321E-02
          281         2  0.84405E-03
```

Fig. 7. A typical run of "Calculate observed scores." The scores are followed by their observed number of occurrences expressed both absolutely and as a percentage of the total number of points.

3. Position the crosshair on the top right of the segment to be aligned, and hit the space bar on the keyboard. The bell will ring.
4. Define "Penalty for starting each gap."
5. Define "Penalty for each residue in gap." A noticeable time will elapse before the alignment is displayed on the screen. A typical alignment is shown in Fig. 8.

```
Align the sequences
Aligning region 1 to  461
   with region 1 to  514
Working
    V     1         11        21        31        41        51
          MA--TGKIVQ VIGA------ VVDVEFPQDA VPRVYDALEV QNG------N ERLVL-----
          *      *    *         **         * *    *     *       * *
          MQLNSTEISE LIKQRIAQFN VVSEAHNEGT IVSVSDGVIR IHGLADCMQG EMISLPGNRY
    H     1         11        21        31        41        51
    V     61        71        81        91        101       111
          EVQQQLGGGI VRTIAMGSSD GLRRGLDVKD LEHPIEVPVG KATLGRIMNV LGEPVDMKGE
          *     *  **  *  * **     *****   *** *  ** * * **
          AIALNLERDS VGAVVMGPYA DLAEGMKVKC TGRILEVPVG RGLLGRVVNT LGAPIDGKGP
    H     61        71        81        91        101       111
    V     121       131       141       151       161       171
          IGEEERWAIH RAAPSYEELS NSQELLETGI KVIDLMCPFA KGGKVGLFGG AGVGKTVNMM
          *     **    * *         ** * * **       *    * *        *  ***
          LDHDGFSAVE AIAPGVIERQ SVDQPVQTGY KAVDSMIPIG RGQRELIIGD RQTGKTALAI
    H     121       131       141       151       161       171
    V     181       191       201       211       221       231
          ELIRNIAIEH SGYS-VFAGV GERTREGNDF YHEMTDSNVI DKVSLVYGQM NEPPGNRLRV
          *   *  **          *                             *       *
          DAI--INQRD SGIKCIYVAI GQKASTISNV VRKLEEHGAL ANTIVVVATA SESAALQYLA
    H     181       191       201       211       221       231
    V     241       251       261       271       281       291
          ALTGLTMAEK FRDEGRDVLL FVDNIYRYTL AGTEVSALLG RMPSAVGYQP TLAEEMGVLQ
          *  *  *** *  *  *          *      * **  *   *
          RMPVALMGEY FRDRGEDALI IYDDLSKQAV AYRQISLLLR RPPGREAFPG DVFYLHSRLL
    H     241       251       261       271       281       291
    V     301       311       321       331       341       351
          ERITST---- ---------- -KTGSITSVQ AVYVPADDLT DPSPATTFAH LDATVVLSRQ
          **         ****  *    * *  *         *       *
          ERAARVNAEY VEAFTKGEVK GKTGSLTALP IIETQAGDVS AFVPTNVISI TDGQIFLETN
    H     301       311       321       331       341       351
    V     361       371       381       391       401       411
          IASLGIYPAV DPLDSTSRQL DPLVVGQEHY DTAR----GV QSILQRYQEL KDIIAILGMD
          ** ***   *  * **        * *                *     *  **
          LFNAGIRPAV NPGISVSR-- ---VGGAAQT KIMKKLSGGI RTALAQYREL AAFSQFAS--
    H     361       371       381       391       401       411
    V     421       431       441       451       461       471
          ELSEEDKLVV ARARKIQRFL SQ----PFFV AE----VFTG SPGKYVSLKD --TIRGFKGI
          *            * *  *     * *     *          *  * *
          DLDDATRKQL DHGQKVTELL KQKQYAPMSV AQQSLVLFAA ERG-YLADVE LSKIGSFEAA
    H     421       431       441       451       461       471
    V     481       491       501       511       521
          MEG--EYDHL P-EQAFYMVG SIEEAVE--- --------KA KKL*
          **  *  * *     *           *                *
          LLAYVDRDHA PLMQEINQTG GYNDEIEGKL KGILDSFKAT QSW*
    H     481       491       501       511       521
Conservation  22.5%
Number of padding characters inserted     63 and    10
```

Fig. 8. A typical output from "Align the sequences." The horizontal and vertical sequences are labeled H and V.

6. Reject "Keep alignment." If the alignment is "kept," the padded sequences from the alignment will replace the original sequences in the active region.

2.8. Comparing a Sequence Against a Library of Sequences

The program SIPL is used for comparing a probe sequence against a whole library of sequences. The searches are very fast and use the "Quick scan" algorithm described above to produce a list of matching sequences sorted in score order. Optionally, this is followed by the production of optimal alignments using the Myers and Miller *(5)* algorithm. The program will search the whole of a library or restrict its search using a list of entry names. The list of entry names can be used either as a list of sequences to search or, conversely, as a list of sequences to exclude from a search.

1. Select SIPL.
2. Select "Personal file."
3. Select "Format."
4. Define "Name of sequence file." This is the name of the file containing the probe sequence.
5. Define "Name of results file."
6. Accept "Display alignments." The alternative will stop after producing a list of the best-matching sequences.
7. Define "Minimum library sequence length." This permits the search to skip sequences that are too short to be of interest.
8. Define "Maximum number of scores to list." This is the maximum number of sequences that will be included in the results file.
9. Define "Identity score." This is the minimum number of consecutive sequence characters that will be counted as a match. Only matches of at least this length will be included in the overall score. For proteins, maximum sensitivity is gained using a value of one, but for nucleic acids values of four or six are necessary to achieve reasonable speed.
10. Define "Number of s.d. above mean." This means the number of s.d. above the mean that a diagonal must score in order for it to be scanned using the proportional algorithm.
11. Define "Odd window length." This is the window size for the rescanning of high-scoring diagonals using the proportional algorithm.

12. Define "Proportional score." This is the score used by the proportional algorithm. It depends on the window length and the score matrix.
13. Define "Minimum global score." This is the total score achieved using the proportional algorithm when all the diagonals scoring the defined number of s.d. above the mean are rescanned.
14. Define "Penalty for starting a gap." This is for the alignment algorithm.
15. Define "Penalty for each residue in gap."
16. Select a library to search. The default library will reflect the composition of the probe sequence. That is, a probe sequence that is < 85% acgt will be guessed to be a protein.
17. Select "Search whole library." The alternatives allow the search to be restricted using a list of entry names.

The search will start. A large number of parameters are required, but for normal use, the default value can be taken for them all. A worked example is shown in Fig. 9.

3. Notes

1. The variants on the proportional algorithm are selected by setting parameters using a special menu.
2. For nucleotide sequences, the program also has a function to complement a sequence. If the sequences on one axis are the complement of that on the other, the plots will show possible base pairing.
3. When the crosshair is being employed, in addition, to the standard special keys, the letter m will produce a display showing all the identical sequence characters around the crosshair position. The display is in the form of a matrix.
4. Users should not be misled by the "Quick scan" algorithm. Its function is to perform rapid comparisons. The plots it produces may look quite striking, because they will contain almost no background; however, such plots tell nothing about the significance of the similarities displayed.
5. By using the "Reposition plots" function, users can display several dot matrix plots on the screen at the same time. In this way, plots from several pairs of sequence comparisons can be viewed together.
6. The library search program SIPL is of limited use for searching the nucleic acid libraries, because it does not deal properly with sequences longer than 20,000 characters, but simply truncates them.

```
SIPL (Similarity investigation program (Library)) V3.0 June 1991
Author: Rodger Staden
Compares a probe protein or nucleic acid
sequence against a library of sequences

Select probe sequence
 Select sequence source
 X  1 Personal file
    2 Sequence library
 ? Selection  (1-2) (1) =2
 Select a library
    1 EMBL nucleotide library
 X  2 SWISSPROT protein library
    3 PIR protein library
 ? Selection  (1-3) (2) =
Library is in EMBL format with indexes
 Select a task
 X  1 Get a sequence
    2 Get annotations
    3 Get entry names from accession numbers
    4 Search titles for keywords
    5 Search keyword index for keywords
 ? Selection  (1-5) (1) =
 ? Entry name=bacr$halha
DE   BACTERIORHODOPSIN PRECURSOR (BR) (GENE NAME: BOP).
 Sequence length=   262
 Sequence composition
```

A	C	S	T	P	A	G	N	D	E	Q	B	Z	H
N	0.	14.	19.	12.	30.	26.	3.	10.	11.	4.	0.	0.	0.
%	0.0	5.3	7.3	4.6	11.5	9.9	1.1	3.8	4.2	1.5	0.0	0.0	0.0
W	0.	1219.	1921.	1165.	2132.	1483.	342.	1151.	1420.	513.	0.	0.	0.

A	R	K	M	I	L	V	F	Y	W	-	X	?
N	7.	7.	10.	15.	39.	23.	13.	11.	8.	0.	0.	0.
%	2.7	2.7	3.8	5.7	14.9	8.8	5.0	4.2	3.1	0.0	0.0	0.0
W	1093.	897.	1312.	1697.	4413.	2280.	1913.	1795.	1490.	0.	0.	0.

```
Total molecular weight=   28256.254
 ? Results file=sipl.res
 ? Display alignments (y/n) (y) =
 ? Minimum library sequence length (10-20000) (209) =
 ? Maximum number of scores to list (1-10000) (20) =10
 ? Identity score (1-3) (1) =
 ? Number of sd above mean (0.00-10.00) (3.00) =
 ? Odd window length (1-31) (11) =
 ? Proportional score (1-297) (132) =
 ? Minimum global score (1-69168) (1729) =
 ? Penalty for starting a gap (1-100) (10) =
 ? Penalty for each residue in gap (1-100) (10) =
Select a library
    1 EMBL nucleotide library
 X  2 SWISSPROT protein library
    3 PIR protein library
    4 Personal file in PIR format
```

Fig. 9. A run of SIPL using an entry from a sequence library and a file of entries
to be excluded from the search.

```
? Selection  (1-4) (2) =
Library is in EMBL format with indexes
Select a task
X  1 Search whole library
   2 Search only a list of entries
   3 Search all but a list of entries
? Selection  (1-3) (1) =3
? File of entry names=skip.nam
.
.
.
   21794 entries processed,    25 above cutoff, sorting now
Entries exceeding sd cutoff=  4439
Mean number of diagonals above span cutoff   1.32012
List in score order
  31007 BACA$HALSA DE   ARCHAERHODOPSIN PRECURSOR (AR).
  12177 BACH$NATPH DE   HALORHODOPSIN PRECURSOR (HR) (GENE NAME: HOP).
  10999 BACH$HALSP DE   HALORHODOPSIN PRECURSOR (HR) (GENE NAME: HOP).
   3999 HYAC$ECOLI DE   HYPOTHETICAL 27.6 KD PROTEIN IN HYAB 3'REGION (GENE NAM
   2670 OPS4$DROME DE   OPSIN RH4 (INNER R7 PHOTORECEPTOR CELLS OPSIN) (GENE NA
   2573 PYR1$MESAU DE   CAD PROTEIN (CONTAINS: GLUTAMINE-DEPENDENT CARBAMOYL-PH
   2328 PFLA$ECOLI DE   PYRUVATE FORMATE-LYASE ACTIVATING ENZYME.
   2194 DCOP$CANAL DE   OROTIDINE 5'-PHOSPHATE DECARBOXYLASE (EC 4.1.1.23) (OMP
   2145 BCM1$HUMAN DE   LYMPHOCYTE ACTIVATION MARKER BLAST-1 PRECURSOR (BCM1 SU
   2103 LAG3$HUMAN DE   LAG-3 PROTEIN PRECURSOR (FDC-PROTEIN) (GENE NAME: LAG3
BACA$HALSA DE   ARCHAERHODOPSIN PRECURSOR (AR).
   V    1        11         21         31         41         51
        MLELLPTAVE GVSQAQITGR PEWIWLALGT ALMGLGTLYF LVKGMGVSDP DAKKFYAITT
        *               ** **  ** **   ** ** ** ** *** ** * *
        M-DPIALTAA VGADLLGDGR PETLWLGIGT LLMLIGTFYF IVKGWGVTDK EAREYYSITI
   H    1        11         21         31         41         51
   V   61        71         81         91        101        111
        LVPAIAFTMY LSMLLGYGLT MVPFGGEQNP IYWARYADWL FTTPLLLLDL ALLVDADQGT
        *** **    * *** * *** * *      * * *  ** ****** ********** *** *
        LVPGIASAAY LSMFFGIGLT EVQVGSEMLD IYYARYADWL FTTPLLLLDL ALLAKVDRVS
   H   61        71         81         91        101        111
   V  121       131        141        151        161        171
        ILALVGADGI MIGTGLVGAL TKVYSYRFVW WAISTAAMLY ILYVLFFGFT SKAESMRPEV
        *  *** *   ** ******* *   *  ** *  ** *    *     ***
        IGTLVGVDAL MIVTGLVGAL SHTPLARYTW WLFSTICMIV VLYFLATSLR AAAKERGPEV
   H  121       131        141        151        161        171
   V  181       191        201        211        221        231
        ASTFKVLRNV TVVLWSAYPV VWLIGSEGAG IVPLNIETLL FMVLDVSAKV GFGLILLRSR
        ****  *    *** *** * ** ***   ** * ****** *** *** ******
        ASTFNTLTAL VLVLWTAYPI LWIIGTEGAG VVGLGIETLL FMVLDVTAKV GFGFILLRSR
   H  181       191        201        211        221        231
   V  241       251        261
        AIFGEAEAPE PSAGDGAAAT SD
        ** *  **** ****   *    *
        AILGDTEAPE PSAG-AEASA AD
   H  241       251        261
Conservation  56.1%
Number of padding characters inserted      0 and      2
```

Fig. 9 *(continued)*.

References

1. Staden, R. (1982) An interactive graphics program for comparing and aligning nucleic acid and amino acid sequences. *Nucleic Acids Res.* **10(9),** 2951–2961.
2. McLachlan, A. D. (1971) Test for comparing related amino acid sequences. *J. Mol. Biol.* **61,** 409–424.
3. Schwartz, R. M. and Dayhoff, M. O. (1978) Matrices for detecting distant relationships, in *Atlas of Protein Sequence and Structure*, suppl. 5, vol. 3, National Biomedical Research Foundation, Washington, DC, pp. 353–358.
4. Lipman, D. J. and Pearson, W. R. (1985) Rapid and sensitive protein similarity searches. *Science* **227,** 1435–1441.
5. Myers, E. W. and Miller, W. (1988) Optimal alignments in linear space. *Comput. Applic. Biosci.* **4,** 11–17.

CHAPTER 15

Staden Plus*

John H. McVey

1. Introduction

The Staden-Plus software is an old menu-driven version of Rodger Staden's software for manipulation of DNA and protein sequences that will run on an IBM compatible PC. This program is of use to those workers who do not have access to the newer versions of these programs, which are described in greater detail in Chapters 15–27 of this book. The PC version is very user-friendly, owing in part to the menus used to take the user through the programs and also because help is available at all levels by pressing <F10>. It is helpful, however, to have an idea of the overall structure of the program (Fig. 1). The user enters the program through the master menu after selecting a working directory. From the master menu, the user can select the following programs: SEQIN, sequence entry; DBAUTO, automatic screening and contig assembly; DBUTIL, sequencing project maintenance; ANALYSEQ, DNA sequence analysis; ANALYSEP, protein sequence analysis; DIAGON, sequence comparison; GENAX, GenBank databank access; ANCIL, ancillary programs (Chapters 18–25). Each of these programs has a menu offering a number of further options. In order to access another program, the user must return to the master menu; <esc> takes you back up through the structure to the master menu.

*Editors' Note: To the best of our knowledge, the "Staden Plus" software package is no longer commercially available. This chapter is included since many laboratories possess this software and may still wish to use it. The reader is referred to other chapters in this volume that describe more recent versions of Staden software (Chapters 2–14).

From: *Methods in Molecular Biology, Vol. 25: Computer Analysis of Sequence Data, Part II*
Edited by: A. M. Griffin and H. G. Griffin Copyright ©1994 Humana Press Inc., Totowa, NJ

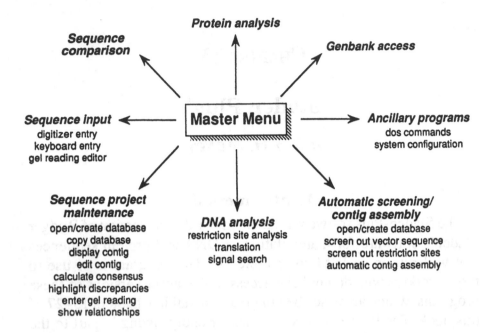

Fig. 1. Staden-Plus program structure. Some of the more commonly used features of the individual programs are shown.

The most useful aspects of these programs concern the storage, assembly, and management of sequences generated by a sequencing project. The program also has the facility to generate a restriction enzyme map of a sequence, to translate the sequence, or to compare one sequence with another. Assembly of the sequences is automated and should take very little time; however, checking and editing can still require a great deal of time and concentration. This chapter will describe how to set up a database, enter sequences, and generate a consensus sequence from which a restriction map and/or translation can be produced.

2. Materials

1. An IBM compatible PC.
2. At least 5 Mbyte of free space on your hard disk.
3. 640k RAM.
4. The 80287 Maths coprocessor.
5. Either the IBM or other compatible enhanced-color display monitor and interface card (EGA) fitted with 256k RAM, or the IBM VGA interface card and IBM 8512 or 8513 color monitor.

6. DOS 3.1 or subsequent version.
7. A dot matrix printer to allow generation of hard copy. Almost all Staden-Plus output can be directed to a printer. Printers, such as the IBM ProPrinter, H. P. QuietJet, and almost any Epson or Epson compatible dot matrix printer, are suitable.
8. A sonic digitizer for entering DNA sequence directly and rapidly from a sequencing gel is not essential, but very highly recommended. The GrafBar GP7 Mark II digitizer is the recommended model, made by:
>   Science Accessories Corp.
>   200 Watson Boulevard
>   P.O. Box 587
>   Stratford, CT 06497
It is available in the UK from:
>   PMS Instruments
>   Waldeck House
>   Reform Road
>   Maidenhead
>   Berks., UK
The SummaGraphics digitizer, model MM1103, is also supported:

SummaGraphics Corp.	SummaGraphics Ltd.
60 Silvermine Road	Ringway House
Seamore, CT 06483	Kelvin Road
	Newbury RG13 2DB, UK

9. Light box.

3. Methods

3.1. Starting Up the Software

1. Type <Asp-go>. The program renames the existing autoexec.bat and congig.sys files, and replaces them with two new files that will start Staden-Plus when the computer is rebooted.
2. To reboot the computer, press <ctrl><alt><del> keys simultaneously.
3. The program will ask if the current directory, the root directory (/), is satisfactory. Type the name of an alternative subdirectory name. This directory must already exist. Create a subdirectory for the sequencing project by typing <mkdir name> before starting the program.
4. The Master Menu should now be displayed.

3.2. Sequence Input

1. <F4> to select the Sequence Input Menu.
2. <F1> to enter sequence using a digitizer.
3. <y> to confirm that the digitizer is a Grafbar on port Com1.

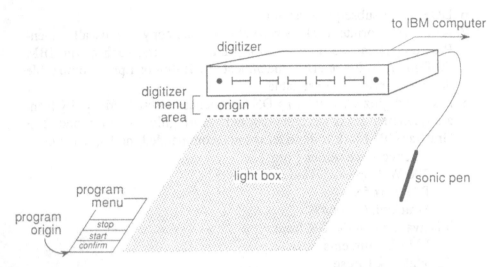

Fig. 2. Digitizer setup.

4. Type a <file of file names>; this is a file name that allows the program to process the gel readings in a batch manner.
5. Using the digitizer stylus, hit the device origin, the program origin, and start on the program menu (Fig. 2).
6. Hit <confirm> or <reset>.
7. If you hit <reset>, then digitize your lane order using the menu, and <confirm>.
8. Hit <start>, and then the center of each of the four lanes of the sequencing reaction you wish to enter. This should be at the point you wish to start reading, where you drew the line (*see* Section 4).
9. Hit <confirm>; separation is ok.
10. Hit <start>, and then read the gel hitting the center of each band.
11. When you want to transfer to the long gel run, hit <reset>.
12. Redefine the center of each lane at the point you wish to start reading.
13. Hit <confirm>, <start>, and begin reading as before.
14. Hit <stop> when you want to complete this gel entry.
15. Type <name of the gel reading>; you have up to a maximum of 10 characters for the file name.
16. Do you want to enter another gel reading? If you type <y>, then you will be returned to step 5.
17. Continue until you have entered all the gel readings. Then type <n> in response to the question in step 16. This returns you to the Sequence Input Menu.
18. <esc> returns the user to the Master Menu.

19. Note the file of file names and the gel file names you used, and label the autoradiograms. This is important since you may have to check the sequence during the subsequent editing process.

3.3. Automatic Screening and Contig Assembly

1. Select <F6>, Automatic Screening and Contig Assembly from the Master Menu.
2. From this General Control Menu, press <F1> to create a database. In subsequent entry of gel readings, you will only have to open a database, <F2>.
3. You will be asked to enter the size for the new database. The default is 50, allowing the entry of about 40 gel readings. This is usually large enough, the database can be increased at a later stage if necessary by copying the database. Hit <return> to accept 50.
4. You will be asked for a name for the database; this must be six characters long.
5. The program will create the database files and return you to the General Control Menu. Press <F3> to screen out vector sequences. If you do not have access to the vector sequence, then proceed to either step 10, if you want to screen out sequences containing a specific restriction endonuclease site, or step 15, if you wish to omit the screening process completely.
6. Type the <file of file names> you used when entering your gel readings.
7. You will be asked for a file name for the passed gels. It may be less confusing to give the file of file names with a different file extension. For example, if your file of file names was <gels1>, type <gels1.vp>.
8. Type the name of the vector sequence file. The complete nucleotide sequence of most of the commercially available vectors can be obtained on disk from the suppliers. The sequence file, however, may have to be edited to change it to the Staden sequence format this can be done easily using a word processor to insert <--- ABCDEF.123 -----> in front of the sequence. The angle brackets and the dashes are important and only the letters and numbers may be changed. Save this file as a DOS file.
9. You will be asked if the minimum match = 12. Hit return to accept this default. The program will then screen all the files for vector sequences and write the files that pass to the file of passed file names, <gels1.vp>.
10. The program will return you to the General Control Menu. Press <F4> to run the restriction site screening program.
11. Type the file of passed file names from the vector screening program, <gels1.vp>.
12. Type a new file of file names to which the program can write the names of the passed gel readings. For example, <gels1.vsp>.

13. You will be asked for the name of the recognition sequence file. Create a file called <rsite> before starting the Staden-Plus program. This file should only contain, in upper-case letters, the recognition sequence of the restriction sites you want to screen out. There should only be one recognition sequence per line.
14. The program will carry out the screening process and write the passed gel names to the file of file names.
15. The program will return you to the General Control Menu. Press <F2> to open a database. Type the name of the database. The copy number will be the default (0), then hit return. Press <F5> to start the automatic contig assembly and then <ctrl P> to switch on the printer.
16. Type the file of passed file names, <gels1.vsp>.
17. Type the file of file names to which you want the program to write the filenames of gel readings it fails to enter into the database, for example, <Failed>.
18. The program will suggest default parameters for: minimum match, minimum alignment, maximum pads allowed in the gel and in the contig, maximum percentage mismatch between new gel and contig. Accept all these values by hitting return.
19. Do you want to allow contig joining? <n>.
20. Do you want to examine contig data as they are entered? <y> Note this is not the default.
21. The program will start to enter the gel readings. Press any key as prompted to continue after each reading. You will then be returned to the General Control Menu. Hit <ctrl P> to switch off the printer.

3.4. Database Management

You have created a database and entered your first gel readings. You will now want to look at the contigs in the database, check the gel readings that are inconsistent, check the gels that failed to enter the database, and generate a consensus sequence for each of the contigs in order to obtain a restriction map and/or a translation. The program identifies contigs by the gel reading at the left end. It is therefore necessary to find out the relationships of the gels before being able to display a contig or to calculate a consensus.

1. Hit <esc> to return to the Master Menu, and then <F5> to run the Sequence Project Maintenance program (*see* Fig. 1).
2. Press <F1>, to go to the General Functions Menu and <F2> to open existing database. Type the name of the database. The copy number will be the default (0), then hit return.

3. Hit <esc> to return to the Sequencing Project Maintenance Menu and <F3> to select the Gel Reading Menu.
4. Press <F5> to show relationships. Press <ctrl P> to switch on the printer. Reply yes to the next three questions by typing <y>. You will obtain a printout showing the length of each contig, and the gels that make up the contig, starting from the left.
5. Return to the Sequencing Project Maintenance Menu by hitting <esc>. Select the Contig Handling Menu <F2> and <F1> to display a contig. You will be asked for the number of the gel at the left end of the contig. The contig will be printed out.
6. Hit <esc> to return again to the Sequencing Project Maintenance Menu and <F1> to select the General Functions Menu.
7. Press <F5> to list a text file; type <Failed>, that is the file of file names for the gels that failed to be entered into the database. You will obtain a printout of the file names.
8. Press <F1> to list a new text file. Type the name of the gel that failed to be entered into the database. You will obtain a printout of the sequence. Repeat until you have printed out all the sequences that failed to enter.
9. Press <ctrl P> to switch off the printer, <esc> to return to the Sequencing Project Maintenance Menu, and <esc> to return to the Master Menu. You will be asked to confirm this by typing <y>.
10. Stop! Look at the data: Check the contigs for discrepancies, go back to the autoradiographs of the sequencing gels, and decide which entry is correct. Check the printout of the contig assembly to find out why the failed gels were not entered into the database. If necessary, check the sequence of the failed gels against the autoradiographs.
11. If you want to edit the sequence of the failed gels and rerun the contig assembly, then press <F4> to obtain the Sequence Input Menu and <F2> to select keyboard entry of sequences.
12. Type <n> for nucleotide sequence and <y> to edit an existing file.
13. Type the name of the file you wish to edit <SeqX>. You will see your sequence displayed on screen, and you can edit it. Press <F8> to exit and save your changes. Confirm this by typing <y>.
14. You will be asked for the name of file to edit. If you have more sequences to edit, type the file name and continue as above. If you have finished editing, then press <esc> and <esc> again to return to the Master Menu.

This screen editor can only be used to edit sequences that have not been entered into the database. To edit a contig or a gel reading, you must use the Sequence Project Maintenance program.

15. The Automatic Screening and Contig Assembly requires the gel file names to be contained in a file of file names even if there is only one gel! To create a new file of file names, press <F8> to go to the Ancillary Program and <F8> to use Dos Commands.

16. Type: <copy con gels2>
             <gel1><retn>
             <gel2><retn>
             <gelx><retn>
             <ctrl Z>

    This creates a file called gels2, which contains the names of the files gel1, gel2, ...gelx.

17. Type <Exit> to return to the Staden-Plus programs, and <esc> to return to the Master Menu. To rerun the Automatic Contig Assembly, carry out the process described in Section 3.3.; however, this time hit <F2> to open a database rather than create one.

18. Select <F5>, Sequence Project Maintenance from the Master Menu.

19. Hit <F1> to obtain the General Functions Menu and <F2> to open an existing database. Type the name of your database and <retn> to accept the default copy number of the database (0).

20. Hit <esc> to return to the Sequencing Project Maintenance Menu. Press <F2> to go to the Contig Handling Menu, and select <F7> to calculate a consensus. You will be asked to type a file name for the consensus sequence.

21. Do you want a consensus for all of the contigs? If you answer <y>, the program will write all the consensus sequences to one file. If you answer <n>, you will be asked for the number of the gel at the left end of the contig. You then have the option to add another contig consensus or to write further consensus sequences to a different file.

22. Once you have generated your file containing the consensus sequence, return to the Master Menu by hitting <esc>, <esc>, <esc>, and confirm that you wish to stop the program.

3.5. DNA Analysis

1. Select <F1> DNA Analysis from the Master Menu. You will be asked for a sequence name; type the consensus file name.

2. Press <F4> to obtain the Structure Menu, and <F8> to select restriction enzyme search.

3. Type: <F5>, for all enzymes; <n>, not to select the enzymes by name; <0>, to list the restriction sites enzyme by enzyme; <n>, not to plot the output; <Ctrl P>, to switch on the printer; <retn>, to accept the default of inc = 1. The search will move through your sequence one base at a time.

4. You should obtain a printout of each restriction enzyme site found in the sequence and a list of the position of the site(s).
5. Switch off the printer by pressing <Ctrl P>. Hit <esc> to return to the Structure Menu, and <esc> again to return to the Main Menu.
6. Select the Translation and Codon Menu by pressing <F5>, and press <F5> again to translate and list in six phases.
7. Type: <Retn>, to define the line length as 60; <y> or <n>, to select either three-letter or single-letter amino acid codes; <Ctrl P>, to switch on the printer; <Retn>, to start translation at nucleotide 1; <retn>, to stop translation at the end of your sequence.
8. You should obtain a printout of the sequence with a translation in all possible reading frames.
9. Press <Ctrl P> to switch off the printer. To return to the Master Menu, hit <esc>, <esc>, <esc>, and confirm that you want to stop the program.

3.6. Exiting the Software

1. To exit the software at any time, return to the Master Menu from wherever you are in the program by pressing <esc>.
2. Press <esc> again to leave Staden-Plus.
3. Press <ctrl><alt><del> simultaneously to restore your normal computer configuration.

4. Notes

1. Before starting to input the sequences, tape your autoradiograph to the light box. Read the beginning of the sequence, identify the end of the vector and the start of the cloned sequence, and draw a line across all four lanes at this point. Also, identify the point of overlap between the short and long gel runs.
2. If the program fails to respond to commands and you have switched on the printer, check that it is on line.
3. Keep a log of all the files of file names and their contents, the gel names, and the DNA and primer used in each of the sequencing reactions.
4. Always have the digitizer switched on while running the program. Otherwise, the program will fail to find the security device, and the program will stop.

CHAPTER 16

DNA Strider

A Macintosh Program for Handling Protein and Nucleic Acid Sequences

Susan E. Douglas

1. Introduction

In 1988, version 1.0 of DNA Strider by Christian Marck was described in *Nucleic Acids Research (1)*. It was the first DNA analysis program designed specifically for the Macintosh. Written in C language and compiled with the Apple MPW C compiler, the program ran extremely efficiently and was able to analyze sequences of up to 32,500 characters very quickly. It combined standard features of Macintosh programs with the classical types of analyses that molecular biologists require for handling sequence data. This program was distributed free of charge to anyone who requested it.

One year later, Strider 1.1 was released. This version had a number of new features and some major improvements over the previous version. These included the option to save analyses as "Text" or "Pict" files for subsequent modifications in word processing or graphics programs, various changes to the Page Setup, the ability to read sequence data using different formats and the choice of 8 different genetic codes and several sets of start codons for translations. A financial contribution to the author to help offset the cost of the developments was requested.

From: *Methods in Molecular Biology, Vol. 25: Computer Analysis of Sequence Data, Part II*
Edited by: A. M. Griffin and H. G. Griffin Copyright ©1994 Humana Press Inc., Totowa, NJ

A new version is expected early in 1992 that will include dot matrix analyses and searches.* Given the high quality of analyses and ease of use, Macintosh users are eagerly awaiting its release. Brief descriptions of the analyses performed by Strider 1.1 and the interactions of this program with other DNA sequence analysis programs are described in this chapter. For a more detailed description of Strider 1.1, the reader is referred to the manual.

2. Materials

DNA Strider 1.1 runs on all Macintoshes so far produced by Apple, except the early 128 and 512 computers and the Mac XL. It requires System 6.0.2 or higher, and the new version will also support System 7.0. The program occupies 167 kbyte and supports most ImageWriter and LaserWriter printers, including the LaserWriter II SC, NT, and NTX. However, some of the Page Setup options are not available in Draft mode of Imagewriters.

3. Methods

3.1. Running the Program

3.1.1. Organization of Files and Menus

DNA Strider uses the Macintosh Hierarchical File System (HFS) and must contain the file named RELibrary (containing the names and sites of restriction endonucleases) in the same HFS folder as the program. It can be run using either the Finder or Multifinder. Standard Macintosh menus (here indicated in **bold** print) and key equivalents are used. There are also some hierarchical menus that are accessed by sliding the cursor sideways into a submenu. In addition, some menus are followed by (...), which means there is an optional dialog box that appears if the Option key is held down while the menu is selected. Certain commands are disabled (dimmed on the screen) when their use is inappropriate, e.g., performing a hydropathy plot on a nucleotide sequence or constructing a restriction map of a protein sequence. At the bottom of the **Conv, Enz,** and **AA** menus are commands listed in bold face. Thee are reminder tables for the DNA alphabet, the genetic code, physical data on amino acids, and descriptions of restriction endonuclease recognition sites.

---

*Christian Marck, Service de Biochimie et de Génétique Moléculaire, Bat. 142, Centre d'Etudes de Saclay, 91191 GIF-SUR-YVETTE CEDEX, France

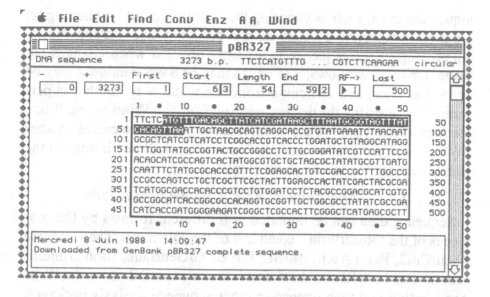

Fig. 1. The DNA Strider 1.1 screen showing the sequence worksheet (SWS) window. The SWS is shown beneath seven menus. The title of the sequence file and the length, topology (circular or linear), and first and last 12 nucleotides of the sequence appear at the top of the SWS. Within the sequence box, a short segment that corresponds to an open reading frame is highlighted. Above the sequence box is a row of counters displaying information about sequence positions. Below the sequence box is the comment box.

3.1.2. Starting the Program

DNA Strider is started by a double-click on the program icon or on any DNA Strider sequence file. An empty sequence worksheet window (SWS) or one containing the selected sequence file is then displayed. Data entry, editing, and analysis are performed in the SWS. Subsequent modification of analysis output can be performed by copying it to a word processing or graphics package.

3.1.3. Sequence Worksheet (SWS) Window

The SWS (Fig. 1) contains a title, which is the name of the sequence file itself; information about the sequence, such as its length, topology, and type; five counters that display various data about the sequence positions; and two boxes, one for the sequence with a vertical scroll bar, and one for comments of up to 220 characters. The boxes are selected by clicking within them, and information is typed in or pasted from the clipboard. Standard Macintosh editing techniques are used, such as choosing the insertion point by clicking on the mouse, selecting by clicking and dragging, and cut/copy/paste. Numbering of the sequence is instantly updated when changes are made. Up to 12 windows, 6 of which may be SWS windows, may be opened at once, and switching from one window to another is performed by selecting the desired window from the **Wind** menu. When the limit of 12 windows are open, it is not possible to create another SWS or perform another analysis (commands will be dimmed in the menus) until some windows are closed.

3.1.4. Importing and Exporting Sequence Data

Sequence data can be imported in 6 different formats by the sub-menus of the "Read/Write" command of the **File** menu. These include Ascii/Citi2, Plain Ascii, EMBL, UWGCG/GenBank, Bionet/Intelligenetics, and NBRF/Dayhoff. Data can be exported using the Write Ascii option and then opened in other sequence analysis packages. Interaction with other packages is described later.

3.1.5. Printing

The "Print Window" command in the **File** menu is used to print any analysis performed by DNA Strider. If more condensed output than that seen on the screen is required, the enlargement (in the LaserWriter Page Setup dialog box) is reduced to below 85%. Similarly, if enlarged output is required, the appropriate adjustment is made in the LaserWriter Page Setup dialog box. It is recommended that Font Substitution be turned off in the LaserWriter Page Setup dialog box, since DNA Strider performs its own font substitution. The size of graphic output can also be modified using the scaling/smoothing dialog on many of the **AA** menu commands. If further modifications need to be made to obtain publication-quality figures, the output may be exported to the Clipboard using the "Copy TEXT..." or "Copy PICT..." commands in the **Edit** menu, or saved as a file using the "Save TEXT..." or "Save PICT..." in the **File** menu (*see* Note 1).

3.2. Editing Sequence Data
3.2.1. File Menu

Most of the commands in the **File** Menu are standard Macintosh commands and are self-explanatory. However, the "New" and "Read/Write" commands allow data to be entered, read, or written as DNA, degenerate DNA, RNA, or protein codes. When the "Read" command is chosen, a dialog box appears that allows the user to select one of the six formats described above. The "Autosave" command can be selected when entering very long sequences from the keyboard and the disk copy of the file will be automatically updated every 30 s.

3.2.2. *Edit Menu*

Again, many of these commands ("Cut", "Copy," "Paste," "Clear") are standard Macintosh commands. There are additional commands that allow portions of the sequence to be selected, and there is provision for choosing whether characters are upper case or lower case. A sequence may be locked to prevent inadvertent editing. If editing is to be performed, a dialog appears that gives the user the opportunity to unlock editing. The "Undo" command does not yet work.

3.2.3. *Find Menu*

Most of the commands in this menu pertain to moving to particular locations in the sequence (the beginning, the end, a sequence position, a particular sequence up to 32 characters long). An especially useful command is "Find ORF," which looks for a start codon, proceeds to an in-frame stop codon, and highlights the entire open reading frame (*see* Fig. 1). The protein sequence may be displayed by using the "Protein 5'→3'" command in the **Conv** menu. Using toggled commands, searches can be performed downward or upward (5'→3' or 3'→5'), and can be sensitive to upper or lower case letters and to the coding phase of the sequence.

3.2.4. *Conv Menu*

The commands in this menu allow the user to convert sequences between DNA, degenerate DNA, RNA, Protein 5'→3', or Protein 3'→5' codes. This can occur over the entire sequence or over just a selected portion. In addition, a DNA sequence can be converted to the antiparallel, reverse, or complementary sequence. The origin of a sequence can be changed and renumbered using negative base numbers, and a linear sequence may be converted to a circular one using a toggled command.

3.3. *Analyzing Sequence Data*
3.3.1. General

When analyzing data in the SWS, a new window appears with the results of the analysis. The contents of this window may be saved or printed. Windows will be stacked on the screen and listed under the

	Bgl I	Drd I	Hha I	Nde I	Sau96 I
Aat II	Bgl II	Dsa I	HinC II	Nhe I	Sca I
Acc I	BsaA I	Eae I	HinD III	Nla III	ScrF I
Afl II	Bsm I	Eag I	Hinf I	Nla IV	Sec I
Afl III	BsmA I	Ear I	HinP I	Not I	SfaN I
Aha II	Bsp1286 I	Eco47 III	Hpa I	Nru I	Sfi I
Alu I	BspH I	Eco57 I	Hpa II	Nsi I	Sma I
Alw I	BspM I	EcoN I	Hph I	Nsp7524 I	SnaB I
AlwN I	BspM II	EcoO109 I	Kpn I	NspB II	Spe I
Apa I	Bsr I	EcoR I	Mae I	NspH I	Sph I
ApaL I	BssH II	EcoR II	Mae II	PaeR7 I	Spl I
Ase I	BstB I	EcoR V	Mae III	PflM I	Ssp I
Asp718	BstE II	Esp I	Mbo I	Ple I	Stu I
Ava I	BstN I	Fnu4H I	Mbo II	Pml I	Sty I
Ava II	BstU I	Fok I	Mlu I	PpuM I	Taq I
Avr II	BstX I	Fsp I	Mme I	Pst I	Tth111 I
BamH I	BstY I	Gdi II	Mnl I	Pvu I	Tth111 II
Ban I	Bsu36 I	Gsu I	Msc I	Pvu II	Xba I
Ban II	Cfr10 I	Hae I	Mse I	Rsa I	Xca I
Bbe I	Cla I	Hae II	Msp I	Rsr II	Xho I
Bbv I	Dde I	Hae III	Nae I	Sac I	Xcm I
Bbv II	Dpn I	Hga I	Nar I	Sac II	Xma I

◉ Unique   ○ Double   ○ Triple   ○ None   [OK]   [Cancel]

Fig. 2. The enzyme chooser. The names of the restriction enzymes in RELibrary are listed alphabetically. Enzymes with unique, double and triple sites in the sequence can be selected using the buttons. By depressing the command key, one can select a number of different enzymes.

Wind menu, from which they may be selected. As mentioned previously, specific analyses can only be performed on the appropriate type of sequence—nucleotide or protein—and the disabled menus will be dimmed.

3.3.2. *Enz Menu*

The "List" command lists the sequence in eight blocks of ten nucleotides, unless the condensed format of ten blocks of ten nucleotides is chosen (by specifying 85% reduction in the LaserWriter page setup). By depressing the option key with the "Graphic map" command, the Enzyme Chooser dialog appears (Fig. 2), and the enzymes used in various restriction endonuclease analyses can be selected. Restriction maps can be drawn, which contain all the sites in the sequence (by default), or only certain selected ones. It is possible to choose enzymes that appear once, twice, or three times in the sequence, and it is also possible to select several specific enzymes at once by depressing the command (or shift) key when in the Enzyme Chooser.

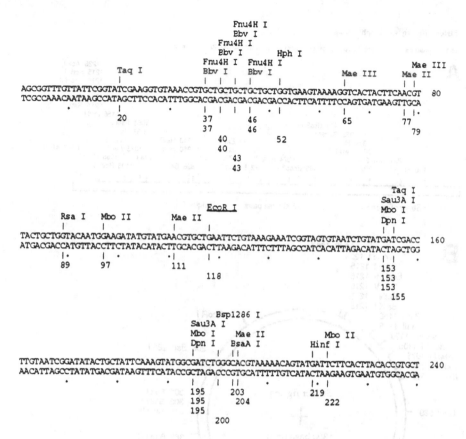

Fig. 3. A multiline restriction map. The names of the restriction sites are shown above the sequence, and their positions below the sequence. Unique sites are underlined (*Eco*RI).

Although there are approx 130 enzyme sites in the library, it is possible to add more sites to the RELibrary, as they become available. The multiline restriction map (Fig. 3) shows the positions of all RELibrary restriction sites in the DNA sequence and underlines the unique sites. Tabulations of site usage, all sites present, and all absent sites may also be obtained. In addition to the multiline restriction map, both linear (Fig. 4A) and circular (Fig. 4B) maps can be drawn, depending on whether they are designated as linear or circular molecules in the **Conv** menu, and selected enzymes are indicated by a dot. The sizes of restriction fragments resulting from digestion with a given restriction enzyme or combination of restriction enzymes can be predicted using the "Digestion" command with the Option key.

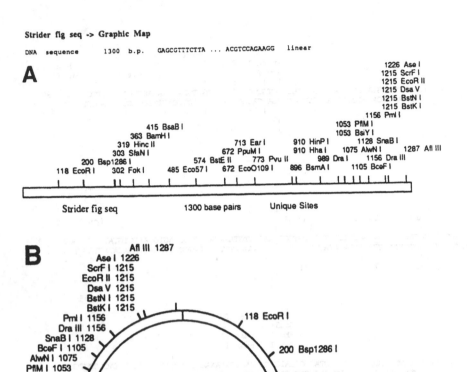

Fig. 4. A graphic restriction map. A graphic representation of a linear (**A**) or circular (**B**) sequence, showing unique restriction sites. The *Eco*RI site has been selected and is indicated by a dot.

3.3.3. AA Menu

The commands in the **AA** menu are involved in various translation analyses of DNA sequences and analyses of protein sequences. The "Translation" command may be performed in one, three, or six phases (Fig. 5) and a graphic representation of the translation where start codons are indicated by small bars and stop codons are indicated by

```
1/1                           31/11
   arg phe val ile arg tyr arg arg cys lys pro cys cys cys cys cys trp OPA ser lys
   ala val cys tyr ser val ser lys val OCH thr val leu leu leu leu leu val lys OCH
   ser gly leu leu phe gly ile glu gly val asn arg ala ala ala ala ala gly glu val
   AGC GGT TTG TTA TTC GGT ATC GAA GGT GTA AAC CGT GCT GCT GCT GCT GCT GGT GAA GTA
   TCG CCA AAC AAT AAG CCA TAG CTT CCA CAT TTG GCA CGA CGA CGA CGA CGA CGA CCA CTT CAT
   ala thr gln OCH glu thr asp phe thr tyr val thr ser ser ser ser ser thr phe tyr
   arg asn thr ile arg tyr arg leu his leu gly his gln gln gln gln gln his leu leu
   pro lys asn asn pro ile ser pro thr phe arg ala ala ala ala ala pro ser thr phe

61/21                         91/31
   arg ser leu leu gln arg tyr cys trp tyr asn gly arg tyr val OPA thr cys OPA ile
   lys val thr thr ser thr leu leu leu val gln trp lys ile cys met asn val leu asn
   lys gly his tyr phe asn val thr ala gly thr met glu asp met tyr glu arg ala glu
   AAA GGT CAC TAC TTC AAC GTT ACT GCT GGT ACA ATG GAA GAT ATG TAT GAA CGT GCT GAA
   TTT CCA GTG ATG AAG TTG CAA TGA CGA CCA TGT TAC CTT CTA TAC ATA CTT GCA CGA CTT
   phe thr val val glu val asn ser ser thr cys his phe ile his ile phe thr ser phe
   leu asp ser ser OPA arg OCH gln gln tyr leu pro leu tyr thr his val his gln ile
   pro OPA AMB lys leu thr val ala pro val ile ser ser ile tyr ser arg ala ser asn

121/41                        151/51
   leu OCH arg asn arg AMB cys asn leu tyr asp arg pro cys asn arg ile tyr cys tyr
   ser val lys lys ser val val OCH ser val OPA ser thr leu OCH ser asp ile leu leu
   phe cys lys glu ile gly ser val ile cys met ile asp leu val ile gly tyr thr ala
   TTC TGT AAA GAA ATC GGT AGT GTA ATC TGT ATG ATC GAC CTT GTA ATC GGA TAT ACT GCT
   AAG ACA TTT CTT TAG CCA TCA CAT TAG ACA TAC TAG CTG GAA CAT TAG CCT ATA TGA CGA
   glu thr phe phe asp thr thr tyr asp thr his asp val lys tyr asp ser ile ser ser
   arg tyr leu phe arg tyr his leu arg tyr ser arg gly gln leu arg ile tyr gln AMB
   gln leu ser ile pro leu thr ile gln ile ile ser arg thr ile pro tyr val ala ile
```

Fig. 5. A six-phase translation. The translation of the nucleotide sequence in the 5'→3' direction in three phases is shown above the sequence, and that in the 3'→5' direction below the sequence.

full bars (Fig. 6) can be obtained using the "ORF map" command. The start codon can be selected to be ATG, GTG, or ATG/GTG/CTG/TTG, and the translation code can be the universal code, the ciliate code, or the mitochondrial code of *S. cerevisiae*, *S. pombe*, vertebrates, *Drosophila*, fungi, or higher plants. Codon usage (Fig. 7) and amino acid usage (Fig. 8) tables can be produced and various analyses related to codon usage bias *(2)* can be produced. Graphic representations of the positions of acidic and basic residues (Fig. 9), cysteines, and histidines (zinc finger motif) (Fig. 10) and hydropathy plots (Fig. 11) according to *(3)*, *(4)*, or *(5)* are also available. Spatial hydropathies can be plotted assuming the whole protein is made of either a single α helix (Fig. 12) or β sheet, or as a function of the helicity angle (hydrophobic moment).

3.4. Interacting with Other Programs
3.4.1. General

Protein sequences exported using the Write Ascii option under the **File** menu may be opened by Microsoft Word and manipulated in that program. Usually, by saving as "Text Only" in Microsoft Word,

Strider fig seq -> 6-phase ORF Map <1>

DNA sequence   1300 b.p.   GAGCGTTTCTTA ... ACGTCCAGAAGG   circular

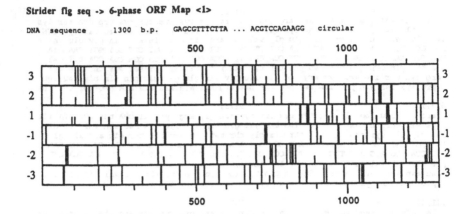

Fig. 6. An ORF map. The open reading frames found by translating the nucleotide sequence in the 5'→3' direction are shown above the sequence, and those found by translating in the 3'→5' direction below the sequence. Start codons (specified by the start chooser in the AA menu) are shown by short lines, and stop codons by full lines.

239 codons

MW : 26149 Dalton   CAI(S.c.) : 0.195   CAI(E.c.) : 0.372

TTT phe F	2	TCT ser S	2	TAT tyr Y	5	TGT cys C	7
TTC phe F	6	TCC ser S	-	TAC tyr Y	2	TGC cys C	-
TTA leu L	8	TCA ser S	2	TAA OCH Z	1	TGA OPA Z	-
TTG leu L	-	TCG ser S	-	TAG AMB Z	-	TGG trp W	4
CTT leu L	9	CCT pro P	3	CAT his H	1	CGT arg R	10
CTC leu L	-	CCC pro P	-	CAC his H	7	CGC arg R	-
CTA leu L	1	CCA pro P	2	CAA gln Q	11	CGA arg R	-
CTG leu L	-	CCG pro P	-	CAG gln Q	-	CGG arg R	-
ATT ile I	8	ACT thr T	6	AAT asn N	2	AGT ser S	3
ATC ile I	6	ACC thr T	-	AAC asn N	10	AGC ser S	1
ATA ile I	1	ACA thr T	10	AAA lys K	7	AGA arg R	-
ATG met M	12	ACG thr T	-	AAG lys K	3	AGG arg R	-
GTT val V	5	GCT ala A	11	GAT asp D	7	GGT gly G	20
GTC val V	-	GCC ala A	-	GAC asp D	7	GGC gly G	-
GTA val V	12	GCA ala A	11	GAA glu E	9	GGA gly G	2
GTG val V	-	GCG ala A	1	GAG glu E	1	GGG gly G	1

Fig. 7. Codon usage table. The calculated molecular weight of the protein, and the codon adaptation indices (CAI) are shown at the top the table. The number of occurrences of specific codons are given in the table.

the information can then be read into other programs, or sent by electronic mail to GenBank or the EMBL Fileserver. Sequences generated in Strider 1.1 can also interact directly with other Macintosh programs, some of which are described below.

Strider aa seq -> A. A. Usage

```
Protein  sequence      239 a.a.    MEDMYERAEFCK ... ADFVETATANKZ

         239 Amino Acids    MW :     26149 Dalton
```

			n	n(%)	MW	MW(%)
A	ala	alanine	23	9.6	1633	6.2
C	cys	cysteine	7	2.9	721	2.8
D	asp	aspartic acid	14	5.9	1610	6.2
E	glu	glutamic acid	10	4.2	1290	4.9
F	phe	phenylalanine	8	3.3	1176	4.5
G	gly	glycine	23	9.6	1311	5.0
H	his	histidine	8	3.3	1096	4.2
I	ile	isoleucine	15	6.3	1696	6.5
K	lys	lysine	10	4.2	1280	4.9
L	leu	leucine	18	7.5	2035	7.8
M	met	methionine	12	5.0	1572	6.0
N	asn	asparagine	12	5.0	1368	5.2
P	pro	proline	5	2.1	485	1.9
Q	gln	glutamine	11	4.6	1408	5.4
R	arg	arginine	10	4.2	1561	6.0
S	ser	serine	8	3.3	696	2.7
T	thr	threonine	16	6.7	1616	6.2
V	val	valine	17	7.1	1684	6.4
W	trp	tryptophan	4	1.7	744	2.8
X	ukw	unknown	-	-		
Y	tyr	tyrosine	7	2.9	1141	4.4
Z	---	STOP	1	0.4		

Fig. 8. Amino acid usage table. The calculated molecular weight of the protein and the number of amino acids it contains are shown at the top of the table. The number of occurrences of a given amino acid and its percentage of the total number of amino acids are given in the table.

3.4.2. FastP

Protein sequences exported using the Write Ascii option under the **File** menu may be read directly into the FastP program for Macintoshes (6). The sequence is selected using the "load search file" command in FastP and is searched against a database. Alignments of matches can then be performed.

3.4.3. PAUP 3.0

Protein or nucleotide sequences exported using the Write Ascii option under the **File** menu may be read directly into the PAUP 3.0

Strider aa seq -> Acid + Basic Map <1>

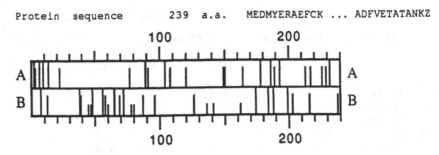

Fig. 9. Acidic + basic map. The positions of the acidic amino acids, aspartic acid (intermediate bar), and glutamic acid (full bar) in a protein sequence are shown in the top panel. The positions of the basic amino acids, histidine (small bar), lysine (intermediate bar), and arginine (full bar) are shown in the bottom panel.

Strider aa seq -> Cys + His Map <1>

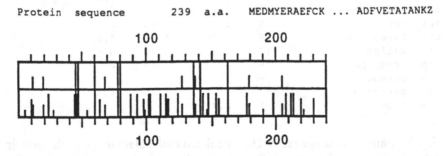

Fig. 10. Cys + His map. The positions of cysteine (small bar) and histidine (full bar) in a protein sequence are shown in the top panel. The positions of amino acids present in zinc finger motifs (tyrosine, cysteine, phenylalanine, leucine, and histidine) are plotted as bars of increasing length on the bottom panel.

for the Macintosh *(7)*. The PAUP command used is "Open" and the "All text files" option is chosen. Phylogenetic analysis may then be performed on manually aligned sequences.

3.4.4. MacClade 2.1

Protein or nucleotide sequences exported using the Write Ascii option under the **File** menu may be read directly into MacClade 2.1 *(8)*. The MacClade "Open File" command is used. As with PAUP, the data in the MacClade file must be manually aligned before phylogenetic analysis is performed.

Strider aa seq -> HW Hydrophilicity <11/1>

Protein sequence   239   a.a.   MEDMYERAEFCK ... ADFVETATANKZ

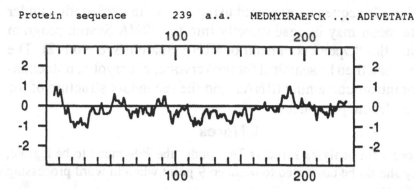

Fig. 11. Hydropathy plot. A Hopp and Wood plot of a protein sequence. Hydrophilic amino acids are shown above the midline and hydrophobic amino acids are shown below.

Strider aa seq -> Alpha Helix Amphiphilicity <11/1>

Protein sequence   239   a.a.   MEDMYERAEFCK ... ADFVETATANKZ

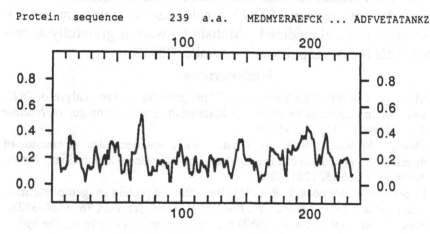

Fig. 12. Amphipathicity plot. Spatial hydropathy of a protein sequence assuming the whole sequence is made of a single alpha helix.

3.4.5. DNA Inspector IIe

Protein or nucleotide sequences exported using the Write Ascii option under the **File** menu may be read directly into the DNA Inspector IIe (Textco). Certain additional routines are available in the DNA Inspector, such as searches for direct and inverted repeats, a graphic representation of base composition, homology matrix analysis, and shotgun sequence alignment.

3.4.6. tRNA Search

Nucleotide sequences exported using the Write Ascii option under the **File** menu may be read directly into the tRNA Search program *(9)* using the "Open" command of tRNA Search *(see* Note 2). The sequence can then be searched for prokaryotic, eukaryotic, mitochondrial, or intron-containing tRNAs, and the secondary structure of the tRNA molecule presented.

4. Notes

1. In order for analyses saved as Text under the **File** menu to be legible, they should be converted to Monaco 9 point when in word processing packages.
2. IBM programs, which have been ported to the Macintosh (such as tRNA Search) may be obtained from Peter Markiewicz, Department of Viral Biology, USAMRIID, Ft. Detrick, MD.

Acknowledgments

The author wishes to thank Christian Marck for information concerning the new version of DNA Strider and for comments on the manuscript. Critical reading by Mohsin Patwary is gratefully appreciated. This is NRCC publication number 33012.

References

1. Marck, C. (1988) "DNA Strider:" a "C" program for the fast analysis of DNA and protein sequences on the Apple Macintosh family of computers. *Nucleic Acids Research* **16,** 1829–1836.
2. Sharp, P. M. and Li, W-H. (1987) The codon adaptation index—a measure of directional synonomous codon usage bias, and its potential applications. *Nucleic Acids Research* **15,** 1281–1295.
3. Hopp, T. P. and Woods, K. R. (1981) Prediction of protein antigenic determinants from amino acid sequences. *Proc. Natl. Acad. Sci. USA* **78,** 3824–3828.
4. Kyte, J. and Doolittle, R. F. (1982) A simple method for displaying the hydrophobic character of a protein. *J. Mol. Biol.* **157,** 105–132.
5. Eisenberg, D., Weiss, R. M., and Terwilliger, T. C. (1984) The hydrophobic moment detects periodicity in protein hydrophobicity. *Proc. Natl. Acad. Sci. USA* **157,** 140–144.
6. Pearson, W. R. and Lipman, D. J. (1988) Improved tools for biological sequence comparison. *Proc. Natl. Acad. Sci. USA* **85,** 2444–2448.
7. Swofford, D. L. (1990) PAUP version 3.0. Illinois Natural History Survey. Champagne, IL, p. 136.
8. Maddison, W. and Maddison, D. (1987) MacClade, version 2.1. Harvard University, p. 55
9. Shortridge, R. D., Pirtle, I. L., and Pirtle, R. M. (1986) IBM microcomputer programs that analyze DNA sequences for tRNA genes. *Computer Applications in the Biosciences* **2,** 13–17.

CHAPTER 17

MacVector: An Integrated Sequence Analysis Program for the Macintosh

Sue A. Olson

1. Introduction

MacVector™, from Laboratory and Research Products, Eastman Chemical Company, New Haven CT, is an integrated comprehensive sequence analysis program. It provides the most commonly used nucleic acid and protein analyses, semiautomatic entry of sequence data from autoradiograms using the optional Gel Reader, and access to the floppy disk formats of the GenBank® nucleic acid database and the NBRF Protein Identification Resource protein database. The program runs on the Apple® Macintosh® computer and is designed to take advantage of the facilities offered by this platform.

1.1. The Macintosh Environment

The Macintosh computer has been a boon for individuals who before this have had difficulty using computers productively. Several factors are responsible for this. First, the Macintosh operating system and programs tend to be easy to use. Instead of requiring the user to learn and remember commands and function key combinations, the Macintosh presents the user with pull-down menus and other graphical objects, such as buttons and check boxes, that can be manipulated by means of a mouse to issue commands and set program options.

Even more important is the high degree of consistency from program to program. Apple publishes interface guidelines that most developers adhere to religiously. Many interface elements, such as dialogs for opening and saving files, window controls, and rudimentary text

From: *Methods in Molecular Biology, Vol. 25: Computer Analysis of Sequence Data, Part II*
Edited by: A. M. Griffin and H. G. Griffin Copyright ©1994 Humana Press Inc., Totowa, NJ

editing operations, are provided in the Macintosh ROM Toolbox, so developers use these instead of reinventing their own solutions. As a result, anyone who has learned how to use one Macintosh program usually feels at home using any other Macintosh program.

Finally, it is relatively easy to transfer data between Macintosh programs. A standard method of copying data from one application and pasting it into another is provided by the operating system, and supported by all Macintosh applications that adhere to the interface guidelines. This makes it easy, for example, to copy a table of analysis results from a sequence analysis program and paste it into a word processor document. Under MultiFinder® or System 7, several programs can be opened simultaneously, so that copy-paste operations can be performed without having to close one application and open another.

1.2. MacVector in the Macintosh Environment

The goal of MacVector's creators was to provide the most commonly used nucleic acid and protein sequence analyses in a form that would be familiar to the researcher who uses a Macintosh word processor or drawing program more frequently than a sequence analysis program. The program adheres to the Macintosh interface guidelines, and takes advantage of the task-switching capabilities of MultiFinder or System 7 to run lengthy analyses, such as database searches, in the background while the user uses another application. Balloon Help™ is supported under System 7.

MacVector is an integrated package, rather than a collection of separate programs linked by a menu system. Many other sequence analysis packages are designed so that the user selects the analysis first and then designates which sequence to use. MacVector uses the convention common to most other Macintosh software: A sequence file is opened first, and then the action to perform on it is selected from a menu.

The program is highly interactive. Many of the analyses store intermediate results in memory and allow the user to apply various "filters" to view different subsets of the results over and over without having to redo the entire analysis. Most analyses allow the user to view the results in three different forms: as a list or table, as a graphical plot, and as an annotated sequence. The mouse can be used to "zoom in" to view small regions of graphical plots in more detail. The outputs can be customized to some extent within the program, and can then be saved as files using one of the two most common Macintosh file types (TEXT for textual data, PICT for graphical data) and imported into word processing or drawing programs for final editing. Selections of data in the results windows can also be copied and pasted into other program's documents.

2. Materials

2.1. Hardware

1. Any Macintosh computer except for the Macintosh Plus and earlier models. (MacVector supports both US and international versions of the Macintosh keyboards.)
2. Minimum memory requirements are 1 Mbyte of RAM if using Finder™ under System 6.0.× or 2 Mbyte of RAM if using MultiFinder or System 7. If the user wishes to have several programs open simultaneously under MultiFinder or System 7, more memory will be needed. If the user plans to analyze very large sequences, more memory may be needed than the minimum requirements.
3. One floppy disk drive and a hard disk. The program and all of its associated data files use about 2 Mbyte of hard disk space.
4. The Eve™ copy-protection device, provided with the MacVector software package.
5. A Macintosh-compatible printer, if a hard copy of results is desired.

2.2. Software

1. Macintosh System Software 6.0.3 or higher.
2. The MacVector application and the MacVector Library file. These must occupy the same folder.
3. The EvE INIT provided with the MacVector software package. This must be located in the System Folder (System 6) or in the Extensions Folder of the System Folder (System 7).

2.3. Data

1. MacVector supports most of the common sequence formats. The native MacVector sequence file type is a binary file. The other supported formats are of type TEXT, and include GenBank and derivatives of it, such as BIONET™ and IBI Pustell text files; CODATA, including NBRF PIR® and DNASTAR variants; Staden; GCG; EMBL and Swiss Prot; and "line" (sequence only) format.

2. MacVector supports the databases provided on the "Entrez™: Sequences" compact disk distributed by the National Center for Biotechnology Information, National Library of Medicine, National Institutes of Health, Bethesda, MD.
3. A number of data files used by different analyses are provided with MacVector. These are in a binary format and can be edited from within MacVector. The files include restriction enzymes, proteolytic agents, nucleic acid and protein subsequences, scoring matrices for nucleic acid and protein comparisons, and mobility standards for use with the Gel Reader. Several codon bias files are also supplied. These can be created by MacVector, but cannot be edited by the user.

2.4. Optional

1. The IBI Gel Reader is a sonic digitizer-based apparatus that can be used for nucleic acid sequence entry and for mobility measurements of nucleic acid or protein fragments.
2. A compact disk reader is required if the user wishes to access the CD version of GenBank. To access the CD, the user will also need the device driver software for a reader and two files named Foreign File Access and ISO 9660 File Access. These files are supplied by the manufacturer of the compact disk reader and should be placed in the System Folder. To inform MacVector that the database is located on a CD rather than on the hard disk, the user needs to run a program called GenBank CD SetUp, which is supplied with MacVector in a folder called GenBank Utilities.
3. If the user wishes to manipulate MacVector's text output files, a word processor or text editor will be needed.
4. To manipulate MacVector's graphical output files, a drawing program capable of opening PICT files will be needed.

3. Methods

3.1. Installing the Eve Device

MacVector uses the Eve device as a form of copy protection. This is a small box that attaches to the computer's Apple Desktop Bus™ (ADB). The device can be daisy-chained with the keyboard or mouse. No ADB device should be attached or removed from the computer while the power is turned on, so turn off the computer before attaching the Eve device.

3.2. Installing the Software

The MacVector software can be installed in two ways. An Installer program is provided on Disk 1 of the MacVector release disks to perform the installation automatically—just double-click on the Installer icon to start the program, click the Install button when the Easy Install dialog box appears, and insert the proper floppy disks when prompted by the install program. Alternatively, the user can install the software by:

1. Dragging the MacVector 4.0 Folder from Disk 2 to the hard disk. The folder can be placed anywhere on the hard disk and can be named whatever the user wishes. It will be referred to as the MacVector Folder.
2. Opening the MacVector Folder by double-clicking on it.
3. Double-clicking on the files MacVector 4.0.sea and MacVector Library.sea to decompress the application and MacVector Library.
4. Dragging the EvE INIT into the System Folder on the hard disk. (This INIT is found on Disk 3 of the version 4.0 release disks.)
5. Dragging any needed data files from the release disks to the hard disk. If the user wishes to follow the tutorials in the user's manual, MacVector Demo folder should be dragged to the hard disk. If the user will be using the CD release of GenBank, the GenBank Utilities folder should be dragged to the hard disk. These do not need to be placed in the MacVector Folder, although it is convenient to do so.
6. Restarting the computer by choosing Restart from the Special menu.

3.3. Starting the Program

MacVector can be started in one of two ways:

1. Open the MacVector Folder and double-click on the MacVector application icon, or on the icon of any files in the folder that were created using the MacVector program.
2. If the user has created a working folder containing MacVector sequence or accessory files, the user should open this folder and double-click on the icon of any files in the folder that were created using the MacVector program.

3.4. An Overview of the MacVector Menus

1. Apple menu: Choose the About MacVector command to see some information about the program. Under System 6, choose Online Help to access MacVector's on-line help system.

2. File menu: To create a new file, choose the New command, and indicate the file type using the submenu. The Open command allows the user to open an existing file. The Print command allows the user to print the contents of the active (topmost) window. If the active window is a sequence, the user can choose to print any or all of the three parts of the sequence file (annotations, features, and the sequence itself).

3. Edit menu: In addition to the familiar Cut, Copy, Paste, Clear, and Select All commands, this menu contains several commands specific to sequence data. Reverse will reverse the order of the residues in a marked block of sequence data. Complement will replace a marked block of nucleotides with its complement, and Reverse & Complement will replace a marked block of nucleotides with its reverse complement. Find is used to locate a specific short sequence of residues in a sequence window, whereas Jump To is used to position the insertion point at a specific residue while editing a sequence.

4. Options menu: Three of the commands in this menu (Format Annotated, Aligned, and Picture Displays) are used to change the way information is presented in analysis results windows. Open Digitizer Port and Define Lanes are used to set up the IBI Gel Reader for reading in sequence data. If the user prefers to enter data from the keyboard, Set Keypad can be used to assign nucleotides to keys of the numeric keypad for one-handed entry. The Show Proofreader command brings up a floating window containing a tape-recorder-like control panel that is used to control voice readback of the sequence in order to proofread the data. The user can choose or create the genetic code assignments to be used in all translations by means of the Modify Genetic Codes command. Make Codon Bias Table is used in conjunction with the GenBank database to create a codon bias table for an organism that can be used in the codon preference analysis.

5. Analyze menu: This menu contains the analysis commands. Analyses that can be performed by MacVector include restriction and proteolytic enzyme analyses, subsequence searches, matrix comparisons of two sequences, open reading frame and codon preference analyses for predicting coding regions, translation, reverse translation (including finding the least ambiguous probes for a protein sequence), base composition analysis, protein analyses (secondary structure predictions and hydrophilicity plots), and mobility measurement of protein or nucleic acid gel fragments using the Gel Reader.

6. Database menu: The user can compare a sequence to the GenBank or NBRF PIR databases by choosing the Align to GenBank or Align to NBRF commands. If the user wishes to compare the sequence to one or

more sequences stored in a folder on the hard disk, Align to Folder should be chosen. The Browse GenBank and Browse NBRF commands allow the user to locate sequences in the databases by organism name, locus name, author name, and so forth. Once a sequence has been found in a database, it can be opened in memory by choosing the Extract to Desktop command, or it can be written to a disk file by choosing Extract to Disk.

7. Windows menu: This menu has commands that allow the user to manipulate the windows on the desktop (Stack, Tile, Close All, and Full Titles). Each window opened in MacVector is listed by name in the menu to allow the user to access windows that have been covered by others. There are also commands that display a list of the one-letter IUPAC codes for nucleotides and amino acids (IUPAC Key) and a list of the graphical symbols MacVector uses for displaying features graphically (Picture Key).

8. Help menu (System 7 only): Under System 7, MacVector's on-line help system as well as Balloon Help can be accessed here.

4. Notes

1. If more than one person is using MacVector, each should have a private working folder to keep sequence files and customized versions of data files (such as restriction enzyme files). To start the program from this folder, just double-click on the icon of any file in that folder that was created by MacVector.

2. MacVector's suggested memory partition size under MultiFinder or System 7 may not be sufficient if the user is analyzing large sequences or performing memory-intensive operations such as database searches or creating codon bias tables. If the user sees a warning of low memory conditions while performing an analysis, or if the line "Warning, window data truncated" appears at the bottom of a results window, the user should increase the partition size. To do this, first exit the application (choose Quit from the File menu). Find the MacVector application icon in the MacVector Folder, and click on it once to highlight it. Choose Get Info from the File menu to open the Info window for MacVector. Type the desired memory partition size in the text entry box in the lower right corner of the Info window.

CHAPTER 18

MacVector: Aligning Sequences

Sue A. Olson

1. Introduction

MacVector™ uses a variation *(1)* of the Wilbur-Lipman-Pearson algorithm *(2–5)* to find a "best" pairwise alignment between a single query sequence in memory and one or more other sequences stored in a folder on disk. The algorithm uses three comparison steps. A very rapid technique called hashing is used to find regions of the two sequences that contain N consecutive matches. The region surrounding each matching nucleus is then scored, using match and mismatch values defined by the user. If this initial score for a matching region exceeds a cutoff score, an optimal alignment is performed, inserting deletions and gaps as necessary to improve the score. The alignment with the best optimized score is saved and reported to the user.

1.1. The Analysis Parameters

The speed and sensitivity of the comparison are controlled by the hash size (the number of consecutive matches that must occur before an alignment is scored). Higher hash values speed up the analysis, but weak similarities between sequences can be missed. Lower hash values result in more sensitive searches, but the comparison process slows down. What constitutes a match in the hash process can be altered by changing the hash code assignments in the scoring matrix file, which is used to score the alignments.

Four empirically derived parameters are used to score the alignments: the match and mismatch scores, the cutoff score, the deletion

From: *Methods in Molecular Biology, Vol. 25: Computer Analysis of Sequence Data, Part II*
Edited by: A. M. Griffin and H. G. Griffin Copyright ©1994 Humana Press Inc., Totowa, NJ

penalty, and the gap penalty. The match and mismatch scores allow the use of criteria other than identity between residues to score the alignments, and permit certain pairings to count as partial matches. This is especially useful when comparing protein sequences. Scoring schemes can be devised that match amino acids according to chemical properties, charge, genetic code, evolutionary replacement, and so on. The scores are stored in a scoring matrix file, such as the DNA matrix and pam250 matrix files supplied with MacVector.

The cutoff score is used to decide whether or not a scored region shows enough similarity to make an optimal alignment worthwhile. The cutoff score is computed from four values that can be changed by the user by editing the scoring matrix file. It is usually best to leave these values at their empirically derived default settings, unless one of the sequences involved in the comparison is very short.

The deletion and gap penalties are used during the optimal alignment. The deletion penalty is the amount subtracted from the alignment score if a single residue gap must be inserted into one of the sequences in order to improve the alignment. The gap penalty is the amount subtracted per residue for extending the gap beyond a single residue. These two parameters interact with the match and mismatch values, as well as with each other. For example, if the maximum mismatch score is −5 and the deletion penalty is 2, deletions will be preferred over mismatches, since they reduce the score less than a mismatch. A high deletion penalty coupled with a low gap penalty will result in few gaps in the alignment, but allows those gaps to be fairly long. A low deletion penalty coupled with a high gap penalty may result in an alignment that contains many single residue deletions. The deletion and gap penalties are also stored in the scoring matrix files and can be edited by the user.

1.2. The Analysis Stage

MacVector's sequence comparison analysis has two stages. In the first stage, the comparison itself is performed using the parameters set by the user. Depending on the parameters and the number of sequences compared, the comparison process can take a long time. The results are stored in memory for use by the second stage, an interactive filter stage.

1.3. The Interactive Filter Stage

The interactive stage allows the user to filter and display the results in different ways. This stage is fairly rapid, since the entire comparison does not have to be redone each time the display requirements change. Results can be viewed as a list of the names and scores of matching sequences, as a linear map display that shows which regions of the sequence set are similar to the query sequence, and as an aligned sequences display that shows the actual alignments.

2. Materials

2.1. Hardware

1. Any Macintosh® except the Macintosh Plus and earlier models.
2. Minimum memory requirements are 1 Mbyte of RAM if using Finder™ under System 6.0.× or 2 Mbyte of RAM if using MultiFinder® or System 7. If very large sequences are typically used, more memory may be needed.
3. One floppy disk drive and a hard disk. The program and its auxiliary files use about 2 Mbyte of hard disk space.
4. The Eve™ copy-protection device, provided with the MacVector software package.
5. A Macintosh-compatible printer if a hard copy of results is desired.

2.2. Software

1. Macintosh System Software 6.0.3 or higher.
2. The MacVector application and the MacVector Library file. These must occupy the same folder.
3. The EvE INIT provided with the MacVector software package. This must be located in the System Folder (System 6) or in the Extensions Folder of the System Folder (System 7).

2.3. Data

1. The sequences to be compared stored in MacVector binary file format or in a TEXT file in a format supported by MacVector. Supported TEXT file formats include GenBank® and derivatives of it, such as BIONET™ and IBI Pustell text files; CODATA, including NBRF PIR® and DNASTAR variants; Staden; GCG; EMBL and Swiss Prot; and "line" format (sequence only). For this tutorial, use the file UBUTA located in the Vector Demo folder inside the MacVector Folder.

2. A scoring matrix file in MacVector's binary format. Several matrices are provided with MacVector and are located in the MacVector Folder: DNA matrix is used for comparisons between two nucleic acid sequences, and pam250 matrix and pam250S matrix are used for comparing protein sequences.

2.4. Optional

1. A word processor or text editor, for manipulation of text output files.
2. A drawing program capable of opening PICT files, for manipulation of graphical output files.

3. Methods

Find the MacVector Folder on the hard disk, and double-click on it to open it. Double-click on the MacVector application icon to launch the program.

3.1. Opening the Files

In order to perform the analysis, a sequence window must be the active (topmost) window. For this tutorial, the protein UBUTA (an α-tubulin) will be compared against a folder containing a number of α- and β-tubulin protein sequences. These sequences are supplied with MacVector in a folder named "Vector Demo" located in the MacVector Folder.

1. Choose the Open command from the File menu.
2. Check the proper file type check box in the dialog box. UBUTA is a MacVector binary protein file, so check the box labeled "Protein."
3. Find the folder name "Vector Demo" in the scrollable list of file and folder names, and double-click on the name to display the list of files it contains.
4. Scroll through the list to find the name UBUTA, and open the file either by double-clicking on its name or by highlighting the name and clicking the Open button.

3.2. Setting Up the Analysis

1. If the sequence window is not the active (topmost) window on the screen, click in it or select its name from the window's menu to activate it.
2. Choose Align to Folder... from the Database menu.
3. Click the Folder to Search button near the top of the dialog box. A file selection dialog box appears. UBUTA will be compared with all protein sequences in Vector Demo, so make sure that the contents of the folder

appear in the scrollable list. Click the Search Here button at the right side of the dialog box.

4. Beneath the Folder to Search button is a check box labeled "GCG." This should only be checked if the sequences in the folder are in the GCG ("Wisconsin") format. Since the sequences in Vector Demo are in MacVector binary format, the box should be unchecked.

5. Make sure the hash size is set to two and that the processing setting is set to align.

6. Make sure the name of the scoring matrix is "pam250 matrix." If the name is "????," tell the program where to find the scoring matrix. Click the button labeled "Scoring Matrix" to bring up a file selection dialog box. The pam250 matrix file is located in the MacVector folder.

7. The check box labeled "align to DNA" should be left unchecked, since a protein will be compared to a set of proteins. (If it is checked, MacVector would compare the protein query sequence with the nucleic acid sequences in Vector Demo.)

8. Click OK to start the analysis.

3.3. The Interactive Stage

When the analysis is complete, the interactive dialog box will appear on the screen. On the top right are the "filter" options that permit the user to subset the results of the analysis. The user can choose to see only a subset of the matches by designating a span in the text input boxes labeled "Entries to show." The outputs could thus be limited to the best five matches by typing "1 to 5" in the boxes. The user could choose to rescore a smaller region of the alignment by designating a region in the "Score Rgn" and "Display Rgn" text entry boxes.

On the bottom right are the display options. The results can be seen as a list of the names of the matching sequences in order of optimized score, horizontal maps of the matching regions, and the aligned sequences themselves.

Once the results are displayed, this interactive dialog box can be returned to as many times as desired without redoing the entire analysis—as long as you keep at least one of the results windows open and active.

1. Check all three output option check boxes, and click OK.
2. After a pause, the three results windows appear. The list window should be on top. Click in the zoom box in the upper right-hand corner of this window to expand it to fill the screen.

3.4. Interpreting the List Display

1. The best matching sequence in the list is UBUTA compared with itself, so its score is the highest possible score that can be achieved in an alignment with UBUTA using the parameters set. The optimized score and initial score are the same, because no gaps were inserted during the optimized alignment. The second sequence in the list, A23035, has lower scores, since there were some mismatches with UBUTA, but again the initial and optimized scores are the same, indicating that no gaps were inserted. The third sequence, A25601, has initial and optimized scores that differ, meaning that a gap had to be inserted into one of the two sequences in order to align them optimally.
2. Do not close any windows.

3.5. Manipulating the Map Window

1. Make the map window the active window. Click in the zoom box in the upper right corner of the window's title bar to expand the window to fill the screen. This window has two panes, each with its own scroll bar. The upper pane contains linear maps of the matching sequences from the folder. The lower pane contains a graphical representation of the query sequence's Features Table entries. UBUTA has no features, so only a thin horizontal line is displayed (Fig. 1).
2. To alter the relative sizes of the two panes, position the cursor over the black bar that separates the scroll bars of the two panes on the right edge of the window. The cursor's shape will change to a double-headed arrow when it is at the right place. Press down on the mouse button, and drag the bar down to reduce the bottom pane to its minimum size.
3. The thick horizontal bars in the upper pane of the map window represent the regions of the sequences in the folder that are similar to UBUTA. Notice that some of the sequences further down in the display only align to a portion of UBUTA. A gap in a bar shows where a gap had to be inserted into the sequence to optimize the alignment with UBUTA (Fig. 2). A vertical line intersecting a bar shows where an insertion had to be placed in the sequence to optimize the alignment. (An insertion in the folder sequence can also be considered to be a gap in UBUTA.)
4. To view a region of the map window in more detail, use the mouse to highlight that region. Zoom into the region between amino acids 250 and 400 of UBUTA, where several gaps and insertions appear in the aligned sequences. Move the cursor into the window. The cursor will change its shape to a magnifying glass. Position it at residue 250, and press down the mouse button. Keeping the button depressed, drag to

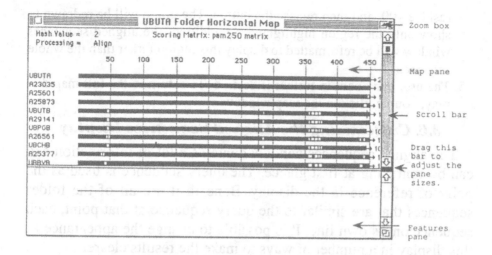

Fig. 1. The map window. Drag the bar that separates the two panes all the way to the bottom of the window to see more of the map display.

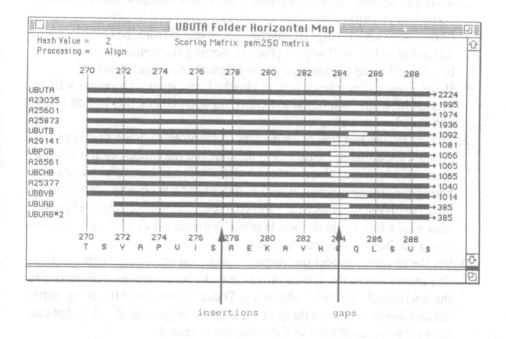

Fig. 2. At higher magnifications, the amino acids of the query sequence UBUTA are displayed at the bottom of the map display.

residue 400, and then release the button. The graph will be redrawn to show only the region highlighted. In addition, the aligned sequences window will be reformatted to display this region rather than the whole sequence.

5. The user can zoom in further if desired. To return to the full map display, double-click in the map window.

3.6. Customizing the Aligned Sequences Display

The aligned sequences window contains a lot of information and can be confusing at first glance. The query sequence is used as the point of reference in the display. Beneath it are all of the folder sequences that are similar to the query sequence at that point, each sequence on its own line. It is possible to change the appearance of this display in a number of ways to make the results clearer.

1. If the map is not currently displaying just the region between residues 250 and 400, zoom into this region now.
2. Make the aligned sequences window the active window, and click in its zoom box to expand it.
3. Choose the Format Aligned Display... command from the Options menu and a dialog box will appear (Fig. 3). Setting the options in this dialog box can alter the appearance of the aligned sequences display.
4. If the Query Line check box is checked, the query sequence will be printed beneath each of the matching regions. Check this option if it is not already checked.
5. The score line is a line of characters of the user's choice that is placed beneath each matching region to indicate the magnitude of the score at each residue. Using three different characters can be useful when comparing protein sequences, where nonidentical amino acids can contribute a positive score. Place the caret character ^ in the box labeled "greater than +1," a hyphen in the box labeled "between −1 & +1," and a lowercase *v* in the box labeled "less than −1."
6. Set the numbering and the vertical alignment to 10. Enter the vertical line character l as the vertical alignment character, and an underscore as the horizontal alignment character. These characters will create vertical and horizontal lines through the aligned sequences display that can help keep the user's place if the display is complex.
7. On the right are two sets of radio buttons that allow the user to define how matching and mismatching residues are to be displayed. Choose upper case for the match character and lower case for the mismatch character.

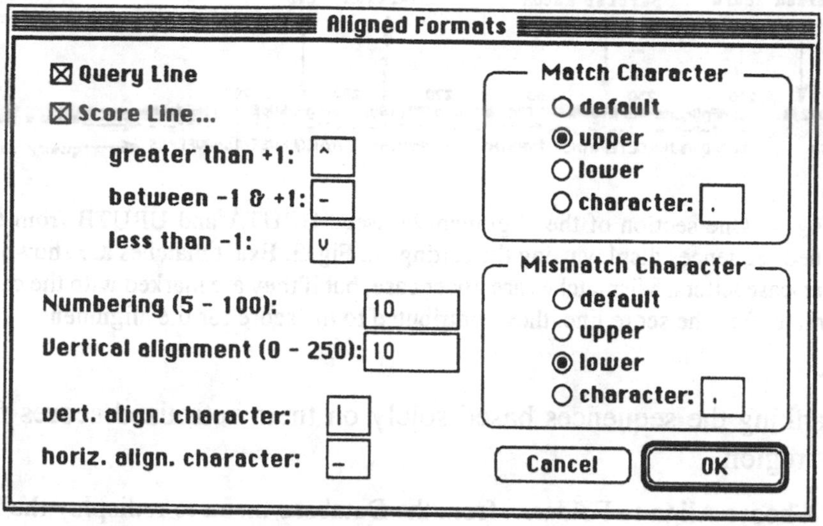

Fig. 3. The initial settings for formatting the aligned sequences display.

8. Once these changes have been made, and the dialog box looks like Fig. 3, click OK, and the aligned sequences window will be redisplayed to reflect the settings. Scroll through the window, and notice how the query line and score line are used. Find the places where gaps and insertions appear to see how they are represented (Fig. 4).

9. Again choose the Format Aligned Display... command from the Options menu. Compress the display somewhat, and make it easier to distinguish matches and mismatches. Uncheck the query line and score line check boxes, and change the match character from upper case to a period by clicking the character radio button and typing a period in the box (if there is not one there already). Click OK and reexamine the aligned sequences to see how these settings affect the display.

10. The user can make as many formatting changes as desired, as long as the aligned sequences window is open. Keep the results windows open, and use one of the filter options to rescore a region.

3.7. Rescoring a Region

While scrolling through the aligned sequences display, the user may notice that the 12th-ranked sequence, UBURAL, has fewer mismatches than some of the other sequences, but has a lower score because it only matches a portion of UBUTA. This can be compensated for to some extent by rescoring the alignments over a certain region and

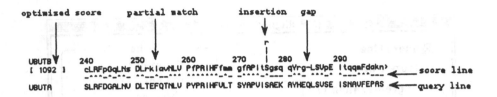

Fig. 4. One section of the alignment between UBUTA and UBUTB from the aligned sequences display using the settings in Fig. 3. Exact matches are shown as upper-case letters. Mismatches are lower case, but if they are marked with the caret character ^ in the score line, they contributed to the score for the alignment.

reranking the sequences based solely on their optimized scores for this region.

1. Choose Align to Folder... from the Database menu to redisplay the interactive dialog box.
2. Change both the "Score Rgn" and "Display Rgn" entries from "1 to 451" to "300 to 451," and click OK.
3. Make the list window active. UBURAL has now moved from 12th place to fourth.

3.8. Saving the Analysis Results

1. To save the contents of a results window to disk, make that window active.
2. Choose Save As... from the File menu. A dialog box will appear that allows the user to type a name for the results file, and choose the disk or folder to save it in. Click OK to save the file.
3. The aligned sequences window is saved as a TEXT file that can be opened within any word processor or text editor. (After the file is opened, change the font to Monaco or Courier, and the font size to nine to maintain proper alignment.) The graphical map window is saved as a PICT file, which may be opened within any drawing program that can read PICT files. Each results window must be saved separately.

3.9. Printing the Analysis Results

1. To print the contents of a results window, make that window active.
2. Choose Page Setup... from the File menu. When the print settings are as desired, click OK. (The sequence format section of this dialog box applies only if the user is printing a sequence directly from the sequence window, and is not applicable to printing results windows.)
3. Choose Print from the File menu to send the data to the printer. Each results window must be printed separately.

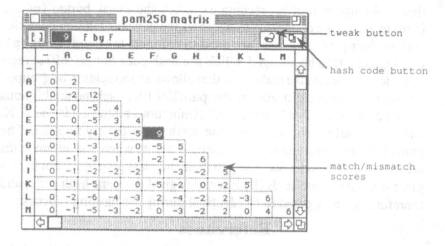

Fig. 5. Click the tweak button of the scoring matrix to change the gap and deletion penalties and the cutoff score. Click the hash code button to change the hash codes for the amino acids. The match and mismatch scores can be changed by editing the scoring matrix itself.

4. Notes

1. The OK button will be dimmed if one or more of the settings in the dialog box are out of range, and the user will not be able to proceed until the offending entry is corrected. Make sure all text entry boxes have valid values. For example, the region to analyze cannot exceed the length of the sequence or the maximum allowable limits for the analysis. At least one of the output options must be checked in the interactive dialog box.
2. There is a limit of 8000 residues for the query sequence. If the sequence is longer than this, it will have to analyzed in sections using the "Region to analyze" text entry boxes. The residue numbers can be typed in by the user, or a region can be selected from the Features Table by accessing the Features Table pop-up menu that appears to the right of the text entry boxes.
3. MacVector analyzes the sequences in the folder in 10,000-residue blocks. Therefore, long sequences may show up more than once in the list of matches if more than one block is similar to the query sequence.
4. The match and mismatch scores can be altered by editing the scoring matrix file. A number of other analysis parameters can also be changed by editing the scoring matrix file. Open the pam250 scoring matrix file (Fig. 5). At the right in the shaded area at the top of the window are two buttons, one with an icon that looks like a mouse and one with an icon

that looks like two offset rectangles. Click the tweak button (mouse icon) to access the dialog box that allows the gap and deletion penalties and the four parameters that are used to compute the cutoff score (p1, p2, p3, and p4) to be changed. Click the hash code button (offset rectangle icon) to access the dialog box that allows amino acids to be grouped during the hashing process. In the pam250 file, each nonambiguous amino acid has its own "hash code." Ambiguous amino acids (B, Z, X, and *) are arbitrarily assigned the hash code for tryptophan. The pam250S file uses a simplified hashing scheme, where the amino acids are assigned hash codes according to their chemical properties. For example, the acidic amino acids (D, E) are assigned the same hash code, and therefore are considered to be identical during the hashing process.

References

1. Pustell, J. M. (1988) Interactive molecular biology computing. *Nucleic Acids Res.* **16,** 1813–1820.
2. Wilbur, W. J. and Lipman, D. J. (1983) Rapid similarity searches of nucleic acid and protein databanks. *Proc. Natl. Acad. Sci. USA* **80,** 726–730.
3. Lipman, D. J. and Pearson, W. R. (1985) Rapid and sensitive protein similarity searches. *Science* **227,** 1435–1440.
4. Pearson, W. R. and Lipman, D. J. (1988) Improved tools for biological sequence comparisons. *Proc. Natl. Acad. Sci. USA* **85,** 2444–2448.
5. Pearson, W. R. (1990) Rapid and sensitive sequence comparison with FASTP and FASTA. *Methods Enzymol.* **183,** 63–98.

CHAPTER 19

MacVector: Sequence Comparisons Using a Matrix Method

Sue A. Olson

1. Introduction

When looking for similarities between two sequences, a matrix comparison is the method of first choice. A matrix analysis is unsurpassed for obtaining an overall picture of how the sequences are related, and it can detect significant features that other methods may miss.

Because it displays results graphically, it takes advantage of the superb pattern recognition capabilities of the human eye to allow detection of even weak similarities between two sequences. Computational alignment methods usually report only one "best" alignment between two sequences; the matrix method will reveal any significant alternative alignments that may exist and can also show the existence of rearrangements. Computational region-finding methods present the researcher with a list of matching regions; the matrix method displays matching regions in the context of the sequence as a whole, making it easy to determine if the regions are repeats or inverted repeats, for example.

The simplest form of matrix analysis (also called a dot plot or dot matrix analysis) can be done by hand if the sequences are not too long. The residues of one sequence are written along the X-axis of a two-dimensional graph, and the residues of the other are written along the Y-axis. A dot is placed at each x, y coordinate where the residues of the two sequences are identical. Regions of similarity stand out as diagonal dotted lines against an amorphous cloud of dots.

From: *Methods in Molecular Biology, Vol. 25: Computer Analysis of Sequence Data, Part II*
Edited by: A. M. Griffin and H. G. Griffin Copyright ©1994 Humana Press Inc., Totowa, NJ

Simple computer programs can be written to automate this process for longer sequences. More sophisticated implementations, such as MacVector™'s Pustell matrix analysis *(1–3)*, have enhancements that increase the signal-to-noise ratio, speed up the comparison process, and allow matching to be done by criteria other than simple identity.

1.1. The Analysis Stage Parameters

The Pustell matrix analysis consists of two stages. In the first stage, the comparison itself is performed. A number of analysis parameters can be altered by the user to increase the usefulness of the analysis. Two of these parameters are the window size and minimum percent match. To reduce the amount of background noise resulting from random insignificant matches, sequences are analyzed in overlapping sections or windows. A window is marked as a region of similarity only if the number of matching residues in the window exceeds a minimum percent. Regions that match weakly can be detected by using long window sizes and low percent match, whereas weak matches can be suppressed by increasing the percent match. The simple dot matrix analysis can be considered to have a window size of one and a minimum percent match of 100.

Changing the value of the hash size parameter affects the speed and sensitivity of the comparison. The process of scoring windows is fairly slow, so the analysis can be sped up if only those windows that have a high likelihood of containing similarities are scored. The technique used to locate such high likelihood windows is called hashing. The program rapidly scans the two sequences until it finds a region where there is an exact match of N consecutive residues. Only then does it pause to score the window surrounding that short exact match. The value N is called the hash size. If this is set to its maximum value of six, the analysis will be rapid, since the program will only score windows that contain at least six consecutive identical residues. Obviously, regions of weak similarity may be missed at this setting. Lower hash values will increase the sensitivity, but at the expense of increasing the time needed to perform the comparison. The simple dot matrix analysis can be considered to have a hash size of one.

A parameter that is unique to the Pustell DNA matrix analysis is the jump parameter. This is used in conjunction with the hash size, and when comparing coding regions, it provides a way of increasing the sensitivity of the comparison without losing the speed of higher hash sizes. When two sequences are compared using a hash setting of six (and a jump setting of one), the program will ignore any windows that do not contain at least six consecutive matching bases. Even if two sequences code for the same protein, they may mismatch at every third base owing to differing codon preferences, and thus, no regions of similarity will be found! If the jump setting is changed to three, the six bases used in the hash step are not derived from six consecutive bases, but from the first base of six consecutive triplets (or codons). Since the first base of each codon is the most conserved, similarities may be found at a jump setting of three that are missed at a jump setting of one.

Another enhancement of the Pustell matrix analysis is the use of scoring matrices. By means of a user-configurable scoring matrix, the program can use criteria other than identity between residues to score the windows. This is most useful when comparing protein sequences. Scoring matrices can be devised that match amino acids according to chemical properties, charge, genetic code, evolutionary distance, and so forth.

1.2. The Interactive Filter Stage

The actual analysis can be a slow process, depending on the lengths of the sequences, the hash size, and the number of similar regions that are found. The results of the comparison are retained in memory for use by the second stage, an interactive "filter" stage during which the results can be filtered and displayed in different ways. In addition to displaying the matrix itself, the user can also see the similar regions as aligned sequences. This interactive stage is fairly rapid, since the entire comparison does not have to be redone each time the display requirements change.

2. Materials

2.1. Hardware

1. Any Macintosh® except for the Macintosh Plus and earlier models.
2. Minimum memory requirements are 1 Mbyte of RAM if using Finder™ under System 6.0.× or 2 Mbyte of RAM if using MultiFinder® or System 7. If very large sequences are typically used, more memory may be needed.

3. One floppy disk drive and a hard disk. The program and its auxiliary files use about 2 Mbyte of hard disk space.
4. The Eve™ copy-protection device, provided with the MacVector software package.
5. A Macintosh-compatible printer if a hard copy of results is desired.

2.2. Software

1. Macintosh System Software 6.0.3 or higher.
2. The MacVector application and the MacVector Library file. These must occupy the same folder.
3. The EvE INIT provided with the MacVector software package. This must be located in the System Folder (System 6) or in the Extensions Folder of the System Folder (System 7).

2.3. Data

1. The sequences to be compared stored in MacVector binary file format or in a TEXT file in a format supported by MacVector. Supported TEXT file formats include GenBank® and derivatives of it, such as BIONET™ and IBI Pustell text files; CODATA, including NBRF PIR® and DNASTAR variants; Staden; GCG; EMBL and Swiss Prot; and line" format (sequence only). For this tutorial, use the file SV4CG located in the Vector Demo folder inside the MacVector Folder.
2. A scoring matrix file in MacVector's binary format. Several matrices are provided with MacVector and are located in the MacVector Folder: DNA matrix is used for comparisons between two nucleic acid sequences, and pam250 matrix and pam250S matrix are used for comparing protein sequences.

2.4. Optional

1. A word processor or text editor, for manipulation of text output files.
2. A drawing program capable of opening PICT files, for manipulation of graphical output files.

3. Methods

Find the MacVector Folder on the hard disk and double-click on it to open it. Double-click on the MacVector application icon to launch the program.

3.1. Opening the Files

In order to perform the analysis, a sequence window must be the active (topmost) window. For this tutorial, SV4CG (the SV40 viral genome) will be compared against itself. This sequence is supplied with MacVector in a folder named "Vector Demo" located in the MacVector Folder.

1. Choose the Open command from the File menu.
2. Check the proper file-type check box in the dialog box. SV4CG is a MacVector nucleic acid binary file, so check the box labeled "Nucleic Acid."
3. Find the Vector Demo folder in the scrollable list of file and folder names, and double-click on the name to display the list of files it contains.
4. Find the name SV4CG, and open the file either by double-clicking on its name or by highlighting the name and clicking the Open button.

3.2. The Analysis Stage

1. Make the sequence window active by clicking in it or by selecting its name from the Windows menu.
2. Choose Pustell DNA Matrix from the Analyze menu.
3. At the top of the dialog box are two scrollable lists containing the names of all nucleic acid sequence files that are currently open in memory. The left-hand list is for designating the sequence to be placed on the X-axis of the graph; the right-hand list is for the Y-axis. Make sure SV4CG is highlighted in both lists.
4. Change the minimum percent match to 55. Make sure the hash size is set to six and that the strand setting is set to both.
5. Make sure the name of the scoring matrix is "DNA matrix." If the name is "????," tell the program where to find the scoring matrix. Click the button labeled "Scoring Matrix" to bring up a file selection dialog box. DNA matrix is located in the MacVector folder.
6. Click OK to start the analysis.

3.3. Filtering the Results of the Analysis

When the analysis stage is complete, the interactive stage dialog box will appear on the screen. On the right are the "filter" options that permit the results of the analysis and the output options to be subset.

1. Check the check box labeled "Matrix map..." and uncheck the check box labeled "Aligned sequences." Click OK.
2. After a pause, a new window appears containing the graphical matrix map. Click in the zoom box in the upper right-hand corner of this window to expand it to fill the screen. Since a sequence is being compared against itself, a long identity diagonal is seen running through the matrix. There will be other diagonal lines showing regions of similarity. At this point, there is no way of determining if some of these diagonals

represent better matches than others. To display this type of information, change the way the diagonals are displayed.

3. Do not close the map window. As long as at least one of the results windows remains open, the user is in the interactive stage of the analysis, and can filter the results or modify the output types without having to repeat the analysis itself.

4. Pull down the Analysis menu again. All of the commands in the menu are dimmed except for Pustell DNA Matrix. Choose this command, and the interactive dialog box will reappear.

5. At the end of the Matrix map... check box line is a pop-up menu, set to "line." Place the pointer over this setting, and press down on the mouse button to reveal the entire pop-up menu (Fig. 1). Change the setting from "line" to "character," and click OK.

6. The matrix map will be redrawn. This time, matching regions are displayed as diagonally arranged characters that indicate the quality of the match at that point. Capital *A* is 100% match, *B* is 98%, and so on, subtracting 2% for each succeeding letter of the alphabet. After *Z* is reached, the lower-case letters are used, starting with *a*. The identity diagonal is composed of *A*s, except near the ends. The other diagonals are composed of letters further down in the alphabet and, thus, indicate regions of weaker similarity.

7. The user can return to the interactive dialog box as many times as desired as long as one of the results windows is the active window. Choose Pustell DNA Matrix from the Analysis menu again to redisplay the interactive dialog box, and look at the output displays in detail.

3.4. The Results Windows

Results of the analysis can be displayed as a graphical matrix map and as an aligned sequence display. These displays can be manipulated and customized to some extent.

1. Make sure the matrix map check box is still checked, and check the check box labeled "Aligned sequences." Use the pop-up menu to change the map back to a line display.

2. Click OK. Two results windows will appear: the matrix map window and an aligned sequence window.

3.5. Manipulating the Map Window

1. Make the map window the active window by clicking in it or by choosing it from the Windows menu. Click in the zoom box in the upper right corner of the window's title bar to expand the window to fill the screen.

2. To view a region of the matrix in more detail, use the mouse to highlight that region. An interesting region is in the upper left corner of the

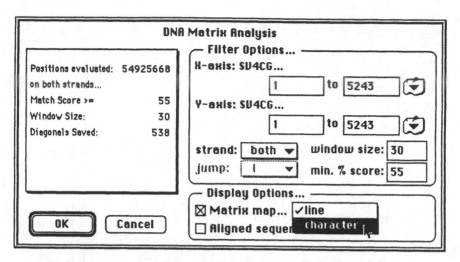

Fig. 1. Using the pop-up menu to change the matrix map from a line display to a character display.

matrix, from bases 1 to 200 on each axis. Move the cursor into the map window. The cursor will change its shape to a magnifying glass. Position it at one corner of the region, and press down the mouse button. Keeping the button depressed, drag to the other corner of the region, and then release the button. The graph will be redrawn to show only the region highlighted (Fig. 2). In addition, the aligned sequences window will be redrawn to display data only for the designated region.

3. Lines that parallel the identity diagonal are regions of similarity between the plus strands of the sequences, whereas lines perpendicular to the main diagonal indicate similarities between the plus strand of the *X*-axis sequence and the minus strand of the *Y*-axis sequence. When comparing a sequence against itself, as is being done here, the matrix display can be used to identify repeats (*1*). They are obvious in this region of the matrix as a series of parallel lines along the main diagonal. Inverted repeats would display as sets of parallel lines perpendicular to the main axis.

4. Zoom in further if desired. To return all windows to their original appearance, double-click in the map window.

3.6. Using the Aligned Sequences Window

The aligned sequences window contains a lot of information and can be confusing at first glance. The *X*-axis sequence, or "query" sequence, is used as the point of reference in the display. Beneath each line of query sequence data are displayed all of the regions of the *Y*-axis sequence that are similar to the query sequence at that point, each

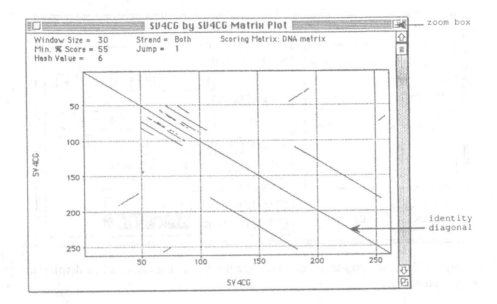

Fig. 2. The matrix map, after zooming into the upper left corner of the map. The long diagonal is the identity diagonal, the short diagonals that parallel it represent a number of direct repeats, and the short lines perpendicular to it indicate a reverse repeat.

matching region on a separate line. The cryptic label at the beginning of each of these lines of matches indicates how the match was found. For example, "jmpl str +" means that the match is on the plus strand of the *Y*-axis sequence using a jump setting of one, whereas "jmp3 str–" means that the match is on the minus strand of the *Y*-axis using a jump setting of three. In addition, an arrow character is placed at the beginning (minus strand matches) or end (plus strand matches) to indicate the direction of the matching sequence.

It is possible to change the appearance of this display in a number of ways to make the results clearer.

1. If the matrix map is not currently displaying just the upper left corner of the complete matrix (bases 1 through about 200 of both axes), zoom into this region now.
2. Make the aligned sequences window the active window, and click in its zoom box to expand it.
3. Choose the Format Aligned Display... command from the Options menu. By setting the options in this dialog box (Fig. 3), the user can alter the appearance of the aligned sequences display.

```
╔══════════════════════════════════════════════════════╗
║▓▓▓▓▓▓▓▓▓▓▓▓▓▓▓ Aligned Formats ▓▓▓▓▓▓▓▓▓▓▓▓▓▓▓║
╠══════════════════════════════════════════════════════╣
║  ⊠ Query Line                    ┌─ Match Character ─┐ ║
║  ⊠ Score Line...                 │  ○ default         │ ║
║      greater than +1: [ | ]      │  ◉ upper           │ ║
║      between –1 & +1: [   ]       │  ○ lower           │ ║
║      less than –1:    [   ]       │  ○ character: [ . ]│ ║
║                                   └────────────────────┘ ║
║  Numbering (5 – 100):   [ 10 ]   ┌─ Mismatch Character ┐ ║
║  Vertical alignment (0 – 250): [ 10 ]  ○ default       │ ║
║                                   │  ○ upper           │ ║
║  vert. align. character: [ | ]    │  ◉ lower           │ ║
║  horiz. align. character: [ _ ]   │  ○ character: [ . ]│ ║
║                                   └────────────────────┘ ║
║                              ( Cancel )  (( OK ))       ║
╚══════════════════════════════════════════════════════╝
```

Fig. 3. The initial settings for formatting the aligned sequences display.

4. If the Query Line check box is checked, the query sequence will be printed beneath each of the matching regions. Check this option if it is not already checked.

5. The score line is a line of characters of your choice that will be placed beneath the matching regions to indicate the quality of the match. Using three different characters can be helpful when comparing protein sequences, where a positive score may be assigned to certain types of mismatch, but for DNA sequences, it is more useful to place the vertical line character | in the box labeled "greater than +1" and leave the other two boxes blank.

6. Set the numbering and the vertical alignment to 10. Enter the vertical line character | as the vertical alignment character and an underscore as the horizontal alignment character. These characters will create vertical and horizontal lines through the aligned sequences display that can help keep the user's place if the display is complex.

7. On the right are two sets of radio buttons that allow the user to define how matching and mismatching residues are to be displayed. Choose upper case for the match character and lower case for the mismatch character.

8. Once these changes have been made, and the dialog box looks like Fig. 3, click OK, and the aligned sequences window will be redisplayed to reflect the settings.

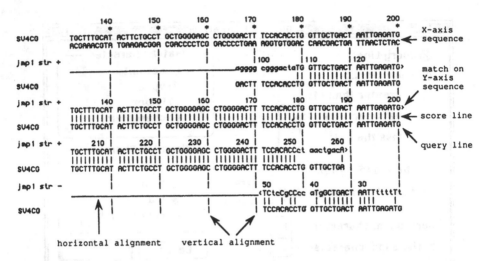

Fig. 4. One "block of the aligned sequences display using the settings in Fig. 3. Notice that arrows at the beginning or end of each of the *Y*-axis matching sequence sections indicate whether the match is on the plus strand or the minus strand.

9. Compare the aligned sequences display (Fig. 4) with the matrix map to get a better idea of where the matching regions come from.
10. Again choose the Format Aligned Display... command from the Options menu. Compress the display somewhat, and make it easier to spot mismatches. Uncheck the query line and score line check boxes, and change the match character from upper case to a period by clicking the character radio button and typing a period in the box (if there is not one there already). Click OK and reexamine your aligned sequences to see how these settings affect the display.
11. As many formatting changes can be made as desired, as long as the aligned sequences window is open.

3.7. Saving the Analysis Results

1. To save the contents of a results window to disk, make that window active.
2. Choose Save As... from the File menu. A dialog box will appear that allows the user to type a name for the results file, and choose the disk or folder to save it in. Click OK to save the file.
3. The aligned sequences window is saved as a TEXT file that can be opened within any word processor or text editor. (After the file is opened, change the font to Monaco or Courier and the font size to nine to maintain proper alignment.) The graphical map window is saved as a PICT file, which may be opened within any drawing program that can read PICT files. Each results window must be saved separately.

3.8. Printing the Analysis Results

1. To print the contents of a results window, make that window active.
2. Choose Page Setup... from the File menu. When the print settings are as desired, click OK. (The sequence format section of this dialog box applies only if the user is printing a sequence directly from the sequence window, and is not applicable to printing results windows.)
3. Choose Print from the File menu to send the data to the printer. Each results window must be printed separately.

4. Notes

1. The OK button will be dimmed if one or more of the settings in the dialog box are out of range, and the user will not be able to proceed until the offending entry is corrected. Make sure all text entry boxes have valid values. For example, the region to analyze cannot exceed the length of the sequence or the maximum allowable limits for the analysis. At least one of the output options must be checked in the interactive dialog box.
2. There is a limit of 8000 residues for the X-axis sequence and 32,000 for the Y-axis sequence. If the sequences are longer than this, they will have to be analyzed in sections using the "Region to analyze" text entry boxes. The residue numbers can be typed in by the user, or a region can be selected from the Features Table by accessing the Features Table pop-up menu that appears to the right of the text entry boxes.
3. If there is too great a size discrepancy between the two sequences, the analysis may not work properly. The longer sequence should probably not be more than 50 times longer than the shorter one.

References

1. Pustell, J. and Kafatos, F. C. (1982) A high speed, high capacity homology matrix: zooming through SV40 and polyoma. *Nucleic Acids Res.* **10,** 4765–4782.
2. Pustell, J. and Kafatos, F. C. (1984) A convenient and adaptable package of computer programs for DNA and protein sequence management, analysis, and homology determination. *Nucleic Acids Res.* **12,** 643–655.
3. Pustell, J. and Kafatos, F. C. (1986) A convenient and adaptable microcomputer environment for DNA and protein sequence manipulation and analysis. *Nucleic Acids Res.* **14,** 479–488.

CHAPTER 20

MacVector: Restriction Enzyme Analysis

Sue A. Olson

1. Introduction

Locating restriction enzyme recognition sites is one of the most commonly used functions of sequence analysis software. MacVector™'s implementation is convenient and flexible, allowing the results of the search to be viewed interactively in many different ways. In addition to finding recognition sites, the program predicts fragment sizes for single digests, double digests, or multiple digests using up to six enzymes.

The restriction enzyme analysis consists of two stages. In the first stage, the search itself is performed. A sequence can be searched with all of the recognition sites in a given file or with selected sites only, and whether the sequence is to be treated as a linear or a circular molecule can be specified. At the conclusion of the search, the results are retained in memory.

The second stage is an interactive "filter" stage in which the results can be filtered and viewed in myriad ways. For example, first only those enzymes that cut within a given region of the sequence can be viewed, then only blunt-cutters can be viewed, and then fragment sizes for single digests can be seen. Results of the analysis can be displayed as a list, as a single linear map showing the sites for all enzymes, or as a collection of linear maps displaying the sites for each enzyme on a separate line, as a circular map, and as an annotated sequence. This interactive stage of the analysis is rapid, since the entire search does not have to be repeated each time the display requirements change.

From: *Methods in Molecular Biology, Vol. 25: Computer Analysis of Sequence Data, Part II*
Edited by: A. M. Griffin and H. G. Griffin Copyright ©1994 Humana Press Inc., Totowa, NJ

2. Materials

2.1. Hardware

1. Any Macintosh® except for the Macintosh Plus and earlier models.
2. Minimum memory requirements are 1 Mbyte of RAM if using Finder™ under System 6.0.× or 2 Mbyte of RAM if using MultiFinder® or System 7. If very large sequences are typically used, more memory may be needed.
3. One floppy disk drive and a hard disk. The program and its auxiliary files use about 2 Mbyte of hard disk space.
4. The Eve™ copy-protection device, provided with the MacVector software package.
5. A Macintosh-compatible printer if a hard copy of results is desired.

2.2. Software

1. Macintosh System Software 6.0.3 or higher.
2. The MacVector application and the MacVector Library file. These must occupy the same folder.
3. The EvE INIT provided with the MacVector software package. This must be located in the System Folder (System 6) or in the Extensions Folder of the System Folder (System 7).

2.3. Data

1. A nucleic sequence stored in MacVector binary file format or in a TEXT file in a format supported by MacVector. Supported TEXT file formats for nucleic acids include GenBank® and derivatives of it, such as BIONET™ and IBI Pustell text files; Staden; GCG; EMBL; and "line" format (sequence only). For this tutorial, use the file SYNPBR322 located in the Vector Demo Folder inside the MacVector Folder.
2. A restriction enzyme data file in MacVector's binary format. A file named "restriction enzyme" is provided with MacVector and is located in the MacVector Folder. It contains data for a large number of commercially available enzymes.

2.4. Optional

1. A word processor or text editor, for manipulation of text output files.
2. A drawing program capable of opening PICT files, for manipulation of graphical output files.

3. Methods

Find the MacVector Folder on the hard disk and double-click on it to open it. Double-click on the MacVector application icon to launch the program.

3.1. Opening the Files

In order to perform the analysis, a nucleic acid sequence window must be the active (topmost) window. For this tutorial, use SYNPBR322, which is supplied with MacVector in a folder named "Vector Demo" located in the MacVector Folder.

1. Choose the Open command from the File menu.
2. Check the proper file-type check box in the dialog box. Beneath the scrolling list of file names is a series of check boxes. SYNPBR322 is a MacVector nucleic acid binary file, so check the box labeled "Nucleic Acid."
3. Find the Vector Demo folder in the scrollable list, and double-click on the name to display the list of files it contains.
4. Find the name SYNPBR322, and open the file either by double-clicking on its name or by highlighting the name and clicking the Open button.

3.2. Selecting Enzymes

1. Choose the Open command from the File menu.
2. Check the check box labeled "Restriction Enzyme" in the dialog box beneath the scrollable list of file names.
3. Position the pointer over the name Vector Demo at the top of the scrollable list, and press the mouse button to reveal the pop-up menu of folder names. Choose the MacVector Folder from the menu.
4. Find the name "restriction enzyme" in the scrollable list of file names. Open this enzyme file by either double-clicking on its name, or by highlighting the name and clicking the Open button.
5. Move the pointer into the window. The shape changes to a pointing hand (Fig. 1). Select six enzymes to use in the analysis: *Bam*H1, *Hae*3, *Hind*3, *Hpa*2, *Mbo*2, and *Pst*I. Position the hand's index finger in front of the name of each enzyme and click. A bullet will appear before the name to indicate that the enzyme is selected. If a mistake is made and the wrong enzyme is selected, click on the bullet to deselect it.

3.3. The Search Stage

1. Make the sequence window active by clicking in it or by selecting its name from the Windows menu.
2. Choose Restriction Enzyme from the Analyze menu.
3. In the center of the dialog box is a pop-up menu whose current setting is all enzymes. Position the pointer over this, and press the mouse button to reveal the pop-up menu (Fig. 2). Choose selected enzymes.
4. Examine the other options in the dialog box to verify that the enzyme file is "restriction enzyme" and the region is "1 to 0."
5. Click OK to start the analysis.

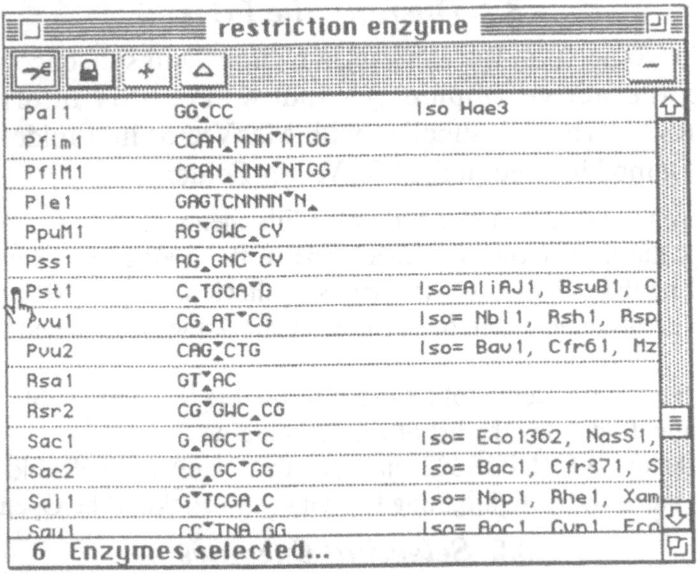

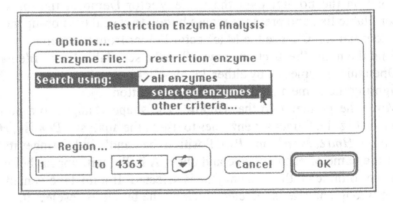

Fig. 1. Selecting the enzyme *Pst*I from the restriction enzyme file.

Fig. 2. Changing the pop-up menu setting from all enzymes to selected enzymes.

3.4. Filtering the Analysis Results

When the analysis stage is complete, the interactive stage dialog box will appear on the screen. On the right is a group of check boxes for display options, whereas on the left are "filter" options that permit the results of the analysis to be subset. The filter options will be looked at first.

1. Check the check box labeled "List cutters by…" and click the OK button.
2. A new window appears containing a list of all of the selected enzymes that cut the sequence. As long as at least one results window remains open, the user is in the interactive stage of the analysis, and can filter the results or modify the output types without having to repeat the search itself.
3. Pull down the Analysis menu again. All of the commands in the menu are dimmed except for Restriction Enzyme. Choose this command, and the interactive dialog box will reappear.
4. Filter the results to show only those enzymes that cut within the tetracycline resistance peptide of pBR322. On the left side of the dialog box, check the check box labeled "cuts only."
5. Now designate the region in the text boxes. Instead of entering the base numbers for the peptide by hand, select the region from the Features Table. Position the pointer over the Features Table icon at the end of the cuts only line, and press down the mouse button to reveal the pop-up menu (Fig. 3). Choose the tetracycline resistance peptide from the menu, and the base numbers for this region will automatically be entered into the text boxes.
6. Click OK. Only the enzyme *Bam*H1 will appear in the list of cutters, since this is the only enzyme in the selected set of six that cuts within the *tet* gene and nowhere else in the sequence.
7. The user can return to the interactive dialog box as many times as desired as long as one of the results windows is the active window.
8. Choose Restriction Enzyme from the Analysis menu once more to bring back the interactive dialog box. Make sure all of the filter options are unchecked, and look at some of the output options in detail.

3.5. Interpreting and Customizing the Results Windows

Results of the analysis can be displayed in three ways: as lists, as graphical maps, and as an annotated sequence. The map and annotated sequence outputs can be manipulated and customized to some extent.

1. Check the check box labeled "Restriction map…." The pop-up menu at the end of this line becomes active, and the user can choose to output a single linear map displaying all enzyme cut sites, separate linear maps for each enzyme, or a circular map. Leave the pop-up on the combined setting.
2. Check the check box labeled "Annotated sequence," and make sure the "List cutters by…" check box is still checked.

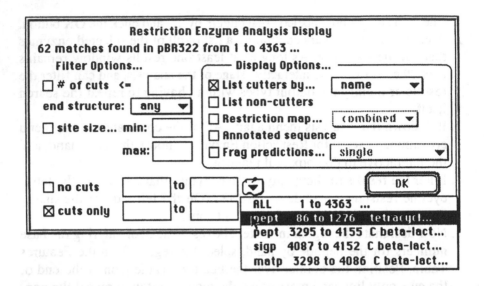

Fig. 3. Using the Features Table pop-up menu to enter the base numbers for the tetracycline resistance gene.

3. Click OK. Three results windows will appear: the list of cutters, a map window, and an annotated sequence window.

3.6. Manipulating the Map Window

1. Make the map window the active window. Click in the zoom box in the upper right corner of the window's title bar to expand the window to fill the screen. This window has two panes, each with its own scroll bar (**Fig. 4**). The upper pane contains a linear map of the cut sites. The lower pane contains a graphical representation of the Features Table entries. (If the sequence has no entries in the Features Table, this pane will contain just a thin horizontal line.)

2. To alter the relative sizes of the two panes, position the cursor over the thick black bar that separates the scroll bars of the two panes on the right edge of the window. The cursor's shape will change to a double-headed arrow when it is in the right place. Press down on the mouse button, and drag the arrow up to see more of the Features Table graphic or down to see more of the analysis results.

3. To view a region of the map window in more detail, use the mouse to highlight that region. Move the cursor into either pane of the window. The cursor will change its shape to a magnifying glass. Position it at the end of the region to be magnified, and press down the mouse button. Keeping the button depressed, drag to the other end of the region, and

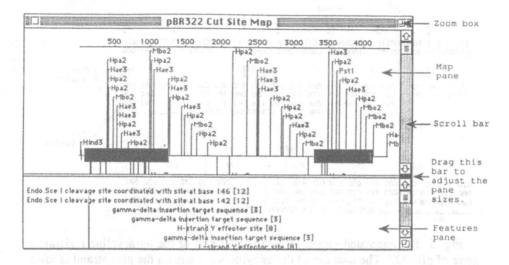

Fig. 4. The cut sites for all of the enzymes are positioned on a single linear restriction map when the map pop-up menu is on the combined setting.

then release the button. The graph will be redrawn to show only the region highlighted. In addition, all of the other results windows will be redrawn to display data only for the designated region. For example, the list of cutters will now show only those enzymes that cut within the region, and the annotated sequence window will display the region rather than the whole sequence.

4. Zoom in further if desired. To return all windows to their original appearance, double-click in either pane of the map window.

3.7. Customizing the Annotated Sequence Window

1. Make the annotated sequence window the active window, and click in its zoom box to expand it.
2. The names of the enzymes appear above the sequence with a vertical line connecting the name with the base after which the enzyme cuts. Certain features from the sequence's Features Table may be displayed along with the sequence, at the proper location (Fig. 5).
3. To alter the appearance of the annotated sequence, choose the Format Annotated Display... command from the Options menu.
4. On the left side of the dialog box are options for changing the way the sequence itself is displayed. The blocking size, the numbering interval, whether the sequence should be displayed as a single- or double-stranded molecule, and so forth, can be changed.

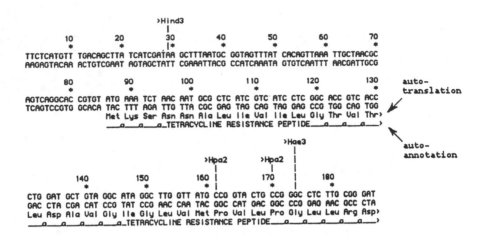

Fig. 5. The annotated sequence for the first part of the tetracycline resistance gene of pBR322. The location of the enzyme's cut site on the plus strand is after the indicated base. Autoannotation of pept features, autotranslation of pept features, and block to phase are turned on. The arrows at the ends of the autoannotation and autotranslation lines indicate that the gene is on the plus strand of the sequence.

5. On the right side of the dialog box are options for changing how the sequence will be annotated. There is a set of check boxes corresponding to Features Table feature types (pept, site, mRNA, and so on). Check those feature types to be annotated along the sequence in the annotated sequence window.

6. Another set of check boxes on the right side allows the user to indicate whether he or she wishes to see the translations of certain translatable features beneath the sequence. The block-to-phase check box will cause the bases to be grouped in codons instead of using the preset blocking size.

7. Change some of the settings, and click OK to apply the changes to the annotated sequence. As many changes can be made as desired, as long as the annotated sequence window is open.

3.8. Saving the Analysis Results

1. To save the contents of a results window to disk, make that window active.

2. Choose Save As... from the File menu. A dialog box will appear that allows a name for the results file to be typed and the disk or folder in which to save it to be chosen. Click OK to save the file.

3. Text windows are saved as TEXT files that can be opened within any word processor or text editor. (After the file is opened, change the font to Monaco or Courier, and the font size to nine to maintain proper align-

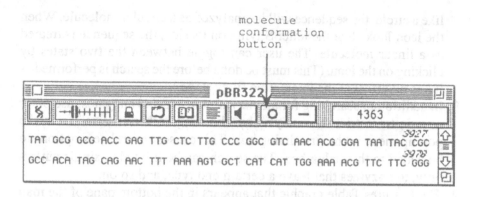

Fig. 6. When the molecule conformation button has a circle icon, the sequence is analyzed as a circular molecule; otherwise it is treated as a linear sequence. Click the button to toggle between circular and linear states.

ment.) The graphical map window is saved as a PICT file, which may be opened within any drawing program that can read PICT files. Each results window must be saved separately.

3.9. Printing the Analysis Results

1. To print the contents of a results window, make that window active.
2. Choose Page Setup... from the File menu. When the print settings are desired, click OK. (The Sequence format section of this dialog box applies only if a sequence is being printed directly from the sequence window, and is not applicable to printing results windows.)
3. Choose Print from the File menu to send the data to the printer. Each results window must be printed separately.

4. Notes

1. If the OK button is dimmed, one or more of the analysis parameters are out of range, and the user will not be able to proceed with the analysis until the offending entry has been corrected. Make sure that the text entry boxes contain valid values. For example, the region to analyze cannot exceed the sequence length. At least one of the output options must be checked in the interactive dialog box.
2. It is important to let MacVector know whether the sequence is a linear or a circular molecule. For example, if pBR322 is analyzed as a linear sequence, the *Eco*R1 site at the beginning of the sequence will not be found. Make the sequence window the active window, and look at the row of buttons in the shaded area at the top of the window (Fig. 6). One of these is used to designate the molecule conformation. When its icon looks

like a circle, the sequence will be analyzed as a circular molecule. When the icon looks like the letter **S** lying on its side, the sequence is treated as a linear molecule. The user can toggle between the two states by clicking on the icon. (This must be done before the search is performed—it has no effect at the interactive filter stage.)

3. In addition to searching for all enzymes or selected enzymes, use other criteria to restrict the analysis by choosing "other criteria" from the pop-up menu in the first dialog box. This causes several new options to be added to the dialog box. The search can be limited to single cutters only, to enzymes that leave a certain end type, and so on.

4. The Features Table graphic that appears in the bottom pane of the map window can also be customized to some extent. Choose Picture Key from the Windows menu to see a legend of what each symbol in the Features Table graphic represents. Choose Format Picture Display... from the Options menu to access the dialog box that allows the user to turn given feature types on and off.

CHAPTER 21

MacVector: Protein Analysis

Sue A. Olson

1. Introduction

One of the goals of molecular biology is to be able to deduce the three-dimensional structure of a protein directly from its amino acid sequence. As a first step toward solving this problem, a number of empirical methods have been developed for predicting a protein's secondary structure and for approximating the surface contours of a protein. Although the current methods are not very accurate (1–4), they often give enough information to be useful for certain applications.

MacVector™'s Protein Analysis Toolbox provides methods for predicting a protein's secondary structure, and for determining which regions of the protein are most likely to be exposed or buried. It also computes the protein's amino acid composition, its molecular weight, and its estimated pI.

1.1. Secondary Structure Prediction Methods

The Toolbox implements two of the most commonly used secondary structure prediction methods: those of Chou and Fasman (5) and of Garnier et al. (6). In addition, the program generates a consensus prediction showing where both methods agree.

Chou and Fasman assigned each of the 20 amino acids to one of four classes (helix former or breaker, β sheet former or breaker) depending on its tendency to appear in a given secondary structure in proteins of known structure. Their method locates clusters of helix- or sheet-forming residues in an amino acid sequence, and applies a set of heuristic rules to determine if these clusters are significant

From: *Methods in Molecular Biology, Vol. 25: Computer Analysis of Sequence Data, Part II*
Edited by: A. M. Griffin and H. G. Griffin Copyright ©1994 Humana Press Inc., Totowa, NJ

enough to nucleate a helix or sheet structure. The same region of a protein may appear to be equally likely to nucleate a helix or sheet; therefore, the likelihoods of helix, β sheet, and turn conformations are plotted on three separate lines.

Garnier et al. used information theory to analyze known protein structures and determine the statistical preferences of amino acids for a given structure. The likelihood that a given residue takes part in any of four conformations (helix, sheet, turn, coil) depends not only on its own statistical preference, but on interactions with its 16 nearest neighbors along the protein backbone.

1.2. Surface Prediction Methods

Four of the Toolbox analyses are designed to predict which regions of a protein are exposed at its surface, and which are buried within the protein itself or within a membrane. Each analysis is based on different parameters.

Hydrophilicity profiles plot the hydrophilicity values of a protein's amino acids along the length of the sequence. The graph is usually smoothed by computing an average value over a user-defined window of 5 to 20 amino acids, and plotting the average above the window's central residue.

A number of scales of hydrophilicity values have been developed for different applications. MacVector provides three commonly used scales. The Kyte-Doolittle scale *(7)* is often used to predict membrane-spanning regions; the Hopp-Woods scale *(8)* was designed to predict the locations of antigenic determinants exposed at the surface of a protein; and the Goldman-Engelman-Steitz (GES) scale *(9)* was developed in order to identify possible transmembrane helices in a protein. The signs of the values in the Kyte-Doolittle and GES scales are the reverse of the published values, so that hydrophilicity rather than hydrophobicity is plotted.

The surface probability profile *(10,11)* is based on statistical information about which amino acid residues are most likely to be found buried or exposed on the surface of proteins whose three-dimensional structures are known. The flexibility profile *(12)* uses normalized crystallographic temperature factors to predict regions of the protein chain that might be relatively flexible and, therefore, more likely to contain antigenic determinants. The antigenic index *(13)* uses a weighted sum of the results of all of the above secondary structure and surface prediction analyses to produce a single composite prediction of the surface contour of a protein.

All of the surface prediction profiles are plotted with the same sense: Values lying above the axis are hydrophilic; values below the axis are hydrophobic.

1.3. Predicting Amphiphilic Regions

Secondary structure elements that are charged or polar on one side and hydrophobic or nonpolar on the other are called amphiphilic or amphipathic structures. Amphiphilic regions often lie at protein-solvent or protein-membrane interfaces, and may be used to form channels through cell membranes. One measure of amphiphilicity is the hydrophobic moment (the periodicity of the hydrophobicity) *(14)*. MacVector's amphiphilicity profile displays a running average of the hydrophobic moment along the length of the sequence using a user-defined window size. A periodicity of 100° is used to predict amphiphilic helices, and 170° to predict amphiphilic strands of β sheet. Regions whose values lie above the axis are predicted to be amphiphilic.

1.4. Estimation of a Protein's pI

The isoelectric point (or pI) of a protein is the pH at which the protein has a net charge of zero. In theory, it can be calculated using the Henderson-Hasselbach equation, and known pK_a values for amino acid side chains and for the amino and carboxy ends of the protein. The pK_a values for amino acid side chains in a protein, however, differ from those of the free amino acids *(15)* and are affected by interactions with neighboring amino acid side chains. Thus, any computation of pI can only be an estimate.

2. Materials
2.1. Hardware

1. Any Macintosh® except for the Macintosh Plus and earlier models.
2. Minimum memory requirements are 1 Mbyte of RAM if using Finder™ under System 6.0.× or 2 Mbyte of RAM if using MultiFinder® or System 7. If very large sequences are typically used, more memory may be needed.
3. One floppy disk drive and a hard disk. The program and its auxiliary files use about 2 Mbyte of hard disk space.

4. The Eve™ copy-protection device, provided with the MacVector software package.
5. A Macintosh-compatible printer if a hard copy of results is desired.

2.2. Software

1. Macintosh System Software 6.0.3 or higher.
2. The MacVector application and the MacVector Library file. These must occupy the same folder.
3. The EvE INIT provided with the MacVector software package. This must be located in the System Folder (System 6) or in the Extensions Folder of the System Folder (System 7).

2.3. Data

A protein sequence stored in MacVector binary file format or in a TEXT file in a format supported by MacVector. Supported TEXT file formats for proteins include GenBank® and derivatives of it, such as BIONET™ and IBI Pustell protein text files; GCG; CODATA and derivatives of it, such as NBRF PIR® and DNASTAR protein text files; EMBL and derivatives of it, such as Swiss Prot; and "line" format (sequence only).

2.4. Optional

1. A word processor or text editor, for manipulation of text output files.
2. A drawing program capable of opening PICT files, for manipulation of graphical output files.

3. Methods

Find the MacVector Folder on the hard disk and double-click on it to open it. Double-click on the MacVector application icon to launch the program.

3.1. Opening a Sequence File

In order to perform the analysis, a protein sequence window must be the active (topmost) window.

1. Choose the Open command from the File menu.
2. Check the proper file-type check box(es) in the dialog box. Beneath the scrollable list of file names is a series of check boxes. Make sure that the check box corresponding to the type of file being opened is checked. For MacVector binary files, check the box labeled "Protein." For TEXT

files, check the box labeled "TEXT," and then choose the proper format from the pop-up menu to the right of the check box. Set the pop-up to "GCG" if the file is in the format used by the GCG ("Wisconsin") VAX programs, 'Staden" if the file is in the format used by Rodger Staden's programs, or "IUPAC" for any other TEXT format.

3. Find the name of your protein file in the scrollable list, and open the file either by double-clicking on its name or by highlighting the name and clicking the Open button.

3.2. Performing the Analysis

1. Choose Protein Analysis Toolbox from the Analyze menu. A dialog box containing analysis settings appears.
2. To view the analysis results graphically, check one or more of the check boxes in the area labeled "Plot... ." For the purposes of this tutorial, check "hydrophilicity," "amphiphilicity," and "secondary structure."
3. To view the analysis results numerically, check one or more of the check boxes in the area labeled "List... ." For the purposes of this tutorial, check "hydrophilicity" and "amphiphilicity."
4. To see the amino acid composition, molecular weight, and estimated pI of the sequence, check the box labeled "AA composition, MW, and pI... ."
5. To see the sequence displayed as an annotated sequence, check the appropriate check box. For the purposes of this tutorial, check it.
6. Set the window sizes for the hydrophilicity and amphiphilicity profiles. Set the hydrophilicity window size to 6 and the amphiphilicity window size to 11.
7. Set the hydrophilicity scale using the pop-up menu. Position the pointer over the current scale setting, and press down the mouse button to reveal the menu. Choose "Hopp-Woods."
8. Indicate the region to analyze. To analyze the whole sequence, leave the values in the text entry boxes at "1 to 0."
9. Click the OK button to start the analysis.

3.3. The Output Displays

Depending on which check boxes were checked, up to four results windows may be seen. If any of the plot check boxes were checked, there will be a map window displaying graphs of the analysis results. All of the other results windows contain textual data, either as lists of values or as an annotated sequence display.

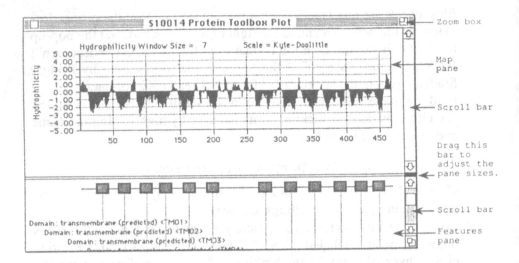

Fig. 1. The map window. Drag the bar that separates the two panes all the way to the bottom of the window to see more of the map display or up to see more of the features graphic.

3.4. Manipulating the Map Window

1. Make the map window the active window. Click in the zoom box in the upper right corner of the window's title bar to expand the window to fill the screen. This window has two panes, each with its own scroll bar (Fig. 1). The upper pane contains the graphs of the analysis results. The lower pane contains a graphical representation of the Features Table entries. (If the user's sequence has no entries in the Features Table, this pane will contain just a thin horizontal line.)
2. To alter the relative sizes of the two panes, position the cursor over the thick black bar that separates the scroll bars of the two panes on the right edge of the window. The cursor's shape will change when it is in the right spot. Press down on the mouse button, and drag the bar up to see more of the Features Table graphic or down to see more of the analysis results.
3. To view a region of the map window in more detail, use the mouse to highlight that region. Move the cursor into the window. The cursor will change its shape to a magnifying glass. Position it at the end of the region desired to be magnified, and press down the mouse button. Keeping the button depressed, drag to the other end of the region, and then release the button. The graph will be redrawn to show only the highlighted region. In addition, all of the other results windows will be

redrawn to display data only for the designated region. For example, the amino acid composition window will now show the composition for that region only rather than for the entire sequence.

4. The user can zoom in further if desired. To return all windows to their original appearance, double-click in the map window.

3.5. Customizing the Annotated Sequence Window

1. Make the annotated sequence window the active window, and click in its zoom box to expand it. Certain features from the sequence's Features Table may be displayed along with the sequence, at the proper location.

2. To alter the appearance of the annotated sequence, choose the Format Annotated Display... command from the Options menu. Most of the options in this dialog box apply only to nucleic acid sequences, but some can be used to customize annotated protein sequences (Fig. 2).

3. Set the line length to 50, the blocking size to 10, the numbering to 10, and the marking to 10. In the boxed area labeled "Characters... ," only the upper-case–lower-case option applies to protein sequences. (Annotated protein sequences are always displayed using the single-letter IUPAC code.) On the left-hand side of the dialog box, only the sites and frag features listed in the "Show... " section apply to protein sequences. Make sure both of these are checked.

4. Click OK to apply the changes to the annotated sequence (Fig. 3). As many formatting changes can be made to the annotated sequence display as desired by repeatedly choosing the Format Annotated Display... command from the Options menu.

3.6. Saving the Analysis Results

1. To save the contents of a results window to disk, make that window active.

2. Choose Save As... from the File menu. A dialog box will appear that allows the user to type a name for the results file, and choose the disk or folder in which it can be saved. Click OK to save the file.

3. Text windows are saved as TEXT files that can be opened within any word processor or text editor. (After the file is opened, change the font to Monaco or Courier and the font size to nine to maintain proper alignment.) The graphical map window is saved as a PICT file, which may be opened within any drawing program that can read PICT files. Each results window must be saved separately.

```
╔════════════════════════════════════════════════════════╗
║ ═══════════════ Annotated Formats ═══════════════      ║
║                                                          ║
║  Line Length (50 - 225): [50  ]    ┌─ Show... ──────┐  ║
║                                     │ ☒ sites ☐ tRNA ☐ pre-msg │
║  Blocking (0 - 50):      [10 ]     │ ☐ pept  ☐ rRNA ☐ mRNA    │
║                                     │ ☐ sigp  ☐ uRNA ☒ frag    │
║  Numbering (5 - 100):    [10 ]     │ ☐ matp  ☐ RNA            │
║                                     └──────────────────────────┘
║  Marking (5 - 100):      [10 ]                               ║
║  ┌─ Characters... ──────────────┐  ┌─ Translate... ─────────┐║
║  Letter case:    ◉ upper ○ lower │ ☐ pept      ☐ matp        │
║  AA code letters: ◉ one  ○ three │ ☐ sigp      ☐ block to phase │
║  Nucleic acid:   ◉ DNA   ○ RNA   └───────────────────────────┘║
║  Strandedness:   ◉ double ○ single ┌ Cancel ┐ ┌═══ OK ═══┐  ║
╚════════════════════════════════════════════════════════╝
```

Fig. 2. The text entry boxes in the upper left corner can be used to alter how an annotated protein sequence is displayed. In the panel labeled "Characters... ," only the Letter case radio buttons affect an annotated protein sequence, whereas in the panel labeled "Show... ," only the sites and frag features can be displayed along an annotated protein sequence.

3.7. Printing the Analysis Results

1. To print the contents of a results window, make that window active.
2. Choose Page Setup... from the File menu. When the print settings are as desired, click OK. (The sequence format section of this dialog box applies only if a sequence is being printed directly from the sequence window, and is not applicable to printing results windows.)
3. Choose Print from the File menu to send the data to the printer. Each results window must be printed separately.

4. Notes

1. If the OK button is dimmed, one or more of the analysis parameters are out of range, and the user will not be able to proceed with the analysis until the offending entry has been corrected. Make sure that at least one check box is checked and that the text entry boxes contain valid values. For example, the window sizes and the region to analyze cannot exceed the sequence length.
2. The window size used for the hydrophilicity analysis will depend on what information the user is trying to obtain. Use smaller window sizes (5 to 7) if looking for small hydrophilic regions that may protrude from the protein surface and thus be antigenic sites. Window sizes of 19 or longer can be used to locate possible membrane-spanning regions in

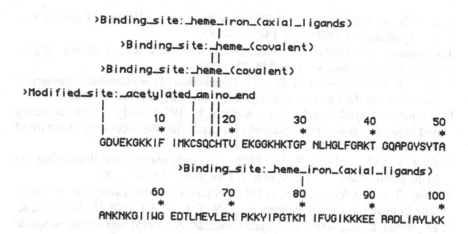

Fig. 3. Part of the annotated protein sequence for human cytochrome c (CCHU), showing the locations of the amino acids that participate in binding the heme prosthetic group.

integral membrane proteins. In general, smaller window sizes yield noisier profiles, whereas larger sizes may smooth the profile too much and cause the user to miss small regions that are highly hydrophilic or hydrophobic. For most globular proteins, window sizes between 6 and 11 are a good compromise.

3. The window size for the amphiphilicity should not be smaller than one "unit" of the structure (three residues for α helix, two residues for β strands). Again, the size to use will depend on the protein and what kind of information the user is trying to obtain. Eisenberg et al. *(14)* suggest a window size of 11; von Heijne *(16)* finds that the optimal window size for surface-seeking peptides ranges from 17 to 26, and for mitochondrial targeting sequences, from 12 to 26.

4. The Features Table graphic that appears in the bottom pane of the map window can also be customized to some extent. Choose Picture Key from the Windows menu to see a legend of what each symbol in the Features Table graphic represents. Choose Format Picture Display... from the Options menu to access the dialog box that allows the feature types to be turned on and off.

References

1. Fasman, G. D. and Gilbert, W. (1990) The prediction of transmembrane protein sequences and their conformation: an evaluation. *Trends Biochem. Sci.* **15,** 89–92.
2. Jahnig, F. (1990) Structure predictions of membrane proteins are not that bad. *Trends Biochem. Sci.* **15,** 93–95.

3. Kabsch, W. and Sander, C. (1983) How good are predictions of protein secondary structure? *FEBS Lett.* **155**, 179–182.
4. Nishikawa, K. (1983) Assessment of secondary-structure prediction of proteins. *Biochim. Biophys. Acta* **748**, 285–299.
5. Chou, P. Y. and Fasman, G. D. (1978) Prediction of the secondary structure of proteins from their amino acid sequence. *Adv. Enzymol.* **47**, 45–148.
6. Garnier, J., Osguthorpe, D. J., and Robson, B. (1978) Analysis of the accuracy and implications of simple methods for predicting the secondary structure of globular proteins. *J. Mol. Biol.* **120**, 97–120.
7. Kyte, J. and Doolittle, R. F. (1982) A simple method for displaying the hydropathic character of a protein. *J. Mol. Biol.* **157**, 105–132.
8. Hopp, T. P. and Woods, K. R. (1981) Prediction of protein antigenic determinants from amino acid sequences. *Proc. Natl. Acad. Sci. USA* **78**, 3824–3828.
9. Engelman, D. M., Steitz, T. A., and Goldman, A. (1986) Identifying nonpolar transbilayer helices in amino acid sequences of membrane proteins. *Ann. Rev. Biophys. Biophys. Chem.* **15**, 321–353.
10. Janin, J., Wodak, S., Levitt, M., and Maigret, B. (1978) Conformation of amino acid side-chains in proteins. *J. Mol. Biol.* **125**, 357–386.
11. Emini, E., Hughes, J. V., Perlow, D. S., and Boger, J. (1985) Induction of hepatitis A virus-neutralizing antibody by a virus-specific synthetic peptide. *J. Virol.* **55**, 836–839.
12. Karplus, P. A. and Schulz, G. E. (1985) Prediction of chain flexibility in proteins. A tool for the selection of peptide antigens. *Naturwiss.* **72**, 212,213.
13. Jameson, B. A. and Wolf, H. (1988) The antigenic index: a novel algorithm for predicting antigenic determinants. *Comput. Applic. Biosciences* **4**, 181–186.
14. Eisenberg, D., Weiss, R. M., and Terwilliger, T. C. (1984) The hydrophobic moment detects periodicity in protein hydrophobicity. *Proc. Natl. Acad. Sci. USA* **81**, 140–144.
15. Wood, W. B., Wilson, J. H., Benbow, R. M., and Hood, L. E. (1974) *Biochemistry. A Problems Approach*. Benjamin, Menlo Park, CA, pp. 17–24.
16. von Heijne, G. (1986) Mitochondrial targeting sequences may form amphiphilic helices. *EMBO J.* **6**, 1335–1342.

Chapter 22

Profile Analysis

Michael Gribskov

1. Introduction

Profile analysis uses information from a group of aligned sequences to create a generalized probe, the profile, for sequence or structural motifs. The profile contains information about the location and type of sequence conservation observed in the aligned sequences, the strength of the sequence conservation, and the observed positions of gaps needed to align the sequences, and is most easily understood as a description of the envelope of sequences capable of forming a given motif. When the sequences are aligned based on their three-dimensional structure, the resulting profile describes the sequences capable of folding into such a structure, and is thus a description of the sequence properties of a structural motif. Standard methods of sequence comparison, such as alignment or database searching, can be performed using a profile instead of a single sequence, and can be thought of as the comparison of the consensus of the motif to a sequence or to a database.

The details of calculating profiles and the modifications that must be made to standard algorithms to use them in sequence alignment procedures have been described in detail *(1),* and will not be repeated here. Instead, this chapter will focus on practical aspects of generating profiles and using the profile analysis approach. Although the profile analysis approach is general and can be used with protein or nucleic acid sequences, only applications to protein sequence analysis will be discussed here.

From: *Methods in Molecular Biology, Vol. 25: Computer Analysis of Sequence Data, Part II*
Edited by: A. M. Griffin and H. G. Griffin Copyright ©1994 Humana Press Inc., Totowa, NJ

The calculation of the profile requires a group of aligned sequences and an amino acid comparison table describing the similarity of each pair of amino acids. The comparison table can be derived from any source, but the default table *(2)* is derived from the mutational distance matrix MDM_{78} *(3)*. The default table differs from the MDM_{78} table primarily by having a constant value of 1.50 for the comparison of identical amino acid residues; the scores for the comparison of nonidentical residues range from 1.43 to –1.25 with a mean of –0.17.

Because profile analysis seeks to extract information from the aligned sequences that cannot be gathered from any single sequence, the selection and weighting of the sequences that are used to generate the profile are quite important. Common sense indicates that sequences that are identical or those that are completely unrelated have no additional information to extract; on the one hand, a group of identical sequences is equivalent to any one of the sequences, whereas a group of completely unrelated sequences contains no real information at all. The most suitable sequences to use in the generation of a profile are therefore those that are similar enough to be unambiguously aligned, but distant enough that patterns of conservation can be discerned. Five to ten sequences are a good number of sequences to use in generating an initial profile, but good results can be obtained with two or even one sequence. Figure 1 shows an alignment of five sequences that will be used as an example in the remainder of the chapter. These sequences are typical of sequences that generate a useful profile: They show conserved and unconserved regions, and sufficient variability to give some idea of the variation allowed within the motif.

A more subtle problem arises because of bias in the source of the sequences on which the profile is based. Typically, one may have several mammalian sequences and several more distantly related sequences from yeast, drosophila, or bacteria. The mammalian sequences will be much more similar to each other than to any of the other sequences, and a profile generated from the sequences will be biased toward the mammalian pattern rather than representing the overall motif. To overcome this problem, a weight is assigned to each sequence when calculating the profile. For instance, the distantly related bacterial, yeast, and drosophila sequences might be given a weight of 1.0, whereas each of three mammalian sequences is given a weight of 0.33. Appropriate weights for the sequences can be calculated by sev-

Fig. 1. S1 repeat motif. Five copies of a repeated sequence motif found in *Escherichia coli* ribosomal protein S1. Repeat 1 comprises residues 105–185, repeat 2 residues 193–272, repeat 3 residues 278–360, repeat 4 residues 365–448, and repeat 5 residues 452–534. The most strongly conserved positions are shown with a black background, and moderately conserved positions are shown with a gray background.

eral methods (4–6) or based on known evolutionary relationships (as in the example above).

Given a set of weights for each sequence, the profile is calculated in a fairly simple way. The value in each column of the profile is the weighted average of the comparison score for each residue at that position in the aligned sequences and the residue represented by the column of the profile. If the residues in the aligned sequences have something in common with the residue represented by the column, the average will be large and positive. If the residues in the aligned sequences are mutually quite unlike the residue represented by the column, the average will be large and negative. And if the residues in the aligned sequences are quite variable, the average will tend to be close to zero, regardless of the column. Each row of the profile is therefore equivalent to the weighted average of the rows in the amino acid comparison table corresponding to the aligned residues at that position (*see* Fig. 2). This procedure has the effect of elucidating from the MDM_{78} table, which is the sum of all ways in which residues are mutationally similar, the property that is conserved at a given position in the aligned sequences.

The profile is a specialized kind of weight matrix with a row for every aligned position (column) in the aligned sequences (Fig. 3).

	A	C	D	E	F	G	H	I	K	L	M	N	P	Q	R	S	T	V	W	Y
I	-2	23	-19	-20	71	-32	-34	150	-23	79	62	-27	-17	-26	-28	-9	16	105	-55	13
L	-12	-80	-55	-32	116	-46	-23	79	-34	150	126	-38	-29	-14	-39	-41	-15	80	47	28
M	-1	-59	-41	-19	48	-31	-33	62	22	126	150	-26	-16	-3	16	-29	-4	63	-25	-10
V	21	25	-18	-19	23	17	-33	105	-22	80	63	-26	7	-25	-27	-8	17	150	-83	-10
Ave	1	-23	-33	-22	64	-23	-31	99	-14	109	100	-29	-14	-17	-19	-22	4	100	-29	5

	A	C	D	E	F	G	H	I	K	L	M	N	P	Q	R	S	T	V	W	Y
C	26	150	-54	-55	-10	15	-13	23	-58	-80	-59	-33	6	-60	-34	72	22	25	-125	103
E	27	-55	103	150	-72	47	36	-20	28	-32	-19	46	13	69	0	19	23	-19	-109	-53
L	-12	-80	-55	-32	116	-46	-23	79	-34	150	126	-38	-29	-14	-39	-41	-15	80	47	28
R	-26	-34	1	0	-53	-34	46	-28	84	-39	16	14	26	36	150	9	-7	-27	143	-59
Ave	4	-5	-1	16	-5	-4	11	14	5	0	16	-3	4	8	19	15	6	15	-11	5

Fig. 2. Generation of the profile by averaging. Two examples of the effect of averaging in generating a profile. In the upper example, four similar residues, isoleucine, leucine, methionine, and valine, are averaged to produce a row in the profile with strong similarity to hydrophobic residues. In the lower example, four dissimilar residues, cysteine, glutamate, leucine, and arginine, are aligned. The resulting profile will show this lack of conservation in having scores for all residues (columns) near to zero.

Each column of the profile contains the score for comparing the residue corresponding to the column to the sequence characters aligned at the corresponding position. These scores are in terms of similarity; a positive score indicates a residue that is in some degree like the aligned sequences, and a negative score indicates one that is dissimilar. A row in which all scores are near zero indicates a position at which there is little sequence conservation, and therefore, any residue can be tolerated. The right-hand two columns of the profile contain weights (given in %) that are applied to gaps that are inserted during the alignment of a sequence and the profile. A gap beginning or crossing that point in the profile receives the gap penalty specified interactively times the gap penalty weight at that point. The gap penalty weights are <100% at all positions where gaps occur in the aligned sequences. Gaps inserted in sequences aligned with the profile will tend to be located in the same regions as they were found in the original aligned sequences.

A profile and a sequence are aligned using a dynamic programming algorithm and the local homology approach of Smith and

Waterman *(7)*. In a standard alignment of two sequences, the score for comparing two positions is found in a scoring table that gives values for comparing each possible pair of residues. When a profile is aligned with the sequence, the score comes from the profile itself and is found in the column of the profile corresponding to the residue in the sequence. The alignment therefore is of the entire group of sequences used to make the profile with a single additional sequence. As with all alignment procedures, there is a systematic correlation of the alignment score with the length of the sequence and profile. When a profile is used as a probe in a database search, the resulting score distribution is therefore normalized to remove these systematic effects by fitting the distribution of scores for sequences unrelated to the profile to an exponential curve

$$< \text{score} > = C \times (1.0 - e^{A \times \text{length} + B}) \qquad (1)$$

where the expected score (< score >) depends on the length of the sequence and the three constants A, B, and C. The alignment scores are then converted to standardized scores (Z).

$$Z = (N - \bar{N})/\sigma_N \qquad (2)$$

where

$$N = \text{score}/< \text{score} > \qquad (3)$$

When a profile is used with the program PROFILESCAN, a further modification of the algorithm *(8)* is used to find multiple regions within the sequence that significantly match with the profile. This allows the detection of repeated motifs within the sequence.

2. Materials

There are currently four programs in the profile analysis package. PROFILEMAKE calculates the profile from a group of aligned sequences, PROFILEGAP aligns a profile with a sequence using local or global homology dynamic programming alignment techniques, PROFILESEARCH searches a database for sequences similar to a profile and evaluates the significance of the result, and PROFILESCAN compares a sequence to a group of statistically validated profiles.

The profile package is written in FORTRAN and is available in two versions. The primary version uses the GCG procedure library

```
(Peptide) PROFILEMAKE v4.40 of: S1_Org.Msf{*}  Length: 86
Sequences: 5  MaxScore: 42.85  October 30, 1991  11:35

           Gap: 1.00              Len: 1.00
      GapRatio: 0.33         LenRatio: 0.10

S1_Org.Msf{Ecorpsa_1}  From: 1  To: 86  Weight: 1.00
S1_Org.Msf{Ecorpsa_2}  From: 1  To: 86  Weight: 1.00
S1_Org.Msf{Ecorpsa_3}  From: 1  To: 86  Weight: 1.00
S1_Org.Msf{Ecorpsa_4}  From: 1  To: 86  Weight: 1.00
S1_Org.Msf{Ecorpsa_5}  From: 1  To: 86  Weight: 1.00
```

Symbol comparison table: ProfilePep.Cmp FileCheck: 1254

Relaxed treatment of non-observed characters
Exponential weighting of characters

Cons	A	B	C	D	E	F	G	H	I	K	L	M	N	P	Q	R	S	T	V	W	Y	Z	Gap	Len
G	70	60	20	70	50	-60	150	-20	-30	-10	-50	-30	40	30	20	-30	60	40	20	-100	-70	30	100	100
A	41	15	-2	20	17	-19	24	-2	6	7	5	11	11	13	11	-6	11	26	11	-39	-16	14	100	100
E	4	19	-21	24	35	-21	4	12	6	28	-6	6	14	4	21	20	7	8	2	-15	-22	29	100	100
V	11	-11	2	-11	-9	17	7	-14	53	-9	45	36	-3	1	-10	-17	-5	22	68	-32	-4	-9	100	100
V	2	-2	2	-2	-2	2	2	-3	11	-2	8	6	-3	-1	-2	-3	-1	2	15	-8	-1	-2	30	30
T	19	19	-5	21	26	-26	20	1	4	26	-10	0	17	14	7	4	16	62	4	-34	-23	19	100	100
G	70	60	20	70	50	-60	150	-20	-30	-10	-50	-30	40	30	20	-30	60	40	20	-100	-70	30	100	100
K	-1	6	-13	3	3	-12	-6	1	19	38	5	19	4	4	8	27	4	8	19	3	-20	7	100	100
V	8	-12	-12	-12	-12	23	1	-18	76	-12	49	37	-18	-1	-14	-18	-6	12	83	-42	-2	-12	100	100
K	11	25	-12	19	17	-28	11	5	-3	43	-13	2	29	9	11	17	14	41	-3	-15	-21	16	100	100
! 11																								

```
N  31  36    5  28  22  -26  38    6  -11   10  -19  -13   43   12   12   -3   32   16   -4  -22  -16   14  100  100
L   0 -11  -20 -12  -8   29 -13   -9   42    4   47   44  -11   -7   -3   -7   -9    4   42   -1    1   -4  100  100
T  21  17    5  21  16  -20  25   -3   17    9   -1    2   12   14    1   -7   13   62   21  -43  -18   11  100  100
D  27  49  -22  64  45  -51  33   16  -10   27  -23  -13   33   10   33    5   14   14   -8  -50  -28   41  100  100
Y -12 -10   15 -18 -15   40 -23    7    7   -5   16    5   -3  -24  -17  -10  -10   -8   -2   39   42  -16   30   30
G  70  60   20  70  50  -60 150  -20  -30  -10  -50  -30   40   30   20  -30   60   40   20 -100  -70   30  100  100
A  44   5   26   7   4  -11  30   -8   14   -9   -7   -3    3   13   -2  -16   21   17   20  -46   -3    0  100  100
F -26 -41   -4 -59 -41   89 -33   -7   45  -41   73   31  -28  -40  -50  -32  -15   -3   14   73   83  -42  100  100
V  12  -8   12  -8  -8   12   9  -15   60    5   38   29  -12    5  -12  -16   -2   25   71  -42   -6   -9  100  100
E  19  31  -15  43  47  -30  36    8    2    7   -9   -5   20    7   23   -6   12   13   11  -54  -24   35  100  100
                                              ...
*  31   0    2  34  38   11  41    7   24   30   34    4   16    8   26   18   21   56    6    5    0
```

Fig. 3. S1 repeat motif profile. The first line of the profile provides documentary information about the profile, such as the version of the program generating the profile, the length of the profile, and the maximum score for the best possible alignment of any sequence and the profile. This is followed by a list of the sequences used to generate the profile and the weights on the sequences. The profile itself begins after the line ending in "..", and continues to the line with the consensus residue "*". The consensus residue for each column in the aligned sequences is shown at the left of the profile and is simply the highest scoring residue in the profile at that position. The last row of the profile, with the consensus residue "*", contains the number of each residue observed in all of the sequences used to generate the profile. This information is used to correct the alignment scores for overall composition. Note that some additional lines not important to the discussion have been omitted from this figure.

(Sequence Analysis Software Package, Genetics Computer Group, Madison, WI), and will run on VAX/VMS or UNIX platforms supported by GCG. This version of the package is distributed with the GCG sequence analysis package, version 6.2 and later, and is also available from the author. A second version, independent of the GCG procedure library, has been ported to many UNIX platforms with only minor modifications. The examples in this chapter are based on the GCG-compatible version of the programs, but the same functionality is available in both versions. It is often useful to plot the score distribution when evaluating the results of the database searches, but a plotter is not essential.

3. Methods

3.1. Method 1: Generating the Profile

This first step in using any of the profile programs is to generate a profile from a group of aligned sequences using the program PROFILEMAKE.

1. Select the sequences to align. It is desirable to have at least three sequences, but not essential. The sequences should be similar enough to be unambiguously alignable. If enough sequences are available, omit those that are difficult to align from the initial profile.
2. Align the sequences using a multiple alignment program (e.g., Chapters 25–28) or by manually assembling a multiple alignment from pairwise alignments. It is usually most convenient to generate the profile from an alignment in interleaved format. Each sequence must be the same length; the sequences should be filled at the ends with gap characters (.) if necessary.
3. Decide on weights for the sequences. In general, sequences that are more than 95% identical should be considered to be identical (i.e., to count as 0.5 sequences). The weights on the other sequences can be estimated by looking for subgroups in the multiple alignment and assigning weights so that each major subgroup has an equal weight (the sum of the weights of its members), by considering the known pattern of phylogenetic similarity between the sequences, or by one of the available computational methods mentioned above *(4–6)*.
4. Generate the profile using the program PROFILEMAKE. The length of the profile should be the same as the lengths of the aligned sequences. Profiles are currently limited to a length of 1000 residues, but this limit can be easily changed by changing a single parameter in the programs.

5. Examine the profile using a text editor. All comparisons to a profile require CPU time proportional to the length of profile. If the profile contains long stretches where all the scores are near to zero, indicating a region in the multiple alignment with little conservation, these regions can be replaced with a single line giving a score of zero for all residues and a suitable reduced weight for the penalties for gaps in that region. A plausible gap penalty weight can be calculated from

$$\text{Weight} = G_R/(1.0 + G_L \times \text{Length}) \qquad (4)$$

where G_R is the maximum desired weight for a short gap, G_L is the ratio of the gap opening penalty to the gap opening penalty that will be used with the comparison programs and Length is the number of lines removed from the profile at that point. Values of $G_R = 0.33$ and $G_L = 0.01$ are typically used to agree with the default gap opening and gap extension penalties of 4.5 and 0.05, respectively. When these values are used, the penalty aligning the known sequences with the profile will be equivalent to one identical match (i.e., -1.5). An insertion of unlimited size with respect to the profile can be allowed by setting the gap penalty weights to 0.0. The profile can also be edited to take into account other known constraints on a given motif. For instance, if a cysteine is required in a certain position to make an essential disulfide bond, all residues other than cysteine should receive a large negative score and cysteine a large positive score. This reasoning would also apply to residues known to be required in the active site of an enzyme.

3.2. Method 2: Database Search

In the database search, the profile is aligned with each sequence in the database using a dynamic programming alignment algorithm. The algorithm has been sped up by omitting all calculations other than that of the optimum score, but is still quit CPU intensive. Using a DEC VAX 8650, a database search requires about 1.9×10^{-3} s/sequence/row in the profile, or about 1.2 CPU h for a complete database search with a 100-residue profile.

1. Generate a profile according to method 1. The time required for the database search will be proportional to the length of the profile, so consider editing the profile to remove long unconserved regions as described above.
2. Compare the profile to each sequence in the database using the program PROFILESEARCH. If there is a limited amount of disk space, it is usually sufficient to have the program report sequences with a stan-

dardized score >4.0 (4σ) or the top 1000 scores. This is done with the command line switches /minlist=4.0 and /listsize=1000. If adequate disk space is available to store the complete results of the database search (about 5400 blocks on VAX/VMS using 22,000 sequences in PIR 27.0), the user will be able to plot the score distribution of the search, which can be useful in evaluating the results.

3. The PROFILESEARCH program automatically normalizes the results of the search and converts the scores to standardized scores (standard normal deviate or Z score). The equations used to normalize the score distribution are reported and should be inspected to ensure that the normalization was successful. In particular, if the correlation of the score distribution and the fitted curve is <90%, it indicates some problem in the normalization.

4. In the normalization process, scores that appear to be significantly similar to the profile are detected and omitted from the calculation of the mean and standard deviation of the score distribution. The number of omitted scores is reported in the normalization section and is an indication of the number of sequences significantly similar to the profile.

5. The results of the database search are reported in order of decreasing standardized score. A standardized score above 6.0 is a strong indication of homology. Scores between 4.0 and 6.0 are unusually high, but must be critically examined. In particular, longer sequences with scores in this region may be proteins in which only one domain is related to the sequences used to generate the profile. The presence of a large amount of unrelated sequence (or duplicated sequence) will cause the score for such a sequence to be reduced during the normalization process. Figure 4 shows the result of a database search using the profile generated from the five sequences shown in Fig. 1. In addition to the ribosomal protein S1 sequences, there is strong similarity to several chloroplast IF-1 sequences, polynucleotide phosphorylase, a *B. subtilis* IF-1 sequence, and weaker similarity to eukaryotic eIF2-α and *E. coli* IF-1 sequences. Further analysis (not shown here) makes it clear that all of these sequences contain the S1 repeat motif, but in the initial search, this cannot be determined by the standardized scores alone.

6. It is often useful to plot the score distribution to aid in detecting significantly similar sequences. This is especially true if the correlation coefficient for the normalization is <90%. Two kinds of plots are useful: a plot of the standardized score vs the sequence length (Fig. 5A) and a plot of the standardized score vs log of the fractional rank of the score (Fig. 5B), where the fractional rank of the top score is the rank of the score

above or below the mean divided by half the number of sequences. The first type of plot requires that all of the scores be present in the output file, and the interesting scores to investigate are those that fall above the large peak of unrelated scores. The second type of plot requires only the top scores, and the interesting sequences are those whose scores fall off of the straight line due to unrelated sequences. Any score that seems to be significantly different from the large group of unrelated scores should be aligned with the profile for further evaluation.

In the example shown in Figs. 4 and 5, it is reasonable to conclude that the chloroplast IF-1, polynucleotide phosphorylase, eIF2-α sequences are significantly similar to the S1 repeat motif described by the profile. It is less clear whether the bacterial IF-1 sequences should be included, but since they are known to be homologous to the chloroplast IF-1, and because the *B. subtilis* sequence appears to be significantly similar, it appears very likely. To test this, the newly detected sequences should be aligned to the original profile (method 3) and a new profile generated (method 1). The database search should then be repeated. If the sequences actually possess the S1 repeat motif, they should all have clearly significant standardized scores, and the standardized scores for the other sequences should not decline substantially from the scores in the original search. Further tests for borderline cases are suggested in method 5.

3.3. Method 3:
Alignment of Sequence(s) and a Profile

The alignment of a sequence and a profile is equivalent to aligning the sequence to an existing group of sequences, without changing the alignment of the existing group.

1. Generate a profile according to method 1.
2. It will often be useful to align a group of sequences to a profile. The sequences might be a group of sequences that there is some reason to believe are similar to the motif in question or, more commonly, the list of most similar sequences found by the program PROFILESEARCH.
3. When aligning a group of sequences with a profile, it is useful to use the command line switch /multout=*filename,* which will cause the alignment of the sequences and the profile to be written out as an interleaved multiple alignment to the file *filename.* It is important to understand that this alignment is not a true multiple alignment, but rather an assembly of each pairwise alignment of a sequence and the profile.

```
(Peptide) PROFILESEARCH of: disk$scratch:[work.gribskov.s1]s1_org.prf;1 Length: 86 to: nbrf:'
Scores are corrected for composition effects

      Gap Weight: 4.50
Gap Length Weight: 0.05
Sequences Examined: 27574
CPU time (seconds): 4710

Profile information:
(Peptide) PROFILEMAKE v4.40 of: S1_Org.Msf{'}  Length: 86
Sequences: 5  MaxScore: 36.76  October 1, 1991  22:00
                        Gap: 1.00           Len: 1.00
                   GapRatio: 0.33      LenRatio: 0.10
  S1_Org.Msf{Ecorpsa_5}  From: 1      To: 86    Weight: 1.00
  S1_Org.Msf{Ecorpsa_4}  From: 1      To: 86    Weight: 1.00

Normalization: October 2, 1991  00:21

Curve fit using 400 length pools
0 of 400 pools were rejected

Normalization equation:

Calc_Score = 9.09 ' ( 1.0 - exp(-0.0074*SeqLen - 0.2815) )

Correlation for curve fit: 0.953

Z score calculation:
Average and standard deviation calculated using 27566 scores
8 of 27577 scores were rejected

Z_Score = ( Score/Calc_Score - 1.002 ) / 0.150

Sequence      Strd ZScore  Orig Length ! Documentation
PIR1:R3EC1      +   16.06    30.55   557 ! Ribosomal protein S1 - Escherichia coli
PIR1:R3ZR1      +   11.23    24.09   568 ! Ribosomal protein S1 - Rhizobium meliloti
PIR2:A23525     +    9.14    12.31    77 ! Initiation factor IF-1 homolog - Spinach chloroplast
PIR2:A26118     +    7.45    19.15   711 ! Polynucleotide phosphorylase (EC 2.7.7.8) alpha chain - Escherichia coli
PIR2:S01562     +    6.34    10.19    78 ! Initiation factor IF-1 - Liverwort (Marchantia polymorpha) chloroplast
PIR2:A05008     +    6.34    10.19    78 ! Hypothetical protein 96 (E.coli infA homolog) - Liverwort
PIR3:A32307     +    6.10     9.67    72 ! 'Initiation factor 1 - Bacillus subtilis
PIR3:A32108     +    5.99    15.85   304 ! 'Initiation factor 2 alpha chain - Yeast (Saccharomyces cerevisiae)
PIR3:PT0051     +    5.96     9.39    69 ! 'Translational initiation factor elF-2 alpha chain - Pig (fragment)
```

PIR2:A26711	+	5.91	15.87	315 !	Initiation factor eIF-2, alpha chain - Rat
PIR2:S06628	+	5.27	10.21	96 !	Glucose dehydrogenase (acceptor) (EC 1.1.99.10) - Drosophila melanogaster
PIR3:F35057	+	5.22	8.17	57 !	*HLA class II histocompatibility antigen, DQ-A beta chain - Baboon (fragment)
PIR1:Q7BPT3	+	5.01	8.58	67 !	Lysis protein - Bacteriophage T3
PIR2:E27397	+	4.96	10.68	114 !	Na+/K+-transporting ATPase (EC 3.6.1.37) alpha chain 5 - Human (fragments)
PIR1:FIRZI	+	4.82	10.28	107 !	Initiation factor IF-1 - Rice chloroplast
PIR2:PL0126	+	4.82	9.51	89 !	H-2 class II histocompatibility antigen, I-A beta chain - Mouse (fragment)
PIR2:S03523	+	4.81	11.11	130 !	T-cell receptor alpha chain precursor V-J region (HD-Mar) - Human (fragment)
PIR1:FIEC1	+	4.81	8.64	71 !	Initiation factor IF-1 - Escherichia coli
PIR1:WJXLAC	+	4.80	8.43	67 !	AC1 homeotic protein - African clawed frog (fragment)
PIR1:CCNC	+	4.80	10.26	107 !	Cytochrome c - Neurospora crassa
PIR1:Q7BPE7	+	4.75	8.39	67 !	Gene 17.5 protein - Bacteriophage T7
PIR2:A27855	+	4.65	8.58	72 !	infA protein - Escherichia coli

Fig. 4. Database search with S1 repeat motif profile. The output from the program PROFILESEARCH is divided into a documentary and a results section. The documentary section is further divided into sections by lines of stars and spaces (* * *). The first documentary section contains information regarding the search: the gap penalties used, number of sequences examined, and elapsed CPU time. The second documentary section is a copy of the first several lines of the documentation section of the profile. The third documentation section details the results of the normalization procedure. Of particular interest are the correlation coefficient for the fit of the calculated scores to the observed data and the number of outlying scores rejected from the calculation of average and standard deviation. The latter number is a rough indication of the number of significantly similar sequences that were found (in this case eight). The results section of the profile lists the sequences in the database most similar to the profile in order of standardized score: the name of the entry, the strand on which the matching sequence was found (this applies only to nucleic acid searches and is always "+" for proteins), the standardized score (Z Score), unnormalized score (Orig), the length of the sequence, and short description of the sequence from the database.

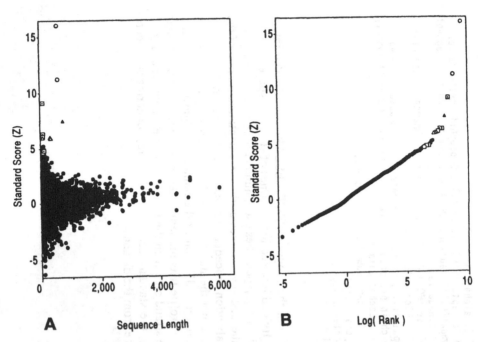

A   Sequence Length   **B**   Log(Rank)

Fig. 5. Plot of database search results. (A) Standardized score vs sequence length. This type of plot is particularly useful when the correlation of the calculated and observed scores is low, and therefore, the standardized scores are not a reliable way to judge the significance of the similarity of a sequence to the profile. (B) Standardized score vs the log of the fractional rank. If the nor normalization procedure is successful, this plot is a good way to visually detect sequences that are significantly similar to the profile. A plot of sequences unrelated to the profile should be roughly linear; sequences similar to the profile should lie above the line. In both panels, the sequences are shown as follows: ribosomal protein S1 (O), polynucleotide phosphorylase (▲), chloroplast IF (⊡), bacterial IF-1 (□), eIF-2α (△), and all other sequences (●).

3.4. Method 4: Quick Evaluation of Interesting Sequences

It is often useful to evaluate the effects of changing the gap penalties or of making changes to the profile, without performing a complete database search for each change. The following method assumes that there is a group of sequences that the user knows or suspects is similar to the motif represented by the profile and wishes to see how the significance changes when various parameters are changed.

1. Generate a profile (method 1), and perform a database search (method 2).
2. Prepare a file containing a list of sequences known or suspected to be similar to the profile. This list should be in the GCG file of sequence

names (FOSN) format, and can be simply the top scores from the database search (e.g., all sequences with standardized scores above 4.0). If attempting to enhance the discrimination between known related and unrelated sequences, include the unrelated sequences desired to discriminate against in this file.

3. Decide on a group of sequences to use as a group of random sequences. There are many ways this could be done. One possibility is to make a file containing every 200th sequence from the initial database search. This should generate a FOSN format file with about 1400 sequences spanning the entire range of standardized score. This FOSN can then be specified as the database using the syntax @*fosn.file*. A second possibility is simply to perform a second database search, but limit the number of sequences *(N)* to 1000–2000 with the command line switch /seqlimit=*N*. This approach will have more bias, since the sequences are not randomly ordered in the database, but is usually good enough for a approximate test. Regardless of the way a random group of sequences is selected, perform the search without normalization using the command line switch /nonorm.

4. Compare the group of suspect sequences to the profile by running PROFILESEARCH and specifying the name of the FOSN as the database using the @*suspectFOSN.file* syntax. Again, use the /nonorm command line switch to turn off the normalization.

5. Edit the output file from the search of the random sequences. Insert the portion of the output file from the search of the suspect sequences beginning at the ".." and continuing to the end of the file. These scores may be inserted anywhere following the ".." in the output file from the search of the random sequences. The user now has a single file with the scores from both the random and suspect sequences.

6. Normalize the scores by running the program PROFILESEARCH with the /nosearch command line switch, specifying the file of combined scores created in step 5 as the one to be normalized. The resulting file will show the effect of the parameter change or change to the profile on both similar and dissimilar sequences. The process must be repeated for every parameter change or change to the profile, but it is an order of magnitude faster than a complete database search.

3.5. Method 5: Testing a Sequence
for Membership in a Homologous Family

When the standard score for the alignment of a sequence and a profile is <6.0, but there is reason to believe that the sequence is related to the motif represented by the profile, additional tests should be performed to try to confirm that the relationship is significant.

The following method assumes that a profile has been generated and a database search performed, and that there are one or more similar sequences with standard scores too low to be conclusive that the user may wish to investigate further.

1. Align the sequence (method 3) to be tested with the original sequences used to generate the profile and generate a new profile (method 1). It is a good sign if the new sequence appears to be most similar to the original sequences in the conserved regions, and that few or no gaps appear in areas that lack gaps in the original sequences.
2. Repeat the database search with the new profile. If the sequence is truly related to the motif in question, the standard score for the sequence should now be higher, usually >6.0, and the standard scores for the original sequences not substantially reduced. If the standard scores for the original sequence are drastically reduced, it suggests that the new sequence is not compatible with the original group.
3. Generate a profile from the new sequence and any obvious homologs, and use it in a database search. If the sequences used to make the original profile are among the top scoring sequences in this search, it suggests that the two profiles represent a common motif.

3.6. Method 6: Validation of Profiles

The process of *validation* is intended to establish that a profile is both sensitive and specific, and to provide the statistical information necessary to determine if any new sequence is significantly similar to the profile. Validated profiles are usually added to the library of profiles used by the program PROFILESCAN.

1. Generate a profile (method 1), and perform a database search (method 2). Examine the results, and confirm that all sequences that should have the motif in question have standard scores above 6.0.
2. Align all of the top scoring sequences with the profile (method 3), and confirm that the alignment is the same as the alignment used to generate the profile. If some of the top scoring sequences appear highly unlikely to contain the motif of interest, try to determine from the alignment what feature of the profile is causing the artifactually high score, and adjust the profile to exclude this sequence more effectively.
3. Return to step 1, and repeat the generation of the profile and database search until all related sequences are detected with standardized scores >6.0, and no unrelated sequences have scores >6.0. Such a profile is both sensitive and specific, and thus validated.

4. Enter the statistics from the normalization section of the output from PROFILESEARCH into the profile library data file *motifs.fil*. The gap and length penalties used for the searches, the three parameters *A, B,* and *C,* used in the normalization, and the mean and standard deviation of the normalized score distribution are recorded in the library data file. In addition, two threshold values are recorded, the *high* and the *interesting* threshold. These values are recorded in terms of the normalized score before conversion to the standard score and can be calculated by

$$\text{threshold} = \text{standard score} \times \sigma + \overline{\text{score}} \qquad (5)$$

where $\overline{\text{score}}$ and σ are the mean and standard deviation for the score distribution. The thresholds are generally set such that the high threshold includes all of the known related sequence, and none of the unrelated sequences, typically a standard score of about 6.0–6.5, and the interesting threshold at a standard score of 4.5 –5.0.

3.7. Method 7:
Scanning for Known Motifs in a Sequence

Once a profile has been validated, it can be entered into the library of motifs and used with the program PROFILESCAN. The purpose of PROFILESCAN is to rapidly compare a sequence or sequences to the motifs in the profile library, and determine immediately if the sequence is significantly similar to any of the motifs. The statistics generated from actual database searches are used to evaluate the significance, and the process of validation ensures that the profiles in the library are both sensitive and specific. A list of the currently available library of validated profiles is shown in Table 1.

1. Run the program PROFILESCAN using either the /interesting or /high command line switch to select the stringency level.
2. Two output files are produced. The summary file (default suffix *.sum*) contains a list of each profile in the library and the number of matching subsequences found in the sequence.
3. The second output file (default suffix *.scan*) shows each alignment of a profile from the library and a subsequence that met the threshold stringency. If the sequence contains a repeated motif, many of the repeats should be detected as separate alignments.

4. Notes

1. A profile can be generated from a single sequence. In older versions of the program PROFILEMAKE (GCG version 6.2 and earlier, or the

Table 1
Validated Profile Library

Profile	Specificity
Ankyrin.prf	Ankyrin repeat motif
Asp.prf	Bacterial glutaminase-asparaginase
Ca-EFHand.prf	Ca binding EF-hand
Chap.prf	Chaperonin
Cy.prf	α-Crystallin—16–27K heat shock protein
Cyclic.prf	Cyclic nucleotide binding structure
CytC4.prf	Cytochrome c
EGF_repeat.prf	EGF repeat motif
Fd.prf	Ferredoxin
Globin.prf	Globins
Grn.prf	Ribonuclease T1-U2
HisJ.prf	LAO-type amino acid binding protein
HMGBox_14.prf	HMG protein conserved motif
Homeo.prf	Homeotic protein
HTH21.prf	Helix-turn–helix DNA binding motif
Ig-constant.prf	Immunoglobulin constant domain
Ig-variable.prf	Immunoglobulin variable domain
Il1.prf	Interleukin-1
JunFos.prf	jun-fos family
Kinase.prf	Ser-Thr protein kinase
Kringle.prf	Kringle motif
Lectin.prf	Hepatic lectin
MAP.prf	MAP repeat motif
Myc.prf	myc oncogene family
NGF.prf	Nerve growth factor
Nr.prf	Pancreatic ribonuclease
Nuc10.prf	Dinucleotide binding structure (Rossman fold)
OmpR.prf	OmpR type regulatory protein
PDGF.prf	Platelet-derived growth factor
Perf.prf	Perforin
POU.prf	POU motif
Pthi.prf	Parathyroid hormone
P450.prf	Cytochrome P-450
Serine.prf	Serine protease
SH3_41.prf	Possible actin binding site
SSRNA.prf	Single-stranded RNA binding motif (pABP/rnp)
S1repeat.prf	Ribosomal protein S1 repeat motif
TPR.prf	TPR repeat motif
ZincI.prf	Zinc finger I
ZincII.prf	Zinc finger II

author's version 4.3 or earlier), characters that do not occur in the aligned sequences are treated as if they occur 0.5 times. When only one sequence is used, the sum of the nonoccurring characters therefore greatly outweighs the single character that occurs. This problem also occurs, to a lesser degree, when less than four sequences are used to generate the profile. The solution is to use the command line switch /stringent when generating a profile from only one sequence, or to increase the weights on each sequence as follows: one sequence—sequence weight of at least 19.0, two sequences—4.75, three sequences—1.4.

2. In order to keep the width of the profile <132 columns, so that the files can be edited with standard text editors, the scores in the profile are multiplied by 100. This allows the decimal point to be omitted and saves one space per column. An entry of 150 in the profile therefore represents a score of 1.5. The programs automatically apply the appropriate corrections when reading the profile.

3. As in any dynamic programming alignment, the scores calculated by PROFILESEARCH and the alignments calculated by PROFILEGAP are sensitive to the gap opening and gap extension penalties (referred to as *gap weight* and *gap length weight* in the GCG versions of the programs). At the least, the user should check that the gap penalties used for a search are capable of reproducing the alignment of the sequences used to generate the profile, and alter the penalties as required. Method 5 provides a way to test the effects of gap penalty changes on the score distribution.

4. Errors in normalization of the results of PROFILESEARCH seem to occur most often with long profiles with a large number of positions with reduced gap penalties. It may be possible to improve the normalization by increasing the gap penalties or by editing the profile. Otherwise, the best option is to plot the score distribution and use it rather than the standardized score to evaluate the significance of the match to the profile.

5. In the validation of profiles, the evaluation of the sensitivity and specificity of a profile toward certain very common motifs can be difficult because of the difficulty of determining whether a sequence possesses the feature of interest. An example of this is the dinucleotide fold motif, which is very common, but difficult to evaluate since the dinucleotide binding properties of proteins are often not specified in the database (or known biochemically). In this case, the *high* and *interesting* thresholds may be set rather arbitrarily, typically to something like 6.5σ and 4.5σ.

6. In the future, the stringency of PROFILESCAN program will be set more flexibly using a command line switch /sigma=n to display all alignments with significance greater than $n\sigma$. Other modifications will allow the scanning of open reading frames from a DNA sequence in order to locate protein coding regions.

Acknowledgments

This research was sponsored by the National Cancer Institute, DHHS, under contract N01-CO-74101 with Advanced BioScience Laboratories. The contents of this publication do not necessarily reflect the views or policies of the Department of Health and Human Services, nor does mention of trade names, commercial products, or organizations imply endorsement by the US government.

References

1. Gribskov, M., Lüthy, R., and Eisenberg, D. (1990) Profile analysis. *Meth. Enzym.* **183,** 146–159.
2. Gribskov, M. and Burgess, R. R. (1986) Sigma factors from *E. coli, B. subtilis,* phage SPO1, and phage T4 are homologous proteins. *Nucleic Acids Res.* **14,** 6745–6763.
3. Schwartz, R. M. and Dayhoff, M. O. (1979) Matrices for detecting distant relationships, in *Atlas of Protein Sequence and Structure,* vol. 5, Supp. 3 (Dayhoff, M. O., ed.), National Biomedical Research Foundation, Washington, DC, pp. 353–358.
4. Felsenstein, J. (1973) Maximum-likelihood estimation of evolutionary trees from continuous characters. *Am. J. Hum. Genet.* **25,** 471–492.
5. Altschul, S. F., Carrol, R. J., and Lipman, D. (1989) Weights for data related by a tree. *J. Mol. Biol.* **207,** 647–653.
6. Sibbald, P. R. and Argos, P. (1989) Weighting aligned protein or nucleic acid sequences to correct for unequal representation. *J. Mol. Biol.* **16,** 813–818.
7. Smith, T. F. and Waterman, M. S. (1981) Comparison of biosequences. *Adv. Appl. Math.* **2,** 482–489.
8. Gribskov, M., Homyak, M., Edenfield, J., and Eisenberg, D. (1988) Profile scanning for three-dimensional structural patterns in protein sequences. *CABIOS* **4,** 61–66.

CHAPTER 23

Prediction of RNA Secondary Structure by Energy Minimization

Michael Zuker

1. Introduction

Secondary structure prediction for RNA is fundamentally different from three-dimensional molecular modeling (1). A secondary structure of an RNA molecule is simply a collection of predicted base pairs subject to a few simple rules. Base pairs can be either G-C or A-U Watson-Crick pairs, or the weaker G-U pair.

Secondary structures can be used in part to explain translational controls in mRNA (2) and replication controls in single-stranded RNA viruses (3). Secondary structure modeling can also be used as a first step to the more intricate process of three-dimensional modeling (4). This could include the modeling of ribosomal RNA or catalytic RNAs, such as group I introns.

When numerous homologous RNAs are available, secondary structure prediction is possible using the comparative sequence analysis method (5). Although largely a labor-intensive method when first used, it is now being made more automatic (6). The comparative approach requires the RNA sequences to be similar enough so that they can be reliably aligned.

When one wishes to fold a single sequence, the usual method is to predict a secondary structure that is optimal in some sense. Recursive algorithms borrowed from the area of sequence alignment have

From: *Methods in Molecular Biology, Vol. 25: Computer Analysis of Sequence Data, Part II*
Edited by: A. M. Griffin and H. G. Griffin Copyright ©1994 Humana Press Inc., Totowa, NJ

been modified and applied to the RNA folding problem by a number of people *(7–13)*. Although some early methods (e.g., *[9]*) maximized the number of base pairs in a secondary structure, the later methods have all attempted to solve the problem by minimizing a computed free energy.

The method described here *(14–16)* uses energy rules developed by Turner and colleagues *(17–19)*. Negative stabilizing energies are assigned to the stacking of base pairs in helical regions and to single bases that stack at the ends of helical regions. Otherwise, destabilizing energies are assigned to bulge, interior, hairpin, and multibranched loops. The energies of base stacking and the destabilizing loops are assumed to be additive in computing the overall energy. Pseudoknots, described in *(20)*, are not permitted in secondary structure calculations.

2. Materials— Computer Equipment and Programs

The "mfold" package by Zuker and Jaeger *(14–16)* computes optimal and suboptimal foldings of an RNA molecule. There are two principal improvements over the older RNAFOLD program *(21)*. The first is the incorporation of up-to-date energy rules, and the second is the program's ability to compute both optimal and suboptimal foldings. There are two forms of output. One is a list of optimal and suboptimal foldings, sorted by energy. The other is a graphics output that depicts all suboptimal foldings in a single image. The graphics output is called the *energy dot plot.*

Mfold was first developed in a VAX/VMS environment. This version is now obsolete, although it is still supported. It will be referred to as the VAX version, the VMS part being understood.

The official version of mfold now runs on UNIX workstations and larger computers. The author's personal version runs on an IRIS workstation (IRIX 4.0.5.) by Silicon Graphics and will be called simply "mfold." It uses the SGI graphics library and is hardware specific. An X11 version of mfold also exists on a number of platforms. It is called "xmfold" on the IRIS, "mfold-dec" on the DEC 3100 workstation (ULTRIX), and "mfold-sun" on the SUN4 workstation or the SUN sparcstation. There are also incomplete versions that run on

CRAY and CONVEX supercomputers, as well as an Apple Macintosh version. A GCG *(22)* compatible version has recently become available.

At the heart of mfold are the two programs "lrna" and "crna." They fold linear RNA and circular RNA molecules, respectively. Computer storage requirements increase as the square of the sequence length. Folding times increase as the cube of the sequence length. On an IRIS 35SG workstation, it takes 14 min to fold 500 bases. Space requirements (both CPU for run time and disk for *save sets*) are roughly $4n^2$ bytes for both lrna and crna. It is usually space limitations that set the upper bound for the maximum size of an RNA that can be folded.

The interactive energy dot plot that is described below requires a graphics interface. In the VAX version, this requires a Tektronix 4105 terminal or terminal emulation, or a Visual 600 series terminal. The VAX version also requires a Tektronix PLOT10 object library. The IRIS mfold version will run on any IRIS workstation, but the user must be working on the color monitor to use the interactive dot plot feature. The X11 versions all require an X11 library to be installed on the workstation, and the program cannot be run from a regular terminal. The SUN version also requires the gnu C compiler. A hard copy of the energy dot plot can be produced with the package if the user has access to the GCG *(22)* graphics program "FIGURE."

The CRAY, CONVEX, and Macintosh versions of the program have been ported without the dot plot. All versions of the program can be run from a nongraphics terminal or in batch mode if the dot plot option is not selected.

The programs are written in FORTRAN using DEC extensions. Some of the UNIX versions have been altered to remove unsupported FORTRAN extensions. Two "include" files, lin.inc and circ.inc, are provided with the source code for lrna and crna, respectively. The value of the "nmax" parameter is the maximum fragment size that can be folded. It can be changed and the programs recompiled using the Makefile that is provided.

The VAX and UNIX workstation versions are available via anonymous ftp at nrcbsa.bio.nrc.ca. Compressed tar files are in the pub directory and are named according to the descriptions above.

3. Methods

3.1. Method 1—
Running the Folding Programs

Both lrna and crna can be run interactively. Menus are displayed, and choices can be made regarding input and output. The first choice is run type.

3.1.1. Run Type Selection

The first menu displayed is:

0 Regular run (default)
1 Save run
2 Continuation run

In a Regular run, the program takes an RNA sequence as input, computes the energy matrices (the fill algorithm), and proceeds to the output section. Because the fill algorithm requires most of the computation time, the computation can be broken into two stages. A Save run computes the energy matrices, writes the results to a large file on disk, and exits. The save file cannot be read by a user. A Continuation run reads a previously computed save file, and proceeds directly to the output stage. When a Regular or Continuation run type is selected, execution continues with run mode selection (Section 3.1.3.). When a Save run is selected, execution continues with output save file name selection (Section 3.1.2.).

3.1.2. Output Save File Name Selection

The user is prompted for the name of a save file into which the energy matrices and other information can be written. Execution continues with sequence file name selection (Section 3.1.6.).

3.1.3. Run Mode Selection

The menu:
Enter run mode

0 Suboptimal plot (default)
1 N best
2 Multiple molecules

is displayed. If the Suboptimal choice is selected, then the *energy dot plot* will be displayed on the screen, and the user will be able to interact with it. Program execution continues with energy dot plot parameter selection (Section 3.1.4.).

In the "N best" mode, the program will compute a selection of optimal and suboptimal foldings sorted by energy. If the "Multiple molecules" mode is selected, the program will run in the "N best" mode for every complete sequence in the input sequence file. This last option can only be selected in the Regular run type. The aim here is to fold many sequences that are contained in a single input file. It is less cumbersome than folding each sequence separately. In the "N best" mode, program execution continues with "N best" parameter selection (Section 3.1.5.), whereas in "Multiple molecules" mode, the user is first prompted for a sequence file name (Section 3.1.6.).

3.1.4. Energy Dot Plot Parameter Selection

The user is prompted for two parameters. The first is the "minimum vector size for plot." This parameter controls the clutter on the display screen. Only helical regions that contain at least this number of consecutive base pairs will be plotted as points. The default is 1; that is, to plot every base pair within a user-selected energy increment from the minimum.

The user is then prompted for the "window size." This parameter can be regarded as one less than the minimum "distance" between any two predicted foldings *(23)*. When a base pair has been selected from the screen and an (otherwise) optimal structure containing that base pair has been computed, the program marks all the base pairs in the computed structure as well as any base pairs that are within a "window size" distance from these base pairs. (In an RNA molecule $r_1, r_2, \ldots, r_n$, the distance between two base pairs r_i-r_j and $r_{i'}$-$r_{j'}$ is defined to be $max\{|i-i'|,|j-j'|\}$.) When the energy dot plot is redrawn, the marked base pairs do not appear. This gives the user a chance to select a base pair that is not "too close" to any base pairs that have already occurred in computed foldings. In addition, subsequent computed foldings will not be output unless they contain at least "window size" unmarked base pairs. The choice of the window parameter is discussed in Section 3.1.5. If one's aim is merely to look at the dot plot and see which structural features are "well defined," with few competing structures, then the value of this parameter is irrelevant.

In a Continuation run, program execution continues with input save file name selection (Section 3.1.7.). Otherwise, in a Regular run, program execution continues with sequence file name selection (Section 3.1.6.).

3.1.5. N Best Parameter Selection

In this case, the program automatically computes a selection of optimal and suboptimal foldings that satisfy certain user-defined conditions. The user is prompted for three parameters. The first is the "percentage for sort." This is an integer, p. All computed foldings will have energies within $p\%$ from the computed minimum free energy. A value of 10% should guarantee that all reasonable secondary structure motifs are found. This value should be increased for short sequences so that the actual energy increment is at least 2–3 kcal/mol. Similarly, p should be decreased for very long sequences, so that the energy increment is not greater than 15–20 kcal/mol, with a recommended range of 10–12 kcal/mol. The default value of p is 0, indicating that only optimal foldings will be computed.

The next prompt is for the "number of tracebacks." This is an upper bound for the number of foldings that will be computed. Although the default is 1, the author strongly recommends that this parameter be set high, to several hundred. Both the "percentage for sort" and "window size" parameters limit the number of foldings that are computed. It is better to let these parameters do their work than to artificially truncate the list of foldings at some arbitrary number.

The third prompt is for the "window size." It has the same meaning as it does in the energy dot plot mode, but in "N best" mode, this parameter ensures that every pair of foldings in the output will be sufficiently different from one another. The default is 0, so that even trivially different foldings might be found. There are no rules, but in Table 1 are listed some recommended window sizes for different sizes of sequences. Selecting a smaller window size will increase the number of foldings that are found, but some of these foldings may be similar. Selecting a larger window size will cut down on the number of computed foldings, but some reasonable folding motifs may be lost.

As in Section 3.1.4., program execution continues with input save file name selection (Section 3.1.7.) in a Continuation run and with sequence file name selection (Section 3.1.6.) in a Regular run. In "Multiple molecules" mode, energy file input (Section 3.1.8.) is next.

3.1.6. Input Sequence File Name Selection

At this point, the user is prompted for the name of a file containing one or more sequences. These sequences can be in the GenBank *(24)*, EMBL *(25)*, PIR *(26)*, IntelliGenetics *(27)*, or GCG *(22)* formats.

Table 1
Suggested "Window Size" Values Depending on Sequence Length[a]

Sequence size	Suggested window size
0–50	2
50–120	3
120–300	5
300–500	7–8
500–800	10–12
800–1200	15
1200–2000	20

[a]The user is encouraged to experiment with this parameter.

Sequences must use upper-case letters. The program recognizes "A," " C," "G," and "U." The letter "T" is treated as "U." In addition, the letters "B," "Z," "H," and " V" are recognized as "A," "C," "G," and "U," respectively. In this case, these bases are flagged by the program as being accessible to single-strand nuclease cleavage. Such a flagged base can base pair only if its 3' neighbor is single stranded. This prevents bases that are accessible to single-strand specific nucleases from being paired in the middle of helical regions. If other letters are used, they will not be allowed to base pairs, nor will they be allowed to contribute to single base stacking energies. The author does not advise the use of other letters. Bases can be prevented from pairing by selecting the "Single Prohibit" option (Section 3.2.6.) in the main menu.

Except in "Multiple molecules" mode, where each complete sequence is read automatically, the program will display the names of the sequences in the file. The user selects a sequence by number, and is then prompted for the 5' and finally the 3' ends of the portion to be folded. Program execution continues with energy file input (Section 3.1.8.).

3.1.7. Input Save File Name Selection

The user is prompted for the name of a save file created previously in a save-type run. After the save file is read in, the user is prompted for a file name for the "continuation dump." This "dump" displays what items have been selected from the main menu and the values of all the "energy parameters." The default file name is /dev/tty (terminal output), but this information can be stored and used to check what options were used in the original save run. The user is then prompted with the question: "Listing of energy files?" A yes to this causes all the energy values for base pairs and loops to be displayed or else appended to the "continuation dump" file. The default is no, but this option can be used to verify what energy files were used in case of confusion. Program execution continues with output file and format selection (Section 3.1.9.).

3.1.8. Energy File Input

Six files containing energy information are needed to run the program. The *dangle* energy file contains energies for single base stacking. The *loop* energy file contains destabilizing energies for internal, bulge, and hairpin loops. The *stack* energy file contains energies for base pair stacking. The *tstack* energy file contains energies for terminal mismatched pairs in interior and hairpin loops. In the future, this file will be split into two, one for interior loops and one for hairpin loops. The *tloop* energy file contains a list of distinguished "tetraloops" *(28)* that the program recognizes and the bonus energies used to bias their occurrence. Finally, the *miscloop* energy file contains some miscellaneous energies related to multibranched loops and asymmetric interior loops. The default names for these files are dangle.dat, loop.dat, stack.dat, tstack.dat. tloop.dat, and miscloop.dat, respectively. These files come with the program, and simulate folding at 37°C. Folding at different temperatures is discussed in Section 3.3.6. After these files are read, program execution continues with output file and format selection (Section 3.1.9.) for a Regular run, and proceeds directly to the main menu (Section 3.1.10.) in a Save run.

3.1.9. Output File and Format Selection

Three different types of folding output formats are available:

1. printer—a rough and inelegant output, but one that can be read and interpreted directly;
2. ct file—an output file that can be used as the input to other programs that draw secondary structures or that analyze a large number of suboptimal foldings; and
3. reg (region) file—an output file that is more condensed than the ct file, but lacks the sequence information. It, too, can be used as the input to other programs.

An example of printer output is given in Fig. 1. Figure 2 contains examples of ct and region output. The first line of a ct file contains three items: the number of bases in the folded fragment, the folding energy, and the sequence label. Subsequent lines contain six fields:

1. The base number, i, within the folded segment. This is called the internal base number.
2. The base identity.
3. The internal base number of the 5' neighbor of base i (0 for the first base).
4. The internal base number of the 3' neighbor of base i (0 for the last base).
5. The internal number of the base to which i is paired. This is 0 if base i is single stranded.
6. The historical numbering of base i within the folded sequence.

The region table format is more compact, giving only the helical regions without base identities. Each line contains five fields:

1. The helix number.
2. The historical numbering, i, of the 5' external base of the helix.
3. The historical numbering, j, of the 3' external base of the helix
4. The number of base pairs, k, in the helix. Thus, in the sequence $r_1, r_2, \ldots, r_n$, the helix contains the k base pairs $r_i\text{-}r_j$, $r_{i+1}\text{-}r_{j-1}$, $\ldots$ $r_{i+k-1}\text{-}r_{j-k+1}$.
5. The energy of the helix in kcal/mol. This does not include single base stacking on the ends that might occur.

The user is prompted on whether or not printer output is desired (default is yes). If "y" is chosen, a further prompt asks if the output should go to the terminal (standard output, default is yes). If "n" is chosen, the user is prompted for a file name. A default based on the sequence label is available. The next prompt is for the number of columns on the printing device or terminal. This is the "record length" of the output file. The default is 80, but other values can be selected.

In a similar way, the user is prompted for ct file and region file output and for file names as required. In a Regular run, program execution will continue with the main menu (Section 3.1.10.). In a Continuation run in "N best" mode, there will be a pause, possibly a long one, while the program computes structures and writes them

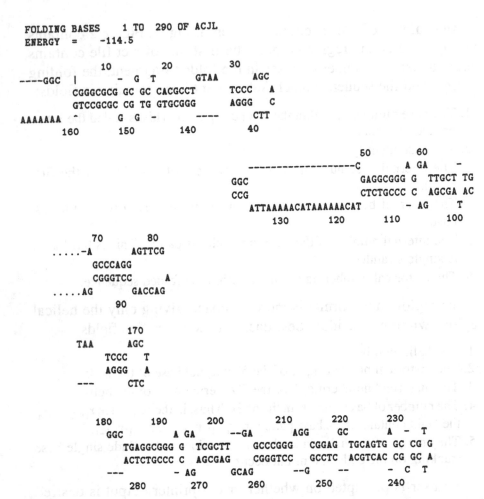

Fig. 1. Printer output of the optimal folding of an Alu consensus sequence *(29)*. Alu sequences can occur within introns and hence, within RNA transcripts. The record length here is 80 columns. Note that the piece from nucleotides 68–97 has two sets of five dots preceding it. This piece is "continued" from the stem above it. The truncation occurs because the entire stem cannot fit into 80 columns. The symbols "|" and "^" point to the base pair C^4–G^{160} that was selected. Base pairs are selected by the user from the energy dot plot in "Suboptimal plot" mode or automatically by the program in "N best" mode, and an optimal structure containing that base pair is computed.

into the output files. The program will print out a message for every structure found and then terminate. In a Continuation run in "Suboptimal plot" mode, program execution continues with the energy dot plot (Section 3.1.11.).

```
--------------- a ---------------                --------------- b ---------------
290 ENERGY =  -114.5      ACJL        (   1)      4     160     8    -16.7
  1 G      0    2    0     1           (   2)     12     151     2     -3.4
  2 G      1    3    0     2           (   3)     15     148     2     -1.9
  3 C      2    4    0     3           (   4)     18     146     7    -13.7
  4 C      3    5  160     4           (   5)     29      44     4     -8.1
  5 G      4    6  159     5           (   6)     45     139     3     -6.3
  6 G      5    7  158     6           (   7)     49     116     8    -15.0
  7 G      6    8  157     7           (   8)     58     108     1      0.0
  8 C      7    9  156     8           (   9)     61     105     5     -7.1
  9 G      8   10  155     9           (  10)     66      99     2     -1.8
... (272 intervening lines) ...        (  11)     69      95     7    -15.6
282 C    281  283  184   282           (  12)    164     179     4     -8.1
283 A    282  284  183   283           (  13)    183     283     9    -16.8
284 A    283  285    0   284           (  14)    193     274     1      0.0
285 A    284  286    0   285           (  15)    196     271     6     -9.9
286 A    285  287    0   286           (  16)    204     261     7    -15.1
287 A    286  288    0   287           (  17)    214     253     5     -8.9
288 A    287  289    0   288           (  18)    221     248     7    -12.6
289 A    288  290    0   289           (  19)    229     241     2     -3.4
290 A    289    0    0   290           (  20)    231     238     2     -2.0
```

Fig. 2. Examples of ct and region files. (a) Partial ct file of the optimal Alu folding presented in Fig. 1. (b) The entire region file for the same Alu folding. Note the 0 energies assigned to the single-base "helices" (numbers 8 and 14). This is because energy assignment in helices is for the stacking of one base pair over another.

3.1.10. Main Menu

At this point, a menu appears offering the user the option of adding auxiliary information to constrain or otherwise alter the folding. This will be dealt with in the next section. If no special features are being selected, or after they have been selected, the user should select item 8 to begin processing. There will be a pause, usually from several minutes to several hours, while the program computes the energy matrices needed for the folding. In a Save run, execution terminates after this long pause, whereas in a Regular run, the program continues on to produce output in "N best" mode, or to the energy dot plot in "Suboptimal plot" mode (Section 3.1.11.). In a UNIX environment, program execution can be pushed into background mode at this stage. It is advisable to run the program in its own window shell. All run types can be run as batch processes, as long as the run mode is not "Suboptimal plot." The auxiliary program, "batgen," discussed in Section 3.3.1., has been designed to facilitate batch runs.

3.1.11. Energy Dot Plot

The energy dot plot displays the superposition of all possible foldings within a user-specified energy increment from the computed minimum. In this triangular plot, a dot in row i and column j represents base pairing between ribonucleotides i and j. In the UNIX versions, different colors are used to represent the different levels of suboptimality. The description of the VAX energy dot plot has already appeared in ref. *16*. The description here will be for the IRIS version that runs only on a Silicon Graphics workstation, and for an X11 version developed for the DECstation 3100 (ULTRIX), but now running on the SGI IRIS, SUN4, and sparcstations, and the DECstation 5000.

In the IRIS-specific version, a nonresizable, nonmovable window that fills the entire screen is created in which the triangular energy dot plot is displayed along with other information. This window can be moved and resized in the X11 version and does not initially fill the screen. Energy values are in kcal/mol, and the base pairs are displayed according to their historical numberings. Energy increments are entered by the user.

Interaction with the program is by means of the mouse and pop-up menus. The pop-up menu can be displayed by pressing the right mouse button. An item can be selected by moving the pointer to the desired selection and releasing the mouse button.

The "Optimal score" that is displayed in the dot plot is the minimum folding energy. The "Energy increment" is the highest deviation in kcal/mol from the minimum energy for which a base pair will be plotted. Thus, all base pairs that are in foldings within this increment from the minimum folding energy will be plotted, consistent with the value of the "minimum vector size for plot" parameter (Section 3.1.4.). The energy increment is initially 0, so that only the optimal foldings appear at first.

3.1.11.1. CHANGING THE ENERGY INCREMENT

The energy increment can be changed by selecting the "Enter new increment" option from the pop-up menu. In the IRIS-specific version, a one-line window will be displayed at the bottom of the screen prompting the user for an energy increment, to be entered in kcal/mol. In the X11 version, the prompt appears in the window shell from which the program is running. After the user enters a number and presses the "return" key, the dot plot will be redrawn using the new energy increment. Base pairs that have already been found in previously computed structures, as well as base pairs within a distance "window size" (Sections 3.1.4–3.1.5) of these base pairs, will not be plotted when the dot plot is redrawn. This allows the user to select base pairing regions different from those that have already been found.

3.1.11.2. SELECTING BASE PAIRS FROM THE DOT PLOT

When the pointer is placed over the dot plot, crosshairs appear. Base pairs can be selected from the dot plot by placing the crosshairs over the desired point and pressing the middle or left mouse button. The historical "coordinates" of the selected base pair will appear in the dot plot on the line beginning with "(i,j) basepair =" and the energy of the optimal structure containing that base pair will be displayed on the following line: " (i,j) score = ". The program will optimize base pair selection by searching the eight points surrounding the selected point, and using the base pair with the minimum energy. This is not necessarily the exact point picked.

3.1.11.3. COMPUTING STRUCTURES

Once a base pair has been selected, the best folding containing that base pair can be computed by selecting the "Compute structure for last (i,j)" (IRIS), or "Compute structure" (X11) option from the pop-up menu. The computed structure will only be sent to the output files if at least "window size" new base pairs occur that are not in or too near base pairs that have already been found in computed structures. The base pairs in and within a distance of "window size" of the computed structure will be marked, and will not appear in the dot plot when it is redrawn.

If one wishes to compute secondary structures in this mode, it is advisable to send them to a file, rather than to the "terminal." In the latter case, the output will be sent to the window that called the program. This window is resized and hidden in the IRIS version, although it can be displayed using the "Toggle textport" option from the pop-up menu.

3.1.11.4. DISPLAYING THE "P-NUM PLOT"

The P-Num plot displays the total number of dots in the ith row and ith column of the dot plot array (ordinate) vs the historical numbering for i (abscissa). For the ith base in the fragment, P-Num(i) is the total number of different base pairs that can be formed with the *i*th base in all possible foldings within the selected energy increment. The P-Num plot is intended to give a rough idea of what parts of the RNA sequence have a "well-determined" structure. It is not as effective as looking at the dot plot, and no "hard copy" option is available for the P-Num plot in the program as distributed. The P-Num plot can be toggled on and off by selecting the "Toggle pnumplot" item from the pop-up menu.

3.1.11.5. Choosing Colors for the Dot Plot

The energy dot plot can be displayed with black dots (default) or with color dots on a white background. To select different colors, choose the "Color" option from the pop-up menu. Slide the pointer to the right along the "Color" entry, and another "rollover" menu will appear. In the IRIS version, the user can select from one to seven different colors from this menu. If more than one color is selected, black is reserved for the "optimal" base pairs and the extra colors display the increasing levels of suboptimality within the user-defined energy increment. The alternative colors are (in order of increasing suboptimality): red, green, yellow, blue, magenta, and cyan. In the X11 version, as soon as more than one color is chosen, the suboptimal points are colored according to a "color wheel" that progresses from red (least suboptimal) to dark blue (most suboptimal). Optimal points are always displayed in black.

3.1.11.6. Creating a "Hard Copy" of the Dot Plot

A "hard copy" of the dot plot can be created by choosing the "Create Plot file" option from the pop-up menu (IRIS version) or the "Save outplot file" option in the X11 version. In the IRIS version, the user is prompted for information at the bottom of the full-screen dot plot window. In the X11 version, these prompts appear in the calling window. The prompts are:

1. "Enter plot file name."
2. "Enter plot label (up to 59 characters)."
3. "Enter tick mark interval."
4. "Enter number of levels."

There are no default values. The "tick marks" display base numbers on the dot plot. The "number of levels" must be between 1 and

9. These correspond to the different colors that will be used in the "hard copy" of the dot plot. The plot file that is created can be read by the "figdot" program (Section 3.3.5.) that produces a device independent plot file for the "FIGURE" program in the GCG *(22)* package. Improvements are being planned for producing better-quality hard copies of the energy dot plot in a less cumbersome way.

3.1.11.7. FINAL DETAILS IN THE DOT PLOT

The IRIS-specific version has a "Redraw screen" option in the pop-up menu. This is lacking in the X11 version, where the screen is automatically redrawn whenever a structure is computed. Program execution is terminated by choosing the "quit" item from the pop-up menu in the IRIS version and the "Exit Program" option in the X11 version.

3.2. Method 2—Using Special Features

The use of auxiliary constraints to force computed foldings to agree with experimental data or preconceived notions was first introduced in ref. *11*. Because the current program computes a selection of optimal and suboptimal structures, and the energy dot plot displays all possible helices that might be contained in close to optimal foldings, there is less of a need to constrain a folding. If an optimal folding conflicts with experimental data, perhaps a suboptimal one will be better. Nevertheless, the current programs, lrna and crna, both have the same ability to use auxiliary information as the earlier *single fold* program *(21)*.

The use of special sequence characters to designate bases susceptible to single-stranded nuclease cleavage has already been discussed in Section 3.1.6. The special features can be selected when the "main menu" (Section 3.1.10.):

1	Energy Parameter	6	Single Prohibit
2	Single Force	7	Double Prohibit
3	Double Force	8	Begin Folding
4	Closed Excision	9	Show current
5	Open Excision	10	Clear current

Enter Choice

is displayed. The choices are selected by number.

3.2.1. Energy Parameter

When this selection is made, a secondary menu appears:

Energy Parameters (10ths kcal/mol)

1	Extra stack energy	[0]
2	Extra bulge energy	[0]
3	Extra loop energy (interior)	[0]
4	Extra loop energy (hairpin)	[0]
5	Extra loop energy (multi)	[46]
6	Multi loop energy/single-stranded base	[4]
7	Maximum size of interior loop	[30]
8	Maximum lopsidedness of an interior loop	[30]
9	Bonus Energy	[−500]
10	Multi loop energy/closing base pair	[1]

Enter Parameter to be changed (<return> for main menu)

This menu displays the user-adjustable "energy" parameters, with their current values within the square brackets. The values shown above are the defaults. The user enters a desired number and is then prompted for a new value for the selected parameter.

The "Extra stack energy" is a user-defined energy that is added to all stacked base pairs. This could be used in conjunction with user-defined energy files containing energy values that are all 0. This would have the effect of giving an equal weight to all base pairs, so that the program would then be optimizing the number of base pairs. Items 2–4 are extra energies that can be assigned to all bulge, interior, and hairpin loops, respectively. Increasing the penalty for bulge, interior, or hairpin loops will tend to produce foldings with fewer such loops.

Energy parameters 5, 6, and 10, denoted by a, b, and c, respectively, are used to assign energies to multibranched loops. A multi-branched loop with n_1 single-stranded base pairs and n_2 helices branching off from it is assigned an energy of $a + b \times n_1 + c \times n_2$. These simple assignments are in lieu of measured quantities or thermodynamic calculations. The default values have been chosen to optimize folding predictions for a number of sequences whose structure is known from phylogenetic comparisons (15).

If an interior loop has m_1 single-stranded bases on one side and m_2 single-stranded bases on the other, then energy parameters 7 and 8 are upper bounds for m_1+m_2 and $|m_1-m_2|$, respectively. These parameters also apply to bulge loops, where $min\{m_1, m_2\} = 0$. The default value for parameter 7 is 30. This should be sufficient for folding at the default temperature of 37°C, but might have to be raised for folding at elevated temperatures where very large interior loops might be expected. Setting this parameter to a smaller number will decrease folding times while increasing the risk of missing the optimal folding. The user can also experiment with the "lopsidedness" parameter to force foldings to have better balanced interior loops.

Energy parameter 9 is the "Bonus Energy" that is used by the program to force certain bases to be double-stranded or to force certain base pairs. The strategy is simple. Base pairs are "strongly encouraged" to form by using large bonus energies. The default is −500, or −50 kcal/mol for each forced base pair. The artificial energies are automatically subtracted from the energies printed in the output files, but the bonus energies do affect the dot plot and the automatic generation of structures in the "N best" mode. These effects are outlined in Section 4.2.

3.2.2. The "Single Force" Option

This option is used to force a group of consecutive bases to be double-stranded. After this item is selected from the main menu, the user is prompted by: "Enter base and length." The historical 5' base number of the segment, i should be entered, followed by the number of bases that are to be double-stranded, k. Thus, bases $r_i, r_{i+1}, \ldots, r_{i+k-1}$ are forced to be double-stranded. Caution should be exercised when using this option.

3.2.3. The "Double Force" Option

This option is used to force a group of consecutive base pairs to form. When this option is selected from the main menu, the user is prompted by: "Enter base pairs and length." If i, j, and k are entered, then the stacked base pair r_i-r_j, r_{i+1}-$r_{j-1}, \ldots, r_{i+k-1}$-$r_{j-k+1}$ are forced.

3.2.4. The "Closed Excision" Option

This option is used to excise a portion of a structure that is either "well-defined" or for which a good secondary structure model already exists. The user is prompted by: "Enter beginning and end" and should enter the historical numbers, i and j, for the base pair that closes off a structural motif that is to be excised. The program folds the remain-

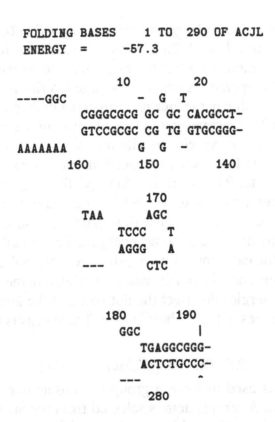

```
            FOLDING BASES    1 TO  290 OF ACJL
            ENERGY  =       -57.3


                         10             20
            ----GGC          -  G  T
                   CGGGCGCG GC GC CACGCCT-
                   GTCCGCGC CG TG GTGCGGG-
            AAAAAAA          G  G  -
                 160        150        140


                         170
            TAA          AGC
                   TCCC    T
                   AGGG    A
                   ---    CTC


                180        190
                GGC          |
                   TGAGGCGGG-
                   ACTCTGCCC-
                   ---         ^
                         280
```

Fig. 3. Printer output of the optimal folding of the Alu consensus sequence described in Fig. 1. Closed excisions have been made at base pairs T^{24}–G^{140} and G^{191}–C^{275}.

der of the sequence, including the designated base pair. An example of a closed excision is given in Fig. 3.

3.2.5. The "Open Excision" Option

In this simple option, the user can excise a portion of the RNA, such as an intervening sequence, and retain the historical numbering of the sequence in the resulting folding. The user enters a beginning base number, i, and an end base number, j, and the program excises bases r_i to r_j inclusive and ligates the ends together. In the closed excision, bases r_i and r_j are not excised. They are forced to pair with one another, and only the intervening stretch is excised.

3.2.6. The "Single Prohibit" Option

This option is used to force a group of consecutive bases to be single-stranded. After this item is selected from the main menu, the user is prompted by: "Enter base and length". The historical 5' base number of the segment, i, should be entered followed by the number of bases that are to be single-stranded, k. Thus, bases $r_i, r_{i+1}, \ldots, r_{i+k-1}$ are forced to be single-stranded.

3.2.7. The "Double Prohibit" Option

This option is used to prevent a group of consecutive base pairs from forming. When this option is selected from the main menu, the user is prompted by "Enter base pair and length". If i, j, and k are entered, then the stacked base pairs $r_i\text{-}r_j, r_{i+1}\text{-}r_{j-1}, \ldots, r_{i+k-1}\text{-}r_{j-k+1}$ are prohibited from forming.

3.2.8. Remaining Options

To exit from the main menu and to begin processing, select item 8 ("Begin Folding") from the main menu. Selection of item 9 ("Show current") will give a list of all force, excision, and prohibit choices that have been made up to that point. These choices can all be erased by selecting menu item 10 ("Clear current").

3.3. Method 3—Additional Programs

Some additional programs are either distributed with the mfold package (Sections 3.3.1.–3.3.6.) or are available separately. Although the description in Sections 3.1. and 3.2. was for the program lrna, the directions for using crna are identical.

3.3.1. Batgen—Command
File Generation for Batch Runs

Both lrna and crna can be run in batch mode as long as the run mode is not "Suboptimal plot." In the UNIX environment, a run can be begun interactively and then pushed into background mode. It is nevertheless desirable to generate command files for batch runs.

The batgen program generates a list of commands that are stored in a user-defined file (default is "fold.com"). The prompts are similar to the lrna prompts, except that the "Suboptimal plot" mode is not offered. A batch run could be initiated from a command file called "fold.com" by issuing the commands:

batch
lrna < fold. com
^D

The "^D" indicates the "EOT" character generated by holding down the "control" key while typing the letter "D."

3.3.2. cttobp—Interconversion Between ct Files and Region Files

The "cttobp" program will take a ct file as input, and produce a region file and "raw" sequence file as output. The "raw" sequence file contains only the sequence with no labels or other characters. Conversely, given a "raw" sequence file and a region table, the program can produce a ct file. In this latter case, the user is prompted for the historical numbering of the first base and for a sequence label. The program is run interactively.

3.3.3. The Distance Program

The "distance" program computes the distance between foldings using the older definition of distance as defined in ref. *30*. The input is a ct file containing multiple foldings. Up to 15 foldings can be processed at one time. The program either gives the distance between the first structure and all subsequent ones in the file, or else it gives the distance between each pair of structures. It is run interactively.

3.3.4. Computing the Energy of a Folding

The program "efn" will compute the energy of a folding from ct file input. It can be used to check the folding energy of a computed structure. It is especially useful for computing the folding energy of a structure computed by other means, such as phylogenetic methods. Also, the energy of a folding computed for one temperature can be reevaluated at another temperature.

The program accepts ct file input. There can be multiple foldings in this file. Energy files, as described in Section 3.1.8. must be provided. The defaults are the same as in Section 3.1.8. The user is given the option of creating printer output and/or region file output (Section 3.1.9).

3.3.5. Creating a "Hard Copy" of an Energy Dot Plot

As described in Section 3.1.11.6., both lrna and crna can be used to create plot files that contain the information encoded in an energy dot plot. The "figdot" program can read up to two such files to create

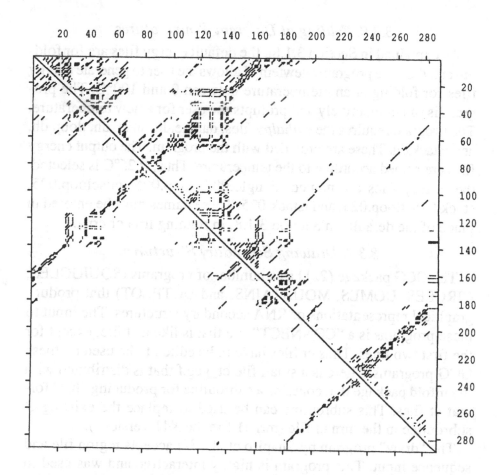

Fig. 4. A double black and white energy dot plot for the Alu sequence. The optimal folding energy is −114.5 kcal/mol. The upper-right dot plot has an 11.4 kcal/mol energy increment (10%), whereas the increment for the lower-left plot is 7.6 kcal/mol.

a *double dot plot* with the second triangular plot rotated through 180° and placed in the lower-left portion of a square. An example is given in Fig. 4. By default, the numbering on the dot plot will be from 1 to the number of bases in the folded sequence. The starting number can be set to another number if desired. Figdot is run interactively, and produces a device independent plot file that can be read by the GCG program "FIGURE." "FIGURE" was used to create a PostScript file for Fig. 4. The creation of the hard copy is cumbersome and inconvenient: lrna → figdot → FIGURE → PostScript printer.

3.3.6. Folding at Different Temperatures

As explained in Section 3.1.8., the default energy files are for folding at 37°C. The program "newtemp" allows the user to generate energy files for folding at any temperature between 0 and 100°C. The program is run interactively and prompts the user for a new temperature. The program requires the *enthalpy* files: dangle.dh, stack.dh, tloop.dh, and tstack.dh. These are provided with the program. The output energy files are named according to the temperature. Thus, if 25°C is selected, the energy files are named: dangle.025, loop.025, miscloop.025, stack.025, tloop.025, and tstack.025. These names must be entered in place of the defaults in Section 3.1.8. in running lrna or crna.

3.3.7. Drawing Secondary Structures

The GCG package (22) has a number of programs (SQUIGGLES, CIRCLES, DOMES, MOUNTAINS, and DOTPLOT) that produce graphical representations of RNA secondary structures. The input to these programs is a "CONNECT" file that is like a ct file, except for the first two lines. Thus ct files have to be edited to be used by these GCG programs. There is a small file ct_gcg.f that is distributed with the mfold package that contains a subroutine for producing GCG format ct files. This subroutine can be used to replace the existing ct subroutine in the mrna.f file (mrna1.f in the X11 versions).

The "draw" program by Shapiro et al. (31) accepts region file and sequence input. This program is highly interactive and was used to produce Fig. 5. It has recently been ported to the Silicon Graphics IRIS platform, where it makes use of the DISSPLA[1] package for producing output files. The "NAVIEW" program by Bruccoleri and Heinrich (32) accepts ct file input and also produces a pleasing output. It has been adapted to run in a general UNIX environment. The output is a device-independent plot file that requires the PLT2 plotting package (33) to produce a plot.

---

[1]CA-DISSPLA (Display Integrated Software System and Plotting Language) proprietary software product of Computer Associates International, Inc., 10505 Sorrento Valley Road, San Diego, CA 92121

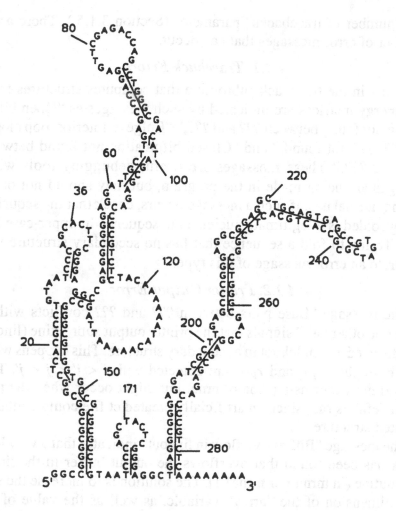

Fig. 5. Output of the "draw" program (Section 3.3.7.) for the optimal folding of the Alu sequence from Fig. 1.

4. Notes

4.1. Error Messages in lrna and crna

The symbol "???" will be used to refer to an indeterminate integer. Normal termination of lrna and crna in "N best" mode is indicated by the message "End reached at traceback ???," which tells the user how many structures were found. This message will not occur if the number of structures found has been artificially truncated by the use of the "number of tracebacks" parameter (Section 3.1.5.). There are a number of error messages that can occur.

4.1.1. Traceback Errors

Errors in the traceback subroutine that computes structures from the energy matrices are indicated by such messages as "Open bifurcation not found between ??? and ???," "Bulge or interior loop closed by ???, ??? not found," and "Closed bifurcation not found between ??? and ???." These messages are useful debugging tools when changes are being made in the program, but they should not occur during normal use. If such a message occurs, check that the sequence being folded is a legitimate nucleic acid sequence in upper-case letters. Trying to fold a sequence that has no secondary structure will generate an error message of this type.

4.1.2. Printer Output Error

The message "Base pairs between ??? and ??? conflicts with at least one other pair" signals that the printer output subroutine (linout) has detected a pseudoknot in a secondary structure. This happens when two base pairs r_i-r_j and $r_{i'}$-$r_{j'}$ are detected with $i < i' < j < j'$. This should not occur using lrna or crna, but might occur when the program "efn" is run, since an artificially created ct file could contain a knotted structure.

The message "Buffer overflow in lineout" indicates that a very long helix has been found that overflows the output buffer in the linout subroutine (in mrna.f or mrna1.f). The solution is to increase the second dimension of the "array" variable, as well as the value of the "amax" variable, to a common number larger than 900.

4.1.3. Other Errors

In a Continuation run, the use of a corrupted save file or a file that is not a save file can lead to the error message "Premature end of save file." This will also happen if a save file from the program "crna" is used for continuation with "lrna."

When a long sequence is being folded in "N best" mode, the number of potential secondary structures might become too large for the sorting buffer, especially if the "window size" parameter (Section 3.1.4.) is small. The error message "More than ??? base pairs in sort at (???,???)" will appear. The first ??? refers to the current value of the "sortmax" parameter defined in the lin.inc and circ.inc files. This parameter should be doubled in size until this error message no longer occurs. If this error is not corrected, there is a risk of missing some valid suboptimal foldings.

4.2. Forcing Base Pairs

The use of bonus energies to force base pair formation creates several problems. The bonus energy value of –500 (–50 kcal/mol) will not create a problem if only a few base pairs are being forced. Note that each closed excision is equivalent to forcing a single base pair formation (the ends of the excised fragment). However, if one wishes to fold the entire 16S rRNA from *E. coli*, for example, forcing all the phylogenetically determined base pairs to form, then there will be a problem. The folding program uses 16-bit integers, and the maximum absolute energy that can be stored is roughly 3200 kcal. This will be exceeded if hundreds of base pairs are forced with a bonus energy of –50 kcal/mol/base pair. In such a case, the solution is to set the bonus energy (Section 3.2.1.) to –2 or –3 kcal/mol—enough so that all the desired base pairs form.

In the energy dot plot, only points corresponding to forced bases or base pairs will appear as long as the energy increment is no larger than the bonus energy. As soon as the bonus energy threshold is passed, the optimal folding and then the suboptimal foldings will be revealed.

In "N best" mode, special care must be taken in choosing the "percentage for sort" parameter (Section 3.1.5.) when base pairs are forced. This is because the internal energies are distorted by the addition of bonus energies. Suppose that p is the desired "percentage for sort" value in a situation where the bonus energy is $-b$, the real optimal energy is $-E$, and $m > 0$ base pairs have been forced. Then the "percentage for sort" number, p', that must be used in this case is given by

$$p' = (pE + 100b)/(E + [m + 1]b). \qquad (1)$$

Note Added in Proof

Since the acceptance of this manuscript, some new features have been added. A "hard copy" of an energy dot plot (Section 3.3.5.) can now be produced more directly using a new program called "dot2ps." The output of the "NAVIEW" program (Section 3.3.7.) can now be turned directly into a PostScript plot file using another new program called "plt2ps."

Acknowledgments

The author wishes to thank Digital Equipment Corporation for the free loan of a DECstation 3100 and the Canadian Institute for Advanced Research for fellowship support. This is NRCC publication no. 34331.

References

1. Major, F., Turcotte, M., Gautheret, D., Lapalme, G., Fillion, E., and Cedergren, R. (1991) The combination of symbolic and numerical computation for three-dimensional modeling of RNA. *Science* **253,** 1255–1260.
2. de Smit, M. H. and van Duin, J. (1990) Control of prokaryotic translation initiation by mRNA secondary structure. *Prog. Nucleic Acids Res. Mol. Biol.* **38,** 1–35.
3. Mills, D. R., Priano, C., Merz, P. A., and Binderow, B. D. (1990) Qβ RNA bacteriophage: mapping cis-acting elements within an RNA genome. *J. Virol.* **64,** 3872–3881.
4. Michel, F. and Westhof, E. (1990) Modelling of the three-dimensional architecture of group I catalytic introns based on comparative sequence analysis. *J. Mol. Biol.* **216,** 585–610.
5. Woese, C. R., Gutell, R. R., Gupta, R., and Noller, H. F. (1983) Detailed analysis of the higher order structure of 16S-like ribosomal ribonucleic acids. *Microbiol. Rev.* **47,** 621–669.
6. Winker, S., Overbeek, R., Woese, C. R., Olsen, G. J., and Pfluger, N. (1990) Structure detection through automated covariance search. *CABIOS* **6,** 365–371.
7. Sankoff, D. (1976) Evolution of secondary structure of 5S ribosomal RNA. Paper presented at *Tenth Numerical Taxonomy Conference,* Lawrence, KS.
8. Waterman, M. S. (1978) Secondary structure of single-stranded nucleic acids, *Studies in Foundations and Combinatorics, Advances in Mathematics, Supplementary Studies* **1,** 167–212.
9. Nussinov, R., Pieczenik, G., Griggs, J. R., and Kleitman, D. J. (1978) Algorithm for loop matchings, *SIAM J. Appl. Math.* **35,** 68–82.
10. Nussinov, R. and Jacobson, A. (1980) Fast algorithm for predicting the secondary structure of single-stranded RNA. *Proc. Natl. Acad. Sci. USA* **77,** 6309–6313.
11. Zuker, M. and Stiegler, P. (1981) Optimal computer folding of large RNA sequences using thermodynamics and auiliary information. *Nucleic Acids Res.* **9,** 133–148.
12. Sankoff, D., Kruskal, J. B., Mainville, S., and Cedergren, R. J. (1984) Fast algorithms to determine RNA secondary structures containing multiple loops, in *Time Warps, String Edits, and Macromolecules: The Theory and Practice of Sequence Comparison* (Sankoff, D. and Kruskal, J. B., eds.), Addison-Wesley, Reading, MA, pp. 93–120.
13. Zuker, M. and Sankoff, D. (1984) RNA secondary structures and their prediction. *Bull. Math. Biol.* **46,** 591–621.

14. Zuker, M. (1989) On finding all suboptimal foldings of an RNA molecule. *Science* **244**, 48–52.

15. Jaeger, J. A., Turner, D. H., and Zuker, M. (1989) Improved predictions of secondary structures for RNA. *Proc. Natl. Acad. Sci. USA* **86**, 7706–7710.

16. Jaeger, J. A., Turner, D. H., and Zuker, M. (1990) Predicting optimal and suboptimal secondary structure for RNA, in *Molecular Evolution: Computer Analysis of Protein and Nucleic Acids Sequences* (Doolittle, R. F., ed.), *Meth. Enzym.* **183**, Academic Press, pp. 281–306.

17. Freier, S. M., Kierzek, R., Jaeger, J. A., Sugimoto, N., Caruthers, M. H., Neilson, T., and Turner, D. H. (1986) Improved free-energy parameters for predictions of RNA duplex stability. *Proc. Natl. Acad. Sci. USA* **83**, 9373–9377.

18. Turner, D. H., Sugimoto, N., Jaeger, J. A., Longfellow, C. E., Freier, S. M., and Kierzek, R. (1987) Improved parameters for prediction of RNA structure. *Cold Spring Harbor Symp. Quant. Biol.* **52**, 123–133.

19. Turner, D. H., Sugimoto, N., and Freier, S. M. (1988) RNA structure prediction. *Annu. Rev. Biophys. Biophys. Chem.* **17**, 167–192.

20. Pleij, C. W. A. and Bosch, L. (1989) RNA pseudoknots: structure, detection and prediction, in *RNA Processing* (Dahlberg, J. E. and Abelson, J. N., eds.), *Meth. Enzym.* **180**, Academic Press, pp. 289–303.

21. Zuker, M. (1989) Computer prediction of RNA structure, in *RNA Processing* (Dahlberg, J. E. and Abelson, J. N., eds.), *Meth. Enzym.* **180**, Academic Press, pp. 262–288.

22. Devereux, J., Haeberli, P., and Smithies, O. (1984) A comprehensive set of sequence analysis programs for the VAX. *Nucleic Acids Res.* **12**, 387–395.

23. Zuker, M., Jaeger, J. A., and Turner, D. H. (1991) A comparison of optimal and suboptimal RNA secondary structures predicted by free energy minimization with structures determined by phylogenetic comparison. *Nucleic Acids Res.* **19**, 2707–2714.

24. Bilofsky, H. S. and Burks, C. (1988) The GenBank genetic sequence data bank. *Nucleic Acids Res.* **16**, 1861–1863.

25. Cameron, G. N. (1988) The EMBL data library. *Nucleic Acids Res.* **16**, 1865–1867.

26. Sidman, K. E., George, D. G., Barker, W. C., and Hunt, L. T. (1988) The protein identification resource. *Nucleic Acids Res.* **16**, 1869–1871.

27. Roode, D., Liebschutz, R., Maulik, S., Friedmann, T., Benton, D., and Kristofferson, D. (1988) New developments at BIONET. *Nucleic Acids Res.* **16**, 1857–1859.

28. Woese, C. R., Winker, S., and Gutell, R. R. (1990) Architecture of ribosomal RNA: constraints on the sequence of "tetra-loops." *Proc. Natl. Acad. Sci. USA* **87**, 8467–8471.

29. Labuda, D. and Striker, G. (1989) Sequence conservation in Alu evolution. *Nucleic Acids Res.* **17**, 2477–2491.

30. Zuker, M. (1989) The use of dynamic programming algorithms in RNA secondary structure prediction, in *Mathematical Methods for DNA Sequences* (Waterman, M. S., ed.), CRC Press, Boca Raton, FL, pp. 159–184.

31. Shapiro, B. A., Maizel, J. V., Jr., Lipkin, L. E., Currey, K. M., and Whitney, C. (1984) Generating non-overlapping displays of nucleic acid secondary structure. *Nucleic Acids Res.* **12,** 75–88.
32. Bruccoleri, R. E. and Heinrich, G. (1988) An improved algorithm for nucleic acid secondary structure display. *CABIOS* **4,** 167–173.
33. Beach, R. C. (1981) The unified graphics system for Fortran 77 Programming Manual. Stanford Linear Accelerator Center Computational Research Group Technical Memo 203, Stanford, CA.

CHAPTER 24

Classification and Function Prediction of Proteins Using Diagnostic Amino Acid Patterns

Aiala Reizer, Milton H. Saier, Jr.,*
and Jonathan Reizer

1. Introduction

Protein sequence similarities offer a convenient means for the classification and identification of protein families and superfamilies. Frequently, proteins descended from a common ancestor preserve their basic three-dimensional conformations even when they have accumulated large numbers of amino acid substitutions and short insertions or deletions. These may prohibit establishment of homology or evolutionary relationships by traditional global sequence alignment means. Limited regions of sequence similarity can also be the result of evolutionary convergence driven by a need for a common function. Regardless of whether divergent or convergent evolution played a role in the appearance of local sequence similarities, these confined regions of similarity can provide insight into structural and functional relationships of proteins that otherwise fail to show significant similarity by global alignment methods.

A useful tool, the PROSITE database, with high predictive value for the classification and identification of proteins has recently been developed by Bairoch *(1)*. This database is a compilation of more than 600 short amino acid sequences that constitute highly specific patterns (signature motifs) found in a wide variety of protein fami-

*Author to whom all correspondence and reprint requests should be addressed.

From: *Methods in Molecular Biology, Vol. 25: Computer Analysis of Sequence Data, Part II*
Edited by: A. M. Griffin and H. G. Griffin Copyright ©1994 Humana Press Inc., Totowa, NJ

lies. These diagnostic sequences include various posttranslational modification sites, topogenic domains, and sequences with specific biological functions or attributes of particular families or groups of proteins. The PROSITE database is accompanied by a PROSITE documentation file that provides:

1. Information regarding the proteins that are known to possess a particular consensus sequence;
2. Indication of the biological function and significance of the signature patterns; and
3. Relevant references.

Altogether, this dictionary of fingerprints or signature sequences provides a useful means for establishing links between the local and global sequence similarities, as well as the classification and prediction of functional properties of proteins. Even when a query protein exhibits "twilight" global similarity with other proteins in the data bank, one can obtain important clues regarding its functional and structural domains based on the existence of a signature pattern.

This chapter will describe (a) two protocols designed for the identification of particular motifs of the PROSITE database in a query protein and (b) an additional protocol that identifies a group of related proteins that share common sequence pattern(s). (*See* Note 1 for additional suggested reading material).

2. Materials

Three protocols designed to search for specific sequence patterns in proteins are described:

2.1. Motifs

This is a search for protein patterns on the VAX computer using the GCG program MOTIFS and the PROSITE database. The following software and data bank are required:

1. The program MOTIFS, which is included in version 7.0 of the Sequence Analysis Software Package of the Genetic Computer Group (GCG) (assembled by John Devereux, Paul Haeberli, and Philip Marquess, and can be obtained from the Genetic Computer Group, University of Wisconsin, Biotechnology Center).
2. VMS operating system.

3. PROSITE database and the PROSITE documentation file (can be obtained from various FTP anonymous file servers, such as the EMBL server, or directly from the author, A. Bairoch, University of Geneva, Switzerland).

The following optional programs may be used (no detailed description of these programs is given in the present chapter, but their use is mentioned where appropriate):

1. A text editor to type in sequences into the computer;
2. Program(s) that extract protein sequence files from protein data banks; and
3. Program(s) that convert protein sequence files to the GCG format.

2.2. ISEARCH

This is a search on the VAX computer for protein sequences that share a common motif. The following software and data banks are required:

1. The DNASYSTEM program ISEARCH, which is the interactive version of the PIR SEARCH program *(2)* (can be obtained from the National Biomedical Research Foundation [NBRF]).
2. VMS operating system; and
3. Protein sequence data bank(s), such as the NBRF protein sequence database or a user-constructed protein library.

2.3. PROSITE (PC/GENE)

This is a Search for motifs on the IBM or compatible personal computer (PC) using the PC/GENE program PROSITE and the PROSITE database. The following hardware, software, and data bank are required:

1. IBM PC XT, PC AT, PS/2, or compatible PC with at least 640 kilobytes (kbyte) of random access memory (RAM).
2. At least 12 Mbyte of free hard disk space to store the programs and at least one additional megabyte of hard disk for user data files.
3. Graphics screen adapter.
4. A printer capable of printing text and graphics is recommended.
5. IBM PC-DOS or MS-DOS operating system, version 3.1 or higher.
6. The program PROSITE, which is included in the PC/GENE, protein and nucleic acid analysis software package (Bairoch Amos, University of Geneva, commercialized by Intelligenetics, Inc.; can be obtained from Intelligenetics, Inc., Mountain View, CA).

7. Additional programs may be needed to convert protein sequence files to a format acceptable by the PC/GENE software (the EMBL or SWISS-PROT format). The PC/GENE package includes the program REFORM, which converts sequence files to EMBL and SWISS-PROT format (no detailed description of the REFORM program is given in the present chapter, but its use is mentioned where appropriate); and

8. The PROSITE database (included in the PC/GENE software package).

3. Methods

3.1. A Search for PROSITE Motifs on the VAX Computer Using the MOTIFS Program

The GCG program MOTIFS detects short sequence patterns, motifs, in a protein sequence (*see also* Note 2). The following method describes the use of the MOTIFS program in searching for patterns defined in the PROSITE database:

1. The protein sequence should be in a GCG format. If necessary, the following GCG programs can be used to convert protein sequence files to a GCG format: REFORMAT (rewrites sequence files to a format that can be read by MOTIFS); FROMSTADEN (converts sequence files from the Staden format); FROMEMBL (converts sequence files from the EMBL format); FROMNBRF (converts sequence files from the NBRF format); FROMIG (converts sequence files from the IntelliGenetic format).

2. To execute the MOTIFS program, the GCG software package must be invoked first. The activation of the GCG software depends on the command files that were set up at the installation. At some installations, command files may have been written to provide the user with a selection menu from which the GCG programs are selected; other installations may have created different user interfaces.

   Regardless of the method used, once the GCG programs have been activated, the MOTIFS program can then be executed. To execute the program in its simplest form, i.e., without any command line parameters, type the key word "MOTIFS" at the VMS $ prompt. The user will be prompted for the name of the protein sequence file(s). This is the name of the file that contains the protein sequence(s) to be analyzed. The user is also prompted for the name of the output file; this is the name of the file in which the results of the analysis are stored.

3. When executed without additional parameters (as described in item 2 above), the output file reports the motif(s) found (the name and the pattern are indicated) and the position of the motif in the query sequence.

```
        Mismatches: 0                    August 10, 1991  01:56  ..

    Bmrbs.Seq  Check: 3188  Length: 389

    !  FROMSTADEN of: Bmrbs.Seq  check: 3188  from: 1  to: 389

    Amidation                      Xg(R,K)(R,K)

                                   Xg(R)(K)

              68: RWVDR      FGRK        IMIVI



    Ck2_Phospho_Site         (S,T)x2(D,E)

                             (S)x(2)(E)

              82: GLLFF       SVSE       FLFGI

                             (T)x(2)(E)

             179: ILSIL       TLRE       PERNP

                             (T)x(2)(D)

             239: HKFGF       TASD       IAIMI
```

Fig. 1. Sample of an output from the GCG MOTIFS program. The name of the pattern, its position, and the pattern amino acid sequence are listed in the output.

(*See* Fig. 1.) Symbols enclosed in parentheses and separated by commas represent alternative amino acid residues; thus, (R,K) indicates R or K, x represents any amino acid residue. Numbers enclosed in curly brackets, {}, indicate the number of times the preceding amino acid must be found. In the above example, (S)x{2}(E), for instance, indicates S followed by any two amino acid residues, followed by E.

The MOTIFS program can be executed with command line parameters which serve a variety of purposes as described below. The following are some of the ommand line parameters that may be of use:

a. The parameter /INFILE = <input sequence file> identifies the name of the protein sequence file. This command line parameter eliminates the interactive prompt for the input file name.

b. The parameter /OUTFILE = <output file name> identifies the name of the output file. This command line parameter eliminates the interactive prompt for the output file name.

c. The parameter /DATA = <pattern file> identifies the name of the protein sequence patterns; the PROSITE database is assumed if this parameter is not provided.

d. The parameter /REFERENCE causes the program to output the PROSITE abstracts, from the PROSITE documentation file, for each pattern found.

e. The parameter /MISMATCH = 1 instructs the program to recognize patterns with one or no mismatches. The default is 0 mismatches, i.e., only exact matches are recognized.

f. The parameter /NAMES outputs a file with the names of the protein sequences that can be used with other GCG programs. When this parameter is used, the output file that lists the patterns and their location is suppressed.

g. The parameter /SHOW lists all the protein sequences searched. When this parameter is not provided, only protein sequences in which patterns were found are shown.

4. A segment of an output file obtained from a run of the MOTIFS program using the /REFERENCE parameter is shown in Fig. 2.

3.2. A Search on the VAX Computer for Protein Sequences That Share a Common Motif Using the ISEARCH Program

ISEARCH compares short sequence patterns with all segments of the same or shorter length in each protein sequence of the database. The ISEARCH program can be executed in an AMBIGUOUS mode. In this mode, patterns of up to 20 alternative amino acids at each given position can be provided. Following is a description of the use of the ISEARCH program in the AMBIGUOUS mode:

1. Issue the command:

    ISEARCH AMBIGUOUS

    at the DCL, $ prompt (the keyword AMBIGUOUS may be truncated up to the first letter, A)

2. When prompted, enter the amino acids (in their one-letter designation) at each position (one at a time). The letter B can be used to represent D or N, the letter Z can be used to represent E or Q, and the letter X can be used to represent any of the 20 amino acids. Up to 20 alternative amino acids can be entered at each position. In this case, the amino acid letter designators are separated with commas.

    For example:

    Position 1: I
    Position 2: V,N,S
    Position 3: I,V
    Position 4: V
    Position 5: X

```
$ MOTIFS /REF

MOTIFS looks for sequence motifs by searching through proteins for the patterns defined in
the PROSITE Dictionary of Protein Sites and Patterns.  MOTIFS can display an abstract of
the current literature on each of the motifs it finds.

MOTIFs from what protein sequence(s) ? BMRBS.SEQ

What should I call the output file (* Bmrbs.Motifs *) ? bmrbs2.motifs

BMRBS.SEQ len:          389 .........
     Total finds:        16
    Total length:       389
Total sequences:          1
CPU time (sec):       23.66
    Output file: "User2:[Reizer]Bmrbs2.Motifs"
MOTIFS from: bmrbs.seq

Mismatches: 0                 August 10, 1991  01:59 ..


Bmrbs.Seq  Check: 3188  Length: 389  !  FROMSTADEN of: Bmrbs.Seq  check: 3188  from: 1
to: 389

Amidation          Xg(R,K)(R,K)
                     Xg(R)(K)
         68: RWVDR      FGRK    IMIVI

********************
* Amidation site *
********************

The precursor of  hormones  and other  active  peptides  which are C-terminally
amidated is always directly followed [1,2] by a glycine residue which provides
the amide group, and  most often by at  least two  consecutive  basic residues
(Arg or Lys) which generally function as an active peptide  precursor cleavage
site.  Although all amino acids can be amidated,  neutral hydrophobic residues
such as Val or Phe are good substrates, while  charged residues such as Asp or
Arg  are much less reactive.  C-terminal  amidation has not  yet been shown to
occur in unicellular organisms or in plants.

-Consensus pattern: x-G-[RK]-[RK]
                    [x is the amidation site]
-Last update: June 1988 / First entry.

[ 1] Kreil G.
     Meth. Enzymol. 106:218-223(1984).
[ 2] Bradbury A.F., Smyth D.G.
     Biosci. Rep. 7:907-916(1987).
^^^^^^^^^^^^^^^^^^^^^^^^^^^^^^^^^^^^^^^^^^^^^^^^^^^^^^^^^^^^^^^^^^^^^^^^^^^^^^^^^^^^^^
```

Fig. 2. Sample of an output obtained from the MOTIFS program using the /REFERENCE parameter. In addition to the pattern, the /REFERENCE parameter retrieves the PROSITE abstract(s) for each pattern found.

Position 6: I,V,N,S

The above example represents the following pattern: I in position 1; V, N, or S in position 2; I or V in position 3; V in position 4; any amino acid in position 5; and I, V, N, or S in position 6.

3. When prompted, enter the name of the protein sequence database, for example, NBRF; the search can be performed on a subset of the database.
4. When prompted, enter the name of the output file in which the search results will be stored. Figure 3 shows an output file obtained from the ISEARCH program using the above pattern and the NBRF protein sequence database. The first column on the left displays the protein sequence code in the database; the second column displays the score, i.e., the number of matched amino acids; the third column displays the amino acid position in the particular protein; and the last column displays the segment of the protein sequence that corresponds to the pattern searched. The sequence identification code can be used to extract the protein sequence from the database using any available query programs, such as the PSQ (Protein Sequence Query) system distributed by the Protein Identification Resource.

3.3. A Search for PROSITE Motifs on the IBM or Compatible Microcomputer Using the PC/GENE Program PROSITE

The PC/GENE program PROSITE is a PC-based program that provides features similar to those found in the VAX-based program MOTIFS. The PROSITE program analyzes a protein sequence, and detects protein patterns or motifs.

The PROSITE program is menu-driven, self explanatory and easy to use. The following description highlights the key features of the PROSITE program:

1. The protein sequence can be entered into the computer using a text editor, such as the PC/GENE program SEQIN. The sequence can also be extracted from a protein sequence data bank. If the sequence is in a format other than EMBL or SWISS-PROT, which are the formats required by PC/GENE, it must be converted. The program REFORM (included in the PC/GENE software package) can be used to convert the protein sequence file to a PC/GENE format.
2. The PC/GENE software is activated by issuing the following command at the DOS prompt:
PCGENE

A user password must also be provided.

3. The PROSITE program is then invoked either by selecting it through a menu-driven process or directly from a list of programs, as described

```
Largest possible score = 6,   Smallest possible score = 1
Number of sequences searched = 7068,   Number of comparisons = 2104196

Test sequence
               IVIVXI
               NV  V
                S  N
                   S
          Score  Residue
O4PSCP      6       218   ISIVAN
DEBYA       6       288   ISIVGS
DEBYA2      6       289   ISIVGS
DECHLM      6        22   ISVVGV
RDBE11      6        72   INVVLS
RDMSD       6        71   INIVLS
RDHY75      6        71   INIVLS
RDBOD       6        71   INIVLS
RDPGD       6        71   INIVLS
RDCHD       6        71   INIVLS
OBSY1       6       462   ISVVGI
XNECGM      6        94   IVVVHN
KIECFA      6       164   ISVVEV
DEECK       6       344   ISVVLI
```

Fig. 3. Sample of an output obtained from the ISEARCH program. The output lists the codes (left column) of all protein sequences that share the common motif. Also listed are the scores (second column from left), and the segment of the protein sequence that corresponds to the pattern searched (last column).

below. To select the PROSITE program from a series of menus, step through the following choices:

   a. Select option **1, Program Selection by Menu**, from the PC/GENE main menu.

   b. Select option **3, Sequence Analysis**, from the Program Selection main menu.

   c. Select option **2, Protein Sequence Analysis**, from the Sequence Analysis menu.

   d. Select option **2, Site Detection Analysis**, from the Protein Sequence Analysis menu.

   e. Select option **2, PROSITE**, from the Site Detection Analysis menu. To select the PROSITE program automatically from a list of programs do the following: (a) Select option **2, Automatic Screen Selection of Programs,** from the PC/GENE main menu. (b) Select the program PROSITE from the list.

4. The features and run options of the PROSITE program are depicted in the following PROSITE menu (Fig. 4).

5. The protein sequence being analyzed is selected via option 1 of the menu.

PC/Gene: the nucleic acid and protein sequence analysis software system.
Locate sites and signatures in a protein sequence.
Version 3.10 / January 1991.

0	EXIT from program.	F10: DOS
1	Select the sequence to be analyzed.	F9 : Gripe
2	Define the content and layout of the output.
3	Scan the sequence for sites and signatures.
4	Display/output the sites & signatures found in the sequence.
5	The 'textbook'.
6	Change output status.
7	Help.

Fig. 4. Features and run options of the PC/GENE PROSITE program. The various features of the program are selected via menu-driven options.

6. The content and format of the output are defined via option 2 of the menu. This option allows the user to define the format of the displayed sites, as well as the general layout of the report.
7. The search for motifs is executed by selecting option 3 from the PROSITE menu.
8. Before displaying the search results, the output destination should be defined (via option 6 of the menu). By default, the search results are displayed on the monitor. The output, however, can be sent to the printer or to a file.
9. The search results can then be obtained via option 4 of the menu. The results are displayed on the monitor, printed on paper, or stored in a file, depending on the output destination that was selected (via option 6).
10. PROSITE documentation text is obtained via option 5 (the "textbook"). The "textbook" entries are displayed on the monitor, printed on paper, or stored in a file, depending on the output destination selected. Various sections of the "textbook" can be extracted: general introduction to the PROSITE dictionary, specific chapters of the "textbook," or chapters corresponding to the sites found in the protein sequence that was analyzed.

4. Notes

1. The use of motifs for the classification and prediction of local tertiary architecture of proteins is gradually emerging as a powerful and subtle tool that undoubtly can achieve notable success. The interested reader may consult the following references for more detailed information regarding this strategy, other computing methods, and catalogs of various motifs *(3–11)*.

2. Earlier version (less than version 7.0) of the GCG software package did not include the MOTIFS programs. In those versions, the GCG program FIND may be used to search for motifs defined in the PROSITE database. However, when using the FIND program, the PROSITE database must be converted to a format recognized by FIND (the PROSITECONV program, written in the Pascal language by Kay Hofmann from the University of Koeln, Germany, can be used for this purpose). Additionally, FIND does not have the capability to output the PROSITE documentation entries, since it merely provides the documentation entry numbers. To browse through the applicable PROSITE abstracts, a hard copy of the PROSITE documentation file must be produced or, alternatively, the program PRODOC (written by Anne Marie Quinn, Salk Institute, La Jolla, CA) can be used to extract documentation entries from the PROSITE file based on entry numbers reported by FIND.

The MOTIFS program combines the functionality of the FIND and PRODOC programs. Therefore, wherever available, the MOTIFS program is recommended, since it is more streamlined and easier to use than the FIND program.

References

1. Bairoch, A. (1991) PROSITE: a dictionary of sites and patterns in proteins. *Nucleic Acids Res.* **19,** 2241–2245.
2. Dayhoff, M. O., Barker, W. C., and Hunt, L. T. (1983) Establishing homologies in protein sequences. *Meth. Enzymol.* **91,** 524–545.
3. Taylor, W. R. and Jones, D. T. (1991) Templates, consensus patterns and motifs. *Curr. Opin. Struct. Biol.* **1,** 327–333.
4. Staden, R. (1988) Methods to define and locate patterns of motifs in sequences. *CABIOS* **4,** 53–60.
5. Staden, R. (1989) Methods for calculating the probabilities of finding patterns in sequences. *CABIOS* **5,** 89–96.
6. Barton, G. J. and Sternberg, M. J. E. (1990) Flexible protein sequence patterns. A sensitive method to detect weak structural similarities. *J. Mol. Biol.* **212,** 389–402.
7. Sibbald, P. R. and Argos, P. (1990) Scrutineer: a computer program that flexibly seeks and describes motifs and profiles in protein sequence databases. *CABIOS* **6,** 279–288.
8. Hodgman, T. C. (1989) The elucidation of protein function by sequence motif analysis. *CABIOS* **5,** 1–13.
9. Sternberg, M. J. E. (1991) PROMOT: a FORTRAN program to scan protein sequences against a library of known motifs. *CABIOS* **7,** 257–260.
10. Seto, Y., Ikeuchi, Y., and Kanehisa, M. (1990) Fragment peptide library for classification and functional prediction of proteins. *Proteins* **8,** 341–351.
11. Argos, P., Vingron, M., and Vogt, G. (1991) Protein sequence comparison: methods and significance. *Prot. Engin.* **4,** 375–383.

CHAPTER 25

CLUSTAL V: Multiple Alignment
of DNA and Protein Sequences

Desmond G. Higgins

1. Introduction

CLUSTAL is a package for performing fast and reliable automatic
multiple alignment of many DNA or protein sequences. It was originally written for IBM-compatible microcomputers (1,2) and was later
reorganized as a single program for VAX mainframes. Recently (3),
the package was completely rewritten as a new program, CLUSTAL V,
which is freely available for a wide variety of computer systems and
which has a number of new features. The main improvements are the
calculation of phylogenetic trees from sequence data sets with a bootstrap option for calculating confidence intervals on the groupings and
the ability to align alignments with each other.

The central feature of CLUSTAL V is the multiple alignment algorithm, which is a variation on the method of Feng and Doolittle (4),
where the alignment is built up progressively, using an initial phylogenetic tree as a guide. This is done in three stages (automatically
done for the user, within the program). First, all sequences are quickly
compared to each other using the algorithm of Wilbur and Lipman
(5). This gives a similarity score (percent identity) between every
pair of sequences. These scores are used to group the sequences in a
crude initial tree using the UPGMA method (6). This tree is referred
to as a dendrogram rather than a phylogenetic tree to emphasize the
fact that it is not reliable as a guide to phylogeny. All of the above is
fast enough to do on a microcomputer, but robust enough to give

From: *Methods in Molecular Biology, Vol. 25: Computer Analysis of Sequence Data, Part II*
Edited by: A. M. Griffin and H. G. Griffin Copyright ©1994 Humana Press Inc., Totowa, NJ

reasonable groupings of sequences except in very difficult cases. The final step is the critical one. Here, the sequences are aligned in larger and larger groups, according to the branching order in the dendrogram. A global optimal alignment method *(7)* is used with penalties for opening and extending each gap and an amino acid weight matrix (the Dayhoff PAM 250 matrix *[8]* is the default).

At each stage in the final multiple alignment, one aligns two alignments. Initially, these are single sequences, but they grow with the addition of new sequences as one goes down the dendrogram. Any gaps that are placed in early alignments remain fixed, and any gaps that are inserted into a group of prealigned sequences are put in all the sequences of the group simultaneously.

The source code of the program is freely available along with documentation on installation and usage, and some make files for various systems. The easiest way of obtaining it is to use the EMBL file server *(9)*. For instructions on how to use the file server, send an electronic mail message to the network address: Netserv@EMBL-Heidelberg.DE with the words: "HELP" and "HELP SOFTWARE" on two lines with no quotes. To obtain the program for an IBM-compatible or Apple Macintosh microcomputer without access to electronic mail, the user should send a formatted diskette to the author, and he or she will receive the full program including an executable image. Please state which machine is being used. The full package consists of C source code, documentation, on-line help file, and make files for various systems. To change the program defaults (e.g., maximum number and length of sequences), edit one small header file and recompile.

Sequence input to the program can be in EMBL/SWISS-PROT *(10)*, FASTA *(11)*, or NBRF/PIR *(12)* formats. The maximum number and lengths of the sequences will depend on the machine being used, but are typically hundreds of sequences of several thousand residues each on a mainframe or work station or 30 sequences of one to two thousand residues on a microcomputer. The alignment output can be in several formats, one of which is compatible with the PHYLIP package *(13)* of Joseph Felsenstein. Usage of the program is generally self-explanatory, being completely menu-driven, and on-line help is available. Alternatively, a full command line interface is offered to facilitate running the program in batch. Defaults are offered for all output file names and parameters.

The phylogenetic trees produced by CLUSTAL V are calculated using the Neighbor-Joining method *(14)*. This is a distance method that works on a matrix of distances calculated from a multiple alignment. Therefore, the sequences must be aligned before a tree is calculated. The distances can be simple percent divergence values (the percentage of residues that differ between the sequences) or can be corrected for multiple substitutions using formulae from ref. *15* for nucleic acid sequences or ref. *16* for proteins. Confidence intervals on the tree can be calculated using a bootstrap procedure similar to that of Felsenstein *(13)*. In simple cases, therefore, CLUSTAL V can be reliably used to produce phylogenetic trees completely automatically from a set of sequences. In more complicated cases, manual adjustment of the alignment may be required or users may prefer to use alternative phylogenetic estimation methods, in which case the alignments may be used as input to the PHYLIP package *(13)*.

2. Materials

CLUSTAL V is written according to the proposed ANSI standard of the C programming language *(17)*. If the user wishes to use the program on a microcomputer, it is not necessary to have a C compiler; an executable file is supplied. However, if the user wishes to change the program defaults, he or she will need to edit the source code and recompile it. It has been compiled and tested on the following systems.

1. IBM-compatible microcomputer running MS DOS version 3.3. The default PC version of the program can align up to 30 sequences of maximum length 1200 residues. This requires 500k of memory, but does not require any special graphics devices or math chips. To recompile the program, a TURBO C compiler is needed.
2. Apple Macintosh with over 800k of memory. The supplied version can align up to 30 sequences of over 200 residues each. The program does not use the standard Macintosh-like interface; it merely opens a window inside which it runs just like on a character-based terminal. To recompile the default version, a THINK C compiler is needed.
3. VAX running VMS. A small command file is supplied for compiling and linking the program. Access to the VAX C compiler is necessary.
4. DECSTATION running Ultrix. A make file is supplied for compiling the program using the Ultrix C compiler.

5. SUN Work station running SUN OS. The native SUN C compiler does not support the ANSI C standard and will not compile CLUSTAL V. The GNU C compiler for SUNs must be used. A make file for GNU C is supplied.

3. Method

CLUSTAL V is normally run interactively. It is possible to use a command line interface, full details of which are described in the documentation. To obtain a very brief summary of the command line options, type CLUSTALV /HELP and press <RETURN> (does not work on the Apple Macintosh). On most systems, enter the program by typing CLUSTALV and pressing the <RETURN> key. The user then gets an opening menu, which is shown in Fig. 1.

Brief on-line help can be obtained by typing H and pressing <RETURN>. To leave the program temporarily in order, for example, to obtain a list of files in the current directory or to edit a file, type S and press the <RETURN> key. This will allow the user to enter a normal operating system command, such as DIR (on PCs or VAX). This feature does not work on Apple Macintoshes. To enter one of the secondary menus or to load sequences, type a number between 1 and 4. All of the secondary menus work in the same way, and help can be obtained or an operating system command entered from them in the same way as described above for the main menu.

3.1. Multiple Alignment of DNA or Protein Sequences

1. All of the sequences to be aligned should be in one file in one of the following formats: EMBL/SWISS-PROT, FASTA, or NBRF/PIR (*see* Section 4. for explanation). If using the GCG *(18)* package to manage sequences, use the TOPIR (formerly the TONBRF command) command of the GCG package to create a file in the appropriate format. Before running CLUSTAL V, create a file with the sequences in one of the allowed formats.
2. Run the program; the opening menu will be seen. Load the sequences by typing 1 and pressing <RETURN>. The user is then asked to give the name of the file that contains his or her sequences. The program attempts to guess the sequence format and sequence type (DNA or protein) that are being used, and if the sequences can be read, will list the names and lengths of all sequences in the file.
3. Enter the multiple alignment menu by typing 2. A menu is then obtained that is shown in Fig. 2.

CLUSTAL V ... Multiple Sequence Alignments

1. Sequence Input From Disc
2. Multiple Alignments
3. Profile Alignments
4. Phylogenetic trees

S. Execute a system command
H. HELP
X. EXIT (leave program)

Your choice:

Fig. 1. The main menu of CLUSTAL V. This menu appears when starting the program.

Multiple Alignment Menu

1. Do complete multiple alignment now
2. Produce dendrogram file only
3. Use old dendrogram file
4. Pairwise alignment parameters
5. Multiple alignment parameters
6. Output format options

S. Execute a system command
H. HELP
or press [RETURN] to go back to main menu

Your choice:

Fig. 2. The multiple alignment menu of CLUSTAL V. This menu is reached from item 2 of the main menu.

4. To do a complete, automatic alignment of all the sequences in the input file, just type 1 and press the <RETURN> key. The user will be offered default output file names for the dendrogram (same as input file with extension: ".DND") and the multiple alignment (same as input file with extension: ".ALN"). Press <RETURN> to accept these. At this stage, the alignment is done using default settings for the gap penalties and weight matrix. The default output format is a self-explanatory, blocked multiple alignment format. To change these defaults, use menu items 4, 5, or 6. The various parameters are explained in Section 4. at the end of this chapter. To use the alignment again at a later date (e.g., to calculate a phylogenetic tree), save it in PIR format.
5. To see the output alignment, use the S command (Execute a system command) to use whatever command is normally used on the system to view text files. This does not work on the Apple Macintosh, where other facilities are available for doing this.

3.2. Phylogenetic Trees

1. Before calculating a tree, the sequences must be aligned. An alignment can be used if the user has just carried one out and the sequences are still loaded in the program. Alternatively, save the sequences in PIR format after aligning them. Files saved in PIR format can be used again as input; gaps are specified using the hyphen character.
2. Go to item 4 from the main menu (phylogenetic trees). A menu that is shown in Fig. 3 will be obtained.
3. If an aligned set of sequences has not already been put into memory, use item 1 and give the name of the file containing an alignment.
4. To calculate a tree using default options, just type 4 (Draw tree now) and press <RETURN>. The user will be prompted for the name of an output file that is the same as the input file, but with the extension: ".NJ", for Neighbor-Joining, the method used to calculate the tree *(14)*.
5. Menu items 2 and 3 allow the user to change the options used to calculate the tree. Item 2 is used to exclude all alignment positions where any sequence has a gap. This is useful in cases where the alignment is unreliable in some positions. It also lessens the biasing effect of the initial dendrogram ("guide tree") if it is incorrect. Further, it means that "like" is compared with "like" in all distance calculations. Menu item 3 allows the user to introduce a correction into the distance calculations to correct for the effect of multiple substitutions. With small distances, this has no effect. With large distances, it compensates for the fact that the number of observed substitutions between two sequences greatly underestimates the actual number that occurred since the sequences diverged.

Phylogenetic Tree Menu

    1. Input an alignment
    2. Exclude positions with gaps?   = OFF
    3. Correct for multiple substitutions? = OFF
    4. Draw tree now
    5. Bootstrap tree

    S. Execute a system command
    H. HELP
or press [RETURN] to go back to main menu

Your choice:

Fig. 3. The phylogenetic tree menu of CLUSTAL V. This menu is reached from item 4 of the main menu. Options 2 and 3 are toggled on or off when chosen.

6. Bootstrap confidence intervals are calculated by using menu item 5. These give a measure of how well supported groupings in the tree are, given the data set and the tree drawing method. The user is asked for a seed number for the random number generator; just enter any integer and press <RETURN>. Remember that the same answer will be obtained if using the same seed number on the same data set. Then the user is asked to enter the number of bootstrap samples to be used. Normally, make this as high as possible, up to about one thousand. This may take over an hour to compute on a microcomputer even for a small number of sequences, but will typically take 5 to 30 min on a work station or mainframe. The output goes to a file with the extension ".NJB".

4. Notes

1. The sequence input format can be any of EMBL/SWISS-PROT *(10)*, PIR *(12)*, or FASTA *(11)* formats. All sequences must be in one file. The only characters that are recognized in sequences are upper- and lower-case letters of the alphabet and the hyphen ("-"), which is used to represent a gap. All other characters, such as digits, spaces, or punctuation marks, are simply ignored. Any letter that is not one of the four nucleotides (a, c, g, or t, u) or one of the 20 amino acids (a, c, d, e, f, g,

h, i, k, l, m, n, p, q, r, s, t, v, w, y) will be treated as unknown. Any gaps that are read on input will remain during the multiple alignment or tree calculating stages. This means that entire alignments can be read in and reused. One of the four output formats (PIR) can be used to write alignments that can be read in again at a later date. Users of the GCG package *(18)* should use the TOPIR command (formerly TONBRF) to prepare their sequences for use by CLUSTAL V.

2. If the user has no idea what the different formats above are, then FASTA format should be used to organize the sequence input file; it is the simplest. The sequences are put one after the other in one file. Each sequence is delimited by a title line that starts with a right-angle bracket (">"). The first 15 characters after the angle bracket are read as the name of the following sequence. The sequence is put on any number of lines after each title line, in free format. The end of each sequence is detected when the end of the file is reached or the next title line is found.

3. The program tries to determine automatically which format is being used, and whether the sequences are DNA or protein. The format is checked by examining the first characters of the first line of the file. If the first character is a right-angle bracket (">"), then FASTA or PIR format is assumed. In order to distinguish between the two, the fourth character is checked. If the fourth character is a semicolon (";"), then PIR format is assumed and the next line is skipped as arbitrary text. EMBL format is assumed if the first two characters in the file are "ID." In order to determine whether DNA or protein sequences are being used, the program counts the number of times one of a, c, g, t, or u occurs. If they occur in over 85% of all positions, then DNA is assumed, otherwise protein. This works in all except the most extreme cases, e.g., very short sequences with very strange residue compositions. It is critical that the program correctly guesses the type of the sequences, so one should be careful.

4. There are four output formats offered for the multiple alignments. They can all be toggled on or off from the menus before alignment, or the user can choose them afterward. The four formats are: CLUSTAL, PIR, GCG MSF, and PHYLIP. CLUSTAL format is self-explanatory with the alignment shown as blocks of aligned sequence. PIR format has the sequences arranged one after the next with the hyphen character ("-") used to represent gaps. This is identical to normal PIR format with the addition of the gap symbol. This format is especially useful, because it can be used to store alignments that can be used as input at a later date, e.g., for calculating trees. The MSF format is new since version 7 of the GCG package *(18)*. The PHYLIP format output can be used by any of

the programs in version 3.4 or later of the PHYLIP package *(13)* that use sequence alignments as input.

5. There are five parameters that control the speed and sensitivity of the fast pairwise alignments that are used to cluster the sequences initially. Usually it is not worth adjusting these parameters in order to make the cluster order more sensitive. The main reason for adjusting them is to increase speed when aligning many (30 or more) or long (500 residues or more) sequences. The parameters are explained in on-line help and include a gap penalty, a word (or "ktuple") size, the number of diagonals to be used, and a window around each top diagonal. The scores that are calculated can be expressed as absolute (number of overlapping words) or percentage (the latter divided by the length of the shorter sequence).

6. The gaps in the final alignment are controlled by two gap penalties: one for opening each gap and another for extending each gap by one residue. Terminal gaps are penalized just like internal ones. The user can choose between three supplied weight matrices for proteins: PAM100, PAM250 *(8)*, or identity. The user can also input his or her own weight matrix from a file. With DNA, the user can choose to weight transitions (purine to purine or pyrimidine to pyrimidine substitutions) more highly than transversions (purine vs pyrimidine changes).

7. The program does not produce graphic output for the phylogenetic trees; they are calculated and described in text files. An example output file is shown in Fig. 4. The estimated distance between every pair of sequences is given followed by a description of the tree topology. The trees are unrooted; they can be rooted manually by choosing the longest branch (correct if there is a good "molecular clock") or by inclusion of an outgroup. An outgroup is a sequence that is certain (on biological grounds) to root outside of the sequences being used. Normally, trees are rooted on biological grounds where there is evidence for the location of the root from a knowledge of the organisms involved or are simply left unrooted.

8. The neighbor-joining method *(14)*, used to calculate the trees, was chosen because it is fast to compute, conceptually simple, and very robust in a wide variety of situations. It does not explicitly assume a molecular clock and performs well under situations where the clock varies greatly in different lineages *(19)*. There are many other methods available, each with its implicit assumptions, strengths, and weaknesses. To learn more about the different methods and the assumptions behind them and try them out, it is highly recommended that the user get the PHYLIP package and read the documentation. Then use CLUSTAL V to generate alignments and any method desired to draw the trees.

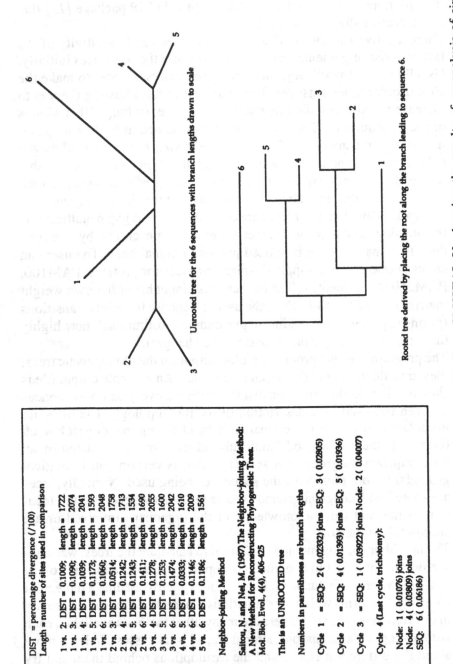

DIST = percentage divergence (/100)
Length = number of sites used in comparison

1 vs. 2: DIST = 0.1009;   length = 1722
1 vs. 3: DIST = 0.1090;   length = 2074
1 vs. 4: DIST = 0.1039;   length = 2041
1 vs. 5: DIST = 0.1173;   length = 1593
1 vs. 6: DIST = 0.1060;   length = 2048
2 vs. 3: DIST = 0.0514;   length = 1758
2 vs. 4: DIST = 0.1242;   length = 1713
2 vs. 5: DIST = 0.1243;   length = 1534
2 vs. 6: DIST = 0.1411;   length = 1690
3 vs. 4: DIST = 0.1278;   length = 2055
3 vs. 5: DIST = 0.1253;   length = 1600
3 vs. 6: DIST = 0.1474;   length = 2042
4 vs. 5: DIST = 0.0333;   length = 1610
4 vs. 6: DIST = 0.1146;   length = 2009
5 vs. 6: DIST = 0.1186;   length = 1561

Neighbor-joining Method

Saitou, N. and Nei, M. (1987) The Neighbor-joining Method:
A New Method for Reconstructing Phylogenetic Trees.
Mol. Biol. Evol., 4(4), 406–425

This is an UNROOTED tree

Numbers in parentheses are branch lengths

Cycle 1   = SEQ: 2 (0.02332) joins SEQ: 3 (0.02805)

Cycle 2   = SEQ: 4 (0.01393) joins SEQ: 5 (0.01936)

Cycle 3   = SEQ: 1 (0.03922) joins Node: 2 (0.04007)

Cycle 4 (Last cycle, trichotomy):

Node: 1 (0.01076) joins
Node: 4 (0.03809) joins
SEQ: 6 (0.06186)

Unrooted tree for the 6 sequences with branch lengths drawn to scale

Rooted tree derived by placing the root along the branch leading to sequence 6.

Fig. 4. An example of a phylogenetic tree output file from CLUSTAL V, showing the results of an analysis of six sequences. The file itself is shown on the left. The unrooted tree that can be derived from it is shown on the right. A rooted tree that is obtained by placing the root along the branch leading to sequence 6 is also shown.

9. Bootstrapping is used to give confidence measures for every grouping in the tree (actually gives confidence levels for every internal branch). The figures are derived by taking a large number (e.g., one thousand) of random samples of sites from the alignment. Each sample is taken with replacement; i.e., from an alignment n residues long, take a random sample of n sites. In any given sample, some sites will not be selected; others will be selected several times. With each sample, calculate a distance matrix and tree in the usual way. This gives one thousand sample trees. The variation in the sample trees gives a measure of how well supported the initial tree is. With each grouping in the initial tree, count how often it occurs in the sample trees. A well-supported grouping will occur in almost all of the sample trees. It is common to express the figures as percentages. Crudely, for a grouping to be significant at the 5 percent level, it should occur in 95 (100 minus 5) percent of the bootstrap samples. Unfortunately, it is not valid to make a significance test of all groupings in a tree at a given level and to treat each test as independent. Either, choose a particular grouping in the tree in advance (before the bootstrap resampling) and test it alone (give its significance), or simply show all the bootstrap figures alongside the tree and let the figures speak for themselves. The latter is very common practice. The reason for taking a very large number of samples is because a different set of bootstrap figures will be obtained with different seed numbers for the random number generator. This variation in the bootstrap figures is confounded with the underlying variation being estimated. Therefore, increase the precision of the bootstrap estimates (reduce the standard error) by taking very many samples.

10. When interpreting bootstrap figures, some words of caution are advised. If a particular grouping is found to be significant, this does not prove biological significance; i.e., it does not prove that the branching pattern actually took place. It does show that the grouping is well supported, given the particular data set and, most importantly, the method used to construct the tree. If a bad method is used, then it can give bad trees, but in a very regular and consistent way. Just be careful not to over-interpret the figures; they are a very useful guide, but that is all. Similarly, if a favorite grouping is not significantly supported, do not be too disappointed and do not be misled into thinking that another method is certain to give better results. The most common reason is that the sequences are too short rather than a defect in the method. The shorter the branch leading to a group of interest, the more sites needed (longer sequences) to resolve it clearly.

References

1. Higgins, D. G. and Sharp, P. M. (1988) CLUSTAL: a package for performing multiple sequence alignments on a microcomputer. *Gene* **73**, 237–244.
2. Higgins, D. G. and Sharp, P. M. (1989) Fast and sensitive multiple sequence alignments on a microcomputer. *Comput. Applic. Biosci.* **5**, 151–153.
3. Higgins, D. G., Bleasby, A. J., and Fuchs, R. (1992) CLUSTAL V: improved software for multiple sequence alignment. *Comput. Applic. Biosci.* **8**, 189–191.
4. Feng, D.-F. and Doolittle, R. F. (1987) Progressive sequence alignment as a prerequisite to correct phylogenetic trees. *J. Mol. Evol.* **25**, 351–360.
5. Wilbur, W. J. and Lipman, D. J. (1983) Rapid similarity searches of nucleic acid and protein data banks. *Proc. Natl. Acad. Sci. USA* **80**, 726–730.
6. Sneath, P. H. A. and Sokal, R. R. (1973) *Numerical Taxonomy.* Freeman, San Francisco, CA.
7. Myers, E. W. and Miller, W. (1988) Optimal alignments in linear space. *Comput. Applic. Biosci.* **4**, 11–17.
8. Dayhoff, M. O., Schwartz, R. M., and Orcutt, B. C. (1978) in *Atlas of Protein Sequence and Structure*, vol. 5, supplement 3 (Dayhoff, M. O., ed.), NBRF, Washington, DC, p. 345.
9. Fuchs, R. (1990) Free molecular biological software available from the EMBL file server. *Comput. Applic. Biosci.* **6**, 120–121.
10. Bairoch, A. and Boeckmann, B. (1991) The SWIS-PROT protein sequence data bank. *Nucleic Acids Res.* **19**, 2247–2248.
11. Pearson, W. R. and Lipman, D. J. (1988) Improved tools for biological sequence comparison. *Proc. Natl. Acad. Sci. USA* **85**, 2444–2448.
12. Barker, W. C., George, D. G., Hunt, L. T., and Garavelli, J. S. (1991) The PIR protein sequence database. *Nucleic Acids Res.* **16**, 1869–1871.
13. Felsenstein, J. (1985) Confidence limits on phylogenies: an approach using the bootstrap. *Evolution* **39**, 783–791.
14. Saitou, N. and Nei, M. (1987) The neighbor-joining method: a new method for reconstructing phylogenetic trees. *Mol. Biol. Evol.* **4**, 406–425.
15. Kimura, M. (1980) A simple method for estimating evolutionary rates of base substitutions through comparative studies of nucleotide sequences. *J. Mol. Evol.* **16**, 111–120.
16. Kimura, M. (1983) *The Neutral Theory of Molecular Evolution.* Cambridge University Press, Cambridge, England.
17. Kernighan, B. W. and Ritchie, D. M. (1988) *The C Programming Language.* 2nd Ed., Prentice Hall, New Jersey.
18. Devereux, J., Haeberli, P., and Smithies, O. (1984) A comprehensive set of sequence analysis programs for the VAX. *Nucleic Acids Res.* **12**, 387–395.
19. Saitou, N. and Imanishi, T. (1989) Relative efficiencies of the Fitch-Margoliash, Maximum Parsimony, Maximum-Likelihood, Minimum-Evolution and Neighbor-Joining Methods of phylogenetic tree construction in obtaining the correct tree. *Mol. Biol. Evol.* **6**, 514–525.

CHAPTER 26

Progressive Multiple Alignment of Protein Sequences and the Construction of Phylogenetic Trees

Aiala Reizer and Jonathan Reizer

1. Introduction

Early attempts to use quantitative measures to provide information on divergent evolution of macromolecules include estimates, such as the degree of crossreactivity of antisera to purified proteins (1,2), the degree of interspecific hybridization of DNA (3), and differences in peptide profile of digested proteins (4,5). The inventions of modern nucleic acid and protein sequencing techniques have led to an avalanche of protein sequence data, which provide a more extensive and quantitative tool for exploration of evolutionary relationships than the aforementioned biological characteristics. Grouping of proteins into families or superfamilies and the reconstruction of protein or organismic history from extant protein sequences can be portrayed in the form of an evolutionary tree. The branching order of such a tree identifies the points of divergence for proteins or organisms that share a common ancestry, whereas the branch lengths reflect the evolutionary distances between present-day proteins and proteins that existed at the point of divergence.

The scheme described here, the progressive sequence alignment method that was devised by Feng and Doolittle (6,7), generates pairwise comparisons of a set of protein sequences; the pair of sequences with the smallest difference score is then used as a starting point for

From: *Methods in Molecular Biology, Vol. 25: Computer Analysis of Sequence Data, Part II*
Edited by: A. M. Griffin and H. G. Griffin Copyright ©1994 Humana Press Inc., Totowa, NJ

alignment of the next most similar sequence. This progressive alignment process is repeated until all protein sequences have been aligned. The comparison scores obtained from the final alignment of all sequences are then used to compute the branch lengths and to determine the topology, i.e., branching order, of the evolutionary tree. Since a gap is never discarded during the progressive alignment process, even if an alignment with some more distantly related sequences might have been improved, the method gives more weight to the more recently diverged sequences than to more distant ones. Other approaches that have been applied to the alignment of a set of protein sequences and to phylogeny reconstruction are described in a recent volume of *Methods in Enzymology (8)*. We have applied the protocol described in this chapter to a vast range of protein sequences *(9–12)*, and found it flexible, easy to use, and reliable in providing phylogenies from protein sequence data.

2. Materials

Hardware: VAX computer; Zeta plotter (optional). Software: (1) Multiple alignment programs, SCORE, PREALIGN, TREE, BLEN TREEPLOT, and MULPUB (can be obtained from R. F. Doolittle, University of California, San Diego) and (2) UNIX (Berkeley) or VMS operating systems. Additionally, the following programs may be required:

1. To extract sequences from the PIR libraries, the PSQ (Protein Sequence Query) program, distributed by the Protein Identification Resource, can be used;
2. To enter or to edit sequences on the computer, an editor, such as the VAX EDT editor, should be used;
3. A program may be required to remove spaces and some other extraneous information or to format a sequence to a fixed line length. The LIGSEQ (LIGate SEQuences), which eliminates blanks and file attributes and formats the sequence to 60 characters/line, is an example of a program that can be used for this purpose; or
4. Programs to convert standard sequences (those entered with an editor) and PIR-formatted sequences (those obtained from the PIR libraries) to a NEWAT format required for the multiple alignment programs. The SEQ2NEWAT and PIR2NEWAT programs (can be obtained from D. W. Smith, University of California, San Diego) can serve such a purpose. The NEWAT program TYPIN may be used instead to type in a sequence directly into the NEWAT format.

3. Methods

1. Before executing the progressive alignment programs (*see* Note 1), the protein sequences, in the proper (NEWAT) format, must exist on the VAX computer. The protein sequences can be either typed in directly on the VAX computer, using a proper editor, such as EDT, extracted from the PIR (Protein Identification Resource) libraries, or obtained from other sources, such as the STADEN programs. Sequences can be also typed in directly on a personal computer and then transmitted, via a communication software, to the VAX computer. In all the above cases, the protein sequences will have to be converted to the NEWAT format. Alternatively, the NEWAT program TYPIN, which allows the user to enter sequences directly into the computer in the NEWAT format, can be used. If sequences are extracted from the PIR libraries, use the COPY command of the PSQ (Protein Sequence Query) system. Sequence files extracted from the PIR libraries contain not only the amino acid letter designator, but also sequence definitions; these files may also contain additional extraneous information, such as file attributes. To remove the definition text, as well as the file attributes, use a text editor or any of the installation's available programs for this purpose, such as the LIGSEQ program.

2. Protein sequence files, which have not been entered with the NEWAT program TYPIN, must be converted to a NEWAT format. The conversion method depends on the source of the sequence and the conversion programs available at the installation. For example, the SEQ2NEWAT program converts standard sequences (those entered into the computer with an editor), whereas the PIR2NEWAT program converts PIR-formatted sequences to a NEWAT format. During this process, the user is prompted to designate a four-letter code to the sequence, which is also used to name the file. As a result, the sequence file name is composed of the four-letter name given by the user with the "AA" extension.

3. The NEWAT-formatted sequences are then combined into a single file. The VMS COPY command can be used for this purpose.

4. The progressive alignment programs, SCORE, PREALIGN, and TREE, are then executed. They must run in a specific order as follows:
   a. SCORE program is executed first. It performs the initial pairwise alignment of the protein sequences. It also provides the initial tree topology and suggests clusters. The topology, that is, the tree's branching order, as well as the proposed clusters can be found in the output file ending with the extension "SC1." The clusters are later used to construct the input files for the PREALIGN program, whereas the branching order is used to construct the input file for the TREE program;

    b. If indeed clusters were proposed by the SCORE program, PREALIGN
is executed next. This program prealigns the protein sequence clusters found by the SCORE program; and

    c. The TREE program is executed last. It performs the final multiple
sequence alignment, and provides the final tree topology and the
branch lengths used in the construction of the phylogenetic tree. (*See
also* Note 2).

5. Execution of the progressive alignment programs.

    a. The SCORE program:

       1. Run the SCORE program by entering SCORE (or RUN SCORE)
at the DCL $ prompt.

       2. When prompted, enter the input file name, which is the file that con-
tains all the protein sequences in the NEWAT format.

       3. When prompted, the user should provide his or her own names for
the output files or accept the default names offered by the program
(default names consist of the input file name with the extension
SC1, SC2, and SCO). The output file with the extension "SC1,"
contains the preliminary branching order and the suggested clusters,
if any. The preliminary branching order is used to construct the input
file for the TREE program, whereas the suggested clusters are used
to construct the input files for the PREALIGN program.

    b. The PREALIGN program: If any clusters were suggested by the
SCORE program, they must be prealigned prior to performing the
final multiple alignment by the TREE program. The following
SCORE output will be used as an example to explain the usage of the
PREALIGN program:

Branching order: (SCTA,SCSA)SCMA)BGCA)GLCA)NACA)GUCA)LLCA
       Clusters: (BGCA,GLCA)NACA)GUCA)LLCA
               (NACA,GUCA)

PREALIGN is executed for each of the clusters, advancing from
the inner loop out. Thus, the (NACA,GUCA) cluster is prealigned
first. Next, the (BGCA,GLCA) cluster is prealigned. Finally, the
(BGCA,GLCA)NACA)GUCA)LLCA cluster is prealigned, using the
prealigned (BGCA,GLCA) and (NACA,GUCA) clusters. Prealignment of clusters is performed as follows:

       1. Combine the NACA and GUCA sequences into a single file using
the DCL's COPY command, and then execute the PREALIGN
program by entering PREALIGN (or RUN PREALIGN) at the
DCL's $ prompt. When prompted, enter the input file name, which
is the file that contains the NACA and GUCA sequences. When
prompted, enter the desired output file names or accept the

program's default file names. The prealigned sequences are in the output file which ends with the extension "PR1."

   2. Similarly, combine the BGCA and GLCA sequence files into a single file, and execute the PREALIGN program.

   3. Finally, prealign the (BGCA,GLCA)NACA)GUCA)LLCA cluster. To create the input file for the PREALIGN program, combine the previously prealigned (BGCA,GLCA) cluster with the previously prealigned (NACA,GUCA) cluster and the LLCA sequence file into a single file.

c. The TREE program:

   1. Construct an input file for the TREE program by combining all the prealigned sequences and those sequences that did not have to be prealigned, in the order suggested by the SCORE program, i.e., the "Branching order." The prealigned sequence files may contain nonstandard attributes that have to be removed before executing the TREE program. The best method that the authors have found for achieving this goal is by using the EDT editor's INCLUDE command when combining sequence file, rather than just using the plain COPY command. This ensures that the extraneous file attributes are not included.

   2. Execute the TREE program. When prompted, enter the name of the input file. When prompted again, the user should provide his or her own names for the output files or accept the program's default names.

      TREE's output file, which ends with the extension "TRE," contains the final multiple alignment and the branch lengths for the construction of the phylogenetic tree.

d. The BLEN program: Occasionally, negative branch lengths may be obtained. To eliminate negative branch lengths, the order of the branches need to be modified and the lengths are recalculated using the BLEN program. To rearrange the order of the sequences in the main branching order (or in any relevant clusters), use TREE's output file, which ends with the extension "ALBR." The capital letters in this file represent the sequences, whereas the alphabetical order represents the branching order. After rearranging the letters to achieve the adjusted topology, execute the BLEN program using the modified .ALBR file as an input; the BLEN program will recalculate the branch lengths. This process may require few trials before the negative branch lengths are completely eliminated.

e. Construction of the phylogenetic tree: The TREEPLOT program, or any other drawing program on the user's personal computer, can be used to draw the phylogenetic tree. Use the branch lengths and the topology of the branches from the TREE or BLEN programs for this purpose. The branch lengths are numbered as r1, r2, r3, and so on, and correspond to the tree branches in a left-to-right and top-to-bottom order.

f. The MULPUB program: The MULPUB program can be used to format the aligned sequences and generate a pretty version of the alignment.

4. Notes

1. The methods for the progressive multiple alignment programs described here are oriented toward the implementation of these programs in the DNASYSTEM under the VAX/VMS environment. These programs are written in the "C" language and run as well under the UNIX system. For the implementation of these programs under the VAX/VMS system, front-end command files have been written to prompt the user for required information and for the handling of output files. The system then generates job files actually to execute the programs. The existence of such job files also enables the user to run the programs either in an interactive mode or in the background as batch files. Under the UNIX operating system, such job files are not required, although script files can be written to provide a "user-friendly" environment.

2. A set of PAPA (Parsimony After Progressive Alignment) programs, which have been recently described *(13)*, can be used for the construction of phylogenetic trees by an independent procedure and for examination of the topology of trees that have been generated by the distance matrix method described in this chapter *(14)*. These programs are also available for the IBM PC and compatibles computers.

References

1. Hafleigh, A. S. and Williams, C. A., Jr. (1966) Antigenic correspondence of serum albumins among the primates. *Science* **151,** 1530–1534.
2. Wilson, A. C., Kaplan, N. O., Levine, L., Pesce, A., Reichlin, M., and Allison, W. S. (1964) Evolution of lactic dehydrogenases. *Fed. Proceed.* **23,** 1258–1266.
3. McCarthy, B. J. and Bolton, E. T. (1963) An approach to the measurement of genetic relatedness among organisms. *Proc. Natl. Acad. Sci. USA* **50,** 156–164.
4. Hill, R. L., Buettner-Janusch, J., and Buettner-Janusch, V. (1963) Evolution of hemoglobin in primates. *Proc. Natl. Acad. Sci. USA* **50,** 885–893.
5. Hill, R. L. and Buettner-Janusch, J. (1964) Evolution of hemoglobin. *Fed. Proceed.* **23,** 1236–1242.
6. Feng, D.-F. and Doolittle, R. F. (1987) Progressive sequence alignment as a prerequisite to correct phylogenetic trees. *J. Mol. Evol.* **25,** 351–360.

7. Feng, D.-F. and Doolittle, R. F. (1990) Progressive alignment and phylogenetic tree construction of protein sequences, in *Methods in Enzymology*, vol. 183 (Doolittle, R. F., ed.), Academic Press, New York, pp. 375–387.

8. Doolittle, R. F. (ed.) (1990) Molecular evolution: Computer analysis of protein and nucleic acid sequences, in *Methods in Enzymology*, vol. 183. Academic Press, New York.

9. Reizer, A., Deutscher, J., Saier, M. H., Jr., and Reizer, J. (1991) Analysis of the gluconate (*gnt*) operon of *Bacillus subtilis. Mol. Microbiol.* **5,** 1081–1089.

10. Reizer, A., Pao, G. M., and Saier, M. H., Jr. (1991) Evolutionary relationships between the permease proteins of the bacterial phosphoenolpyruvate:sugar phosphotransferase system. Construction of phylogenetic trees and possible relatedness to proteins of eukaryotic mitochondria. *J. Mol. Evol.* **33,** 179–193.

11. Wu, L.-F., Reizer, A., Reizer, J., Cai, B., Tomich, J. M., and Saier, M. H., Jr. (1991) Nucleotide sequence of the *fruK* gene of *Rhodobacter capsulatus* encoding fructose-1-phosphate kinase: Evidence for a kinase superfamily including both phosphofructokinases of *E. coli. J. Bacteriol.* **173,** 3117–3127.

12. Vartak, N. B., Reizer, J., Reizer, A., Gripp, J. T., Groisman, E. A., Wu, L.-F., Tomich, J. M., and Saier, M. H., Jr. (1991) Sequence and evolution of the *FruR* protein of *Salmonella typhimurium:* a pleiotropic transcriptional regulatory protein possessing both activator and repressor functions which is homologous to periplasmic sugar binding proteins. *Res. Microbiol.* **142,** 951–963.

13. Doolittle, R.F. and Feng, D.-F. (1990) Nearest neighbor procedure for relating progessively aligned amino acid sequences, in *Methods in Enzymology*, vol. 183 (Doolittle, R. F., ed.), Academic Press, New York, pp. 659–669.

14. Reizer, J., Reizer, A., and Saier, M. H., Jr. (1993) The MIP family of integral membrane channel proteins: sequence comparisons, evolutionary relationships, reconstructed pathway of evolution, and proposed functional differentiation of the two repeated halves of the proteins. *Critical Rev. Biochem. Mol. Biol.* **28,** 235–257.

CHAPTER 27

The AMPS Package for Multiple
Protein Sequence Alignment

Geoffrey J. Barton

1. Introduction

AMPS (Alignment of Multiple Protein Sequences) is a set of programs for the multiple alignment of protein sequences. The programs build on the sequence comparison technique introduced in 1970 by Needleman and Wunsch *(1)*. The Needleman and Wunsch algorithm when supplied with two protein sequences and a matrix of scores for matching each possible pair or amino acids (e.g., the identity matrix, or Dayhoff's mutation data matrix *[2]*) calculates the best alignment between the two sequences and a score for the alignment. Although the Needleman and Wunsch algorithm may be used with virtually any pair of sequences, extending the method to more than three sequences is impractical even with today's fast computers. In order to cope with the alignment of large families of protein sequences, a heuristic approach was therefore developed *(3,4)*.

The protein sequences to be aligned are first compared pairwise using a conventional Needleman and Wunsch algorithm. Thus, for four sequences *A, B, C*, and *D* separate alignments of *A* and *B, A* and *C, A* and *D, B* and *C, B* and *D,* and *C* and *D* are made. For each of these alignments, various statistics are gathered (e.g., % identity), and this information is used to group the sequences using a conventional cluster analysis technique. The information gained from the cluster analysis enables a degree of confidence to be assigned to the alignments. Finally, a multiple sequence alignment is obtained either

From: *Methods in Molecular Biology, Vol. 25: Computer Analysis of Sequence Data, Part II*
Edited by: A. M. Griffin and H. G. Griffin Copyright ©1994 Humana Press Inc., Totowa, NJ

by following the tree produced by cluster analysis, or simply by aligning the sequences in order of their similarity (*see* Sections 5.1. and 5.2.). The result is a multiple sequence alignment that groups the most similar sequences together in the output. In order to assist in the interpretation of the alignment, conservation values *(5)* may be calculated for each position of the alignment to highlight the positions where the amino acids share similar physicochemical properties.

AMPS contains additional functions to allow the position of gaps in the alignment to be restricted *(6)*, and to allow specific residue weights to be assigned. AMPS also has features for flexible pattern matching and data bank scanning *(7)*, however, these aspects will not be discussed in this chapter.

2. Materials

The AMPS package may be supplied for Digital VAX computers running the VMS operating system, Sun Microsystems SPARC work stations, and Silicon Graphics Iris work stations running Unix. No special terminals or printers are required to use the programs, although a printer capable of using small fonts is useful if large multiple sequence alignments are to be studied. The package is not available for IBM PCs or Apple computers.

3. Methods
3.1. General

A multiple sequence alignment is obtained in three stages using the programs MULTALIGN and ORDER.

1. The program MULTALIGN is run in pairwise mode to calculate all pairwise alignments of the sequences, and the results are saved to a file. Randomizations may be carried out at this stage to give standard deviation (SD) scores for the comparisons and provide an estimate of confidence in the alignments *(4,6)*. Normally, only statistics for the alignments are saved; however, optionally, each pairwise sequence alignment may be written to a file if required.
2. The program ORDER is run to perform cluster analysis on the sequence scores obtained from running MULTALIGN in stage 1. An *order file* and *tree file* are written out, and a simple phylogenetic tree is drawn.
3. MULTALIGN is supplied with the *tree file* and *order file* and run in multiple mode, thus generating a multiple sequence alignment.

In order to make use of the programs, the user must first get to know how to use an editor on the computer. On a VAX/VMS system, the user might use the editor *EDT*, whereas under Unix, the most generally available, although awkward to use editor is called *vi*. The system manager should be able to help here. In the following description, the author will assume the user is familiar with the concept of computer files, knows how to use the editor on the computer system, and has found out where the AMPS package is located. All the examples that follow assume that a Unix system (e.g., Sun SPARCstation) is being used. There are small differences in the way files are defined when using a VAX, and these are described in Section 6.

3.1.1. Sequence File

A file containing the sequences to be aligned must first be created. An example file showing some of the possible formats (i.e., the way in which the information is laid out in the file) is illustrated in Fig. 1. The format for each sequence entry has three parts:

1. An identifier code prefixed by the > symbol and of up to nine characters in length.
2. Title line of up to 500 characters in length.
3. The one-letter code amino acid sequence terminated by a * symbol.

All the sequences to be aligned must be put in this format in one file. If the user has access to the NBRF-PIR sequence data base and the program PSQ, then the PSQ COPY/NOTEXT command will create files of a suitable format. MULTALIGN will only read the alphabetic characters from the sequence lines. Valid amino acid codes will be read in either upper or lower case; nonvalid codes (e.g., U) will be read as X (i.e., the unknown amino acid type). All other characters are ignored (e.g. %, 3, 7, blank, and so forth).

3.1.2. Pairscore Matrix File

Several pairscore matrix files are supplied with the program. An example is shown in Fig. 2. The file consists of a title line (Mutation Data Matrix [250 PAM] in Fig. 2), an amino acid index line (ARNDCQEGHILKMFPSTWYVBZX in Fig. 2), and a square matrix where each element represents the score associated with matching

```
>1fb4hv
1fb4h variable domain
EVQLVQSGGGVVQPGRSLRLSCSSSGFIFSSYAMYWVRQAPGKG
LEWVAIIWDDGSDQHYADS
VKGRFTISRNDSKNTLFLQMDSLRPEDTGVYFCARDGGHGFCSSASC*
>1fb4hc
1fb4h constant domain
F G P D Y W G Q G T P V T V S S A S T K G P S V F P L
A P S S K S (T S G G T A A) L G C L V K D Y F P Q P V T
V S W N S G A L T S G V H T F P A V
L Q S S G L Y S L S S V V T V P S S S L G T Q T Y I C N V N H K
P S N T K V D K R V E P K S C*
>1fb4lv
1fb4l variable domain
ESVLTQPPSASGTPGQRVTISCTGTSSNIGSITVNWYQQLPGMAPKLLIY
RDAMRPSGVPTRFSGSKSGTSASLAISGLEAEDESDYYCASWNSSDNSYVFGTGTKVTVL*
>1fb4lc
1fb4lv constant domain
G Q P K A N P T VTLFPPSSEELQANKATLVCLISDFYPGAVTVAWKA
DGSPVKAGVETTKPSKQSNNKYAASSYLSLTPEQWKSHRSYSCQVTHEGSTVEKTVAPTECS*
>3fabhv
Variable domain Sequence from 3fab
VQLEQSGPGLVRPSQT

LS
LTCTVSGTSFDDYYSTWVRQ
PPGRGLEWIGYVFYHGTS
dtdtplrsrv
TMLVNTSKNQFSLRLSSVTAADTAVYYCARN
LIAGCIDVWGQG*

>3fabhc
3fabh constant domain Sequence from 3fab
slvtvssastkgpsvfplapSSKSTSGGTAALGclvkdyfpepvtvswnsgal
tsgvhtfpavlqssglyslssvvtvpssslgtqtyicnvnhkpsntkvdkkvepksc*

>3fablv
3fabl variable domain from 3fab
SVLTQPPSVSGAPGQRVTISCTGSSSNIGAGNHVKWYQQLPGTAPKL
LIFHNNARFSVSKSGSSATLAITGLQAEDEADYYCQSYDRSLRVFGGGTKLTVLRQPKAAPSV*
>3fablc
3fabl constant domain from 3fab
TLFPPSSEELQANKATLVCLISDFYPGAVTVAWKADSSPVKAGVETTTPSKQSNNKYAASS
YLSLTPEQWKSHKSYSCQVTHEGSTVEKTVAPTECS
*
```

Fig. 1. Example sequence data file. A very fexible format is allowed, but all sequences must be stored in a single file in one-letter code.

the pair of amino acids indicated in the row and column. For example, the first row of the matrix in Fig. 2 gives the scores for matching A (Alanine) with each amino acid (e.g., AA=2, AC = -2). The second row gives scores for R (Arginine) vs each amino acid, and so on.

```
Mutation Data Matrix (250 PAMs)
ARNDCQEGHILKMFPSTWYVBZX
  2 -2  0  0 -2  0  0  1 -1 -1 -2 -1 -1 -4  1  1  1 -6 -3  0  0  0  0  0
 -2  6  0 -1 -4  1 -1 -3  2 -2 -3  3  0 -4  0  0 -1  2 -4 -2 -1  0  0  0
  0  0  2  2 -4  1  1  0  2 -2 -3  1 -2 -4 -1  1  0 -4 -2 -2  2  1  0  0
  0 -1  2  4 -5  2  3  1  1 -2 -4  0 -3 -6 -1  0  0 -7 -4 -2  3  3  0  0
 -2 -4 -4 -5 12 -5 -5 -3 -3 -2 -6 -5 -5 -4 -3  0 -2 -8  0 -2 -4 -5  0  0
  0  1  1  2 -5  4  2 -1  3 -2 -2  1 -1 -5  0 -1 -1 -5 -4 -2  1  3  0  0
  0 -1  1  3 -5  2  4  0  1 -2 -3  0 -2 -5 -1  0  0 -7 -4 -2  2  3  0  0
  1 -3  0  1 -3 -1  0  5 -2 -3 -4 -2 -3 -5 -1  1  0 -7 -5 -1  0 -1  0  0
 -1  2  2  1 -3  3  1 -2  6 -2 -2  0 -2 -2  0 -1 -1 -3  0 -2  1  2  0  0
 -1 -2 -2 -2 -2 -2 -2 -3 -2  5  2 -2  2  1 -2 -1  0 -5 -1  4 -2 -2  0  0
 -2 -3 -3 -4 -6 -2 -3 -4 -2  2  6 -3  4  2 -3 -3 -2 -2 -1  2 -3 -3  0  0
 -1  3  1  0 -5  1  0 -2  0 -2 -3  5  0 -5 -1  0  0 -3 -4 -2  1  0  0  0
 -1  0 -2 -3 -5 -1 -2 -3 -2  2  4  0  6  0 -2 -2 -1 -4 -2  2 -2 -2  0  0
 -4 -4 -4 -6 -4 -5 -5 -5 -2  1  2 -5  0  9 -5 -3 -3  0  7 -1 -5 -5  0  0
  1  0 -1 -1 -3  0 -1 -1  0 -2 -3 -1 -2 -5  6  1  0 -6 -5 -1 -1  0  0  0
  1  0  1  0  0 -1  0  1 -1 -1 -3  0 -2 -3  1  2  1 -2 -3 -1  0  0  0  0
  1 -1  0  0 -2 -1  0  0 -1  0 -2  0  1  3 -5 -3  0  0 -1  0  0  0  0  0
 -6  2 -4 -7 -8 -5 -7 -7 -3 -5 -2 -3 -4  0 -6 -2 -5 17  0 -6 -5 -6  0  0
 -3 -4 -2 -4  0 -4 -4 -5  0 -1 -1 -4 -2  7 -5 -3 -3  0 10 -2 -3 -4  0  0
  0 -2 -2 -2 -2 -2 -2 -1 -2  4  2 -2  2 -1 -1 -1  0 -6 -2  4 -2 -2  0  0
  0 -1  2  3 -4  1  2  0  1 -2 -3  1 -2 -5 -1  0  0 -5 -3 -2  2  2  0  0
  0  0  1  3 -5  3  3 -1  2 -2 -3  0 -2 -5  0  0 -1 -6 -4 -2  2  3  0  0
  0  0  0  0  0  0  0  0  0  0  0  0  0  0  0  0  0  0  0  0  0  0  0  0
```

Fig. 2. Example pairscore matrix file. Each element of the matrix gives the score for aligning the corresponding amino acids. (e.g., AA=2).

The pairscore matrix can be customized if desired or matrices generated, perhaps for a specific family of proteins. However, this usually gives little advantage over using the standard Mutation Data Matrix supplied with the program.

3.1.3. MULTALIGN Command File

As the name suggests, the *command file* contains instructions that tell the MULTALIGN program what to do, which sequence file to work with, and where to write the results. The command file consists of a number of lines, each of which consists of a key word followed by an equals sign (=), which may in turn be followed by one or more arguments. The commands may be given in upper- or lower-case or mixed case. For example, the commands:

> OUTPUT_FILE =junk.out
> output_file=junk.out
> OuTpUt_FiLe=junk.out

are all equivalent.

Command files must always start with the OUTPUT_FILE= command, optionally followed by the MODE= command. Most other commands may be specified in any order. There are many possible combinations of commands these will be illustrated with reference to the alignment of the eight immunoglobulin domains shown in Fig. 1.

4. Running a Pairwise Sequence Alignment

If it is assumed that the immunoglobulin domains shown in Fig. 1 are stored in a file called ig.seq, the following command file will instruct MULTALIGN to perform all possible pairwise alignments.

Command File: ig_pairs.com

```
output_file=ig_pairs.out
mode=pairwise
matrix_file=md.mat
pairwise_random=100
gap_penalty=8.0
constant=8
seq_file=ig.seq
```

output_file=ig_pairs.out tells MULTALIGN to write the results to the file ig_pairs.out. A command of this type must always be the first. (Note: On VAX/VMS systems, the output file should not be set to the log file name.) mode=pairwise specifies that pairwise alignments are to be performed. If this command is not included, the program defaults to mode=multiple.

matrix_file=md.mat defines the matrix file to be used in the comparisons. This file is the Dayhoff mutation data matrix or similar file containing values for the substitution of each possible amino acid pair (*see* Fig. 2 for an example).

pairwise_random=100 specifies that 100 randomizations of each sequence pair are to be performed in order to provide statistical SD scores. gap_penalty=8.0 defines the penalty for inserting a gap (amino acids aligned with one or more blanks) in the alignment to the value of 8.0. Experience shows that this value is adequate for virtually all protein comparisons (*see* Section 6.).

constant=8 is a constant added to the matrix_file to remove all negative elements. The reason for this is a little obscure; however, a value of 8 is suitable for the matrix in Fig. 2. seq_file=ig.seq specifies the file containing the sequences to be aligned pairwise.

This command file causes all pairwise comparisons to be performed on the sequences in the file ig.seq. In other words, for the eight sequences, sequence 1 is aligned with 2, 1 with 3, and so on. For *N* sequences, there are $N(N-1)/2$ comparisons performed. In order to execute the command file on a Unix system, simply type:

multalign < ig_pairs.com

For each sequence pair (e.g., 1 and 2), a full Needleman and Wunsch sequence comparison is performed. Then the sequences are shuffled and recompared (in this case 100 times) in order to find the expected distribution of scores that would be obtained if the sequences were unrelated, but have the same length and composition as 1 and 2. Various statistics on the comparisons are then output to the specified output_file (ig_pairs.out).

4.1. Results of a Pairwise Sequence Comparison Run

Figure 3 shows the output file (ig_pairs.out) that results from running MULTALIGN with the commad file ig_pairs.com. This file contains a great deal of useful information that is described in this section. However, if the principal aim is to generate a multiple sequence alignment, then skip to Section 5.

The file first shows the program name, its original source, and limitations of use. Next come the program limits. The important values to note are the Maximum allowed sequence length and Maximum number of sequences. The version of MULTALIGN shown will handle up to 500 sequences of 500 amino acids in length, but a version able to cope with 250 sequences of 1200 length is also distributed. Information regarding the parameters selected in the command file is then shown before listing information about the sequences present in the seq_file.

Index is a number used to refer to the sequence. Here, it is simply the order in the sequence data file (Fig. 1). *Identifier* shows the identifier from the sequence data file. *Length* gives the number of amino acids in the sequence. *Name* shows the title information from each sequence entry. *File order, Score,* and *with* should all be zero, since

```
                    M U L T A L I G N

                            by
                    Geoffrey J. Barton
            Laboratory of Molecular Biophysics
                 The Rex Richards Building
                    South Parks Road
                     OXFORD OX1 3QU
                        LONDON

                       Please cite:
       Barton, G. J. (1990) Methods in Enzymology 183, 403-428.
                   and also as appropriate:
   Multiple alignment: Barton, G. J. and Sternberg, M. J. E. (1987), J. Mol. Biol., 198, 327-337.
   Flexible Patterns: Barton, G. J. and Sternberg, M. J. E. (1990), J. Mol. Biol., 212, 389-402.
Structure Dependent Gap-Penalties: Barton, G. J. and Sternberg, M. J. E. (1987), Prot. Eng., 1, 89-94.

                        THANK YOU


        This copy of MULTALIGN is supplied for use on a single machine
    and must not be copied or redistributed without prior written consent from the author
                   Copyright (c) Geoffrey J Barton (1987)


Multalign Version D1.1  (6.2)
Maximum allowed sequence length  =   500
Maximum pattern length           =   500
Maximum number of sequences      =   500

matrix file read
Mutation Data Matrix (250 PAMs)

Pairwise randomizations defined
Min: 100 Max: 100 Increment:  1
Gap penalty :    8.00

Constant :   8

Sequence file : ig.seq


PAIRWISE mode has been defined

Checking minimal command set

No checking currently performed in this mode

Commands satisfy this mode..

Sequences defined in SEQ file:
```

Index	Identifier	Length	Name	File order	Score	with
1	>1fb4hv	110 1fb4h variable domain		0	0.00	0
2	>1fb4hc	119 1fb4h constant		0	0.00	0
3	>1fb4lv	110 1fb4l variable domain		0	0.00	0
4	>1fb4lc	106 1fb4lv constant domain		0	0.00	0
5	>3fabhv	109 3fabh variable domain Sequence from 3fab		0	0.00	0
6	>3fabhc	110 3fabh constant domain Sequence from 3fab		0	0.00	0
7	>3fablv	110 3fabl variable domain from 3fab		0	0.00	0
8	>3fablc	97 3fabl constant domain from 3fab		0	0.00	0

J	ILEN	JLEN	MATCH	NGAPS	NALIG	NIDENT	%IDENT	NAS	NASAL	RMEAN	STDEV	SCORE	
2	110	119	858.00	5	107	13	12.15	801.87	764.49	863.25	16.31	-0.32	1
3	110	106	959.00	4	103	27	26.21	931.07	900.96	840.96	14.48	8.15	2
4	110	106	841.00	5	101	16	15.84	832.67	793.07	818.16	16.03	1.43	3
5	110	109	1102.00	2	107	44	41.12	1029.91	1014.95	834.67	19.17	13.95	4
6	110	110	847.00	5	105	14	13.33	806.67	768.57	833.91	17.22	0.76	5
7	110	110	879.00	6	95	27	28.42	925.26	874.74	839.07	16.87	2.37	6
8	110	110	814.00	3	97	18	10.56	839.18	814.43	769.83	16.47	2.68	7
3	110	110	871.00	3	107	15	14.02	814.02	799.07	879.00	13.92	-0.57	8
4	119	106	996.00	6	102	36	35.29	976.47	937.25	853.69	14.34	9.93	9
5	119	109	861.00	21	106	21	19.81	812.26	766.98	866.24	16.28	-0.32	10
2	119	110	1258.00	0	110	106	96.36	1234.55	1234.55	883.65	15.68	30.24	11
2	119	107	974.00	4	107	20	18.69	816.82	801.87	876.35	15.09	-0.16	12
8	119	97	925.00	3	93	35	37.63	994.62	960.22	793.66	14.33	9.17	13
4	110	106	863.00	3	100	19	19.00	863.00	839.00	830.00	13.89	2.38	14
5	110	109	933.00	2	103	24	23.30	905.83	890.29	837.88	17.50	5.44	15
6	110	110	866.00	2	107	14	13.08	809.35	794.39	854.26	14.18	0.83	16
7	110	110	1147.00	3	100	73	73.00	1147.01	1123.00	849.66	14.95	19.88	17
8	110	97	841.00	2	97	20	17.53	867.01	850.52	779.99	14.78	4.13	18
5	106	109	817.00	5	100	14	14.00	817.00	777.00	815.07	15.96	0.12	19
6	106	110	996.00	3	102	34	33.33	976.47	952.94	833.30	16.52	9.85	20
7	106	110	840.00	0	99	23	23.23	848.48	816.16	829.63	13.63	0.76	21
8	106	97	1219.00	0	97	94	96.91	1256.70	1256.70	776.07	16.27	27.23	22
5	109	110	846.00	2	107	13	12.15	790.65	775.70	836.06	17.20	0.58	23
6	109	110	883.00	2	107	17	13.08	825.22	810.28	835.95	16.16	2.91	24
7	109	110	784.00	6	92	17	18.48	852.17	800.00	770.52	17.77	0.76	25
8	109	97	872.00	2	107	20	18.69	814.95	792.52	853.36	13.63	1.37	26
8	110	97	927.00	4	93	35	37.63	996.77	962.37	785.51	15.41	9.18	27
8	110	97	826.00	2	97	22	22.68	851.55	810.56	775.98	12.77	3.93	28

0.780

Times: TOTAL: 384.690 USER: 383.910 SYSTEM:

Fig. 3. Example output file from running MULTALIGN in pairwise mode.

these columns are only relevant after multiple sequence alignment (*see* Section 5.1.). There follows a list of the statistics obtained from all the pairwise sequence comparisons.

There are 16 columns headed I, J, ILEN, JLEN, MATCH, NGAPS, NALIG, NIDENT, %IDENT, NAS, NASAL, NRANS, RMEAN, STDEV, SCORE, and blank. I and J show the index numbers of the sequence compared in the row. For example, row one shows the comparison of sequence 1 (1fb4hv) with 2 (1fb4hc). ILEN and JLEN give the lengths of the sequences compared. MATCH is the raw match score obtained for the comparison of the two sequences. This value is not normalized; i.e., it increases as the length of the sequences compared increases. NGAPS is the number of internal gaps in the alignment (overhangs at the ends are not counted). NALIG shows the number of positions at which two amino acids are aligned (i.e., not aligned with a gap). NIDENT gives the number of positions at which identical amino acids are aligned. %IDENT shows the percentage identity (i.e., %IDENT/NALIG). NAS gives a Normalized Alignment score—the match score divided by the number of aligned positions $*$ 100. (MATCH/NALIG) $\times$ 100. NASAL shows an alternative Normalized Alignment score—the match score minus the number of gaps times the gap penalty all divided into the number of aligned positions. (MATCH/(MATCH-NGAPS) *gap_penalty)) $\times$ 100. NRANS records the number of randomizations performed for this pair. RMEAN shows the mean score for the randomizations. STDEV gives the standard deviation of the random scores. SCORE shows the SD score for the alignment, given by the mean random score minus the match score all divided by the standard deviation. (RMEAN-MATCH)/STDEV Column 16 simply numbers each pairwise comparison for reference purposes.

This output file is in the correct format for input by the program ORDER to perform cluster analysis prior to multiple sequence alignment. However, the actual alignments that give the statistics shown have not been output. If these are required, the user must include the command line(s): print horizontal= (for horizontal format) and/or print_vertical= (for vertical format). The alignment output formats are the same as those described in the following sections on multiple sequence alignment. Note, however, that a file containing alignments cannot be read directly by ORDER. (The alignments would first have to be deleted using a text editor.) Alternatively, the command: pair-wise_align_file=name, where name is a file name that does not already exist on the computer, will instruct MULTALIGN to write the sequence alignments to the specified file.

5. Multiple Sequence Alignment

The results of the pairwise sequence comparison run described in the previous section are used to guide the multiple sequence alignment. Two possible strategies for multiple alignment are supported by AMPS, *single order* and *tree*. Both methods require the program ORDER to be run on the results of the pairwise analysis in order to generate an order_file and for *tree* alignment, a tree_file.

5.1. Single-Order Alignment

To run the program ORDER, simply type order and follow the steps shown below. What is typed is shown in **bold**; **return** means press the Return key.

order

```
----------------------
Program O R D E R
----------------------
Processes MULTALIGN pairwise output
Author: Geoff Barton
!!!!!!!!!!!!!!!!!!!!!!!!!!!!!!!!!!!!!!!!!!
This program is supplied for use on a single machine and should not be
copied or redistributed without prior written consent from the author.
!!!!!!!!!!!!!!!!!!!!!!!!!!!!!!!!!!!!!!!!!!
Maximum Number of sequences allowed: 250
Enter pairwise filename: ig_pairs.out
Does the input file have timings? (Def=N) : return
-----------------------------------
number of lines read in = 28
-----------------------------------
ENTER FILENAME FOR COMPARISONS TO EXCLUDE [none] return
Write out prolog format file [N]: return
Options: NGAPS (1), PIDENT (2), NAS (3), NASAL (4), RSCORE (5) ?
Enter choice: 5
-----------------------------
Number items compared 8
-----------------------------
```

Number of identifiers 8

Produce an order file? [Y] **Y**
Enter filename for order **ig_pairs.ord**
Order file written
Reanalyze the data? [Y] **N**
Program ORDER ends

The terminal session shown uses the SD scores calculated from randomization to create an order file. If SD scores have not been calculated, then NAS, NASAL, or PIDENT scores can be used instead.

Having generated the order_file, MULTALIGN may be run to generate a multiple sequence alignment by supplying the program with the following command file:

Command File: ig_mult.com

output_file=ig_mult.out
mode=multiple
matrix_file=md.mat
gap_penalty=8.0
constant=8
seq_file=ig.seq
order_file=ig_pairs.ord
print_vertical=
print_horizontal=
consplot=mz

Note the changes from the file ig_pairs.com. The mode command is now set to multiple, the pairwise_random command is deleted, and four new commands are added: order-file=ig_pairs.ord tells the program to read the order_file generated by the program ORDER. print_vertical= and print_horizontal= instruct MULTALIGN to write out the sequence alignment in both vertical and horizontal formats, whereas consplot=mz asks for a conservation profile *(5)* to be generated for the alignment. MULTALIGN is run using the ig_mult.com file by typing:

multalign < ig_mult.com

Figure 4 (on pages 342 and 343) illustrates the output of the single-order multiple alignment of the eight immunoglobulin domains.

Single-order alignment starts by aligning the most similar pair of sequences, then aligns the next most similar sequence to that pair, then the next most similar sequence to the triple, and so on until all eight sequences are aligned. The *File order, Score,* and *with* columns show how the order of alignment was determined. For example, the sequence with the Index 3 (1fb4lc) is the eighth sequence in the original sequence file ig.seq and shows the greatest similarity of 9.93 SD with sequence 2 (1fb4h).

After the program descriptions, there follow two forms of multiple alignment output. The first is the result of the print_vertical= command, which prints out the sequences in columns. Alongside each aligned position in this alignment, the consplot=mz command has led to the generation of a conservation profile with values between 0 and 1, where 1 equals total identity, and high values show positions with conserved physicochemical properties (*see [5]* for details of conservation values). The conservation values are also shown graphically by a simple graph built up of crosses. Below the vertical format alignment, a more conventional horizontal format alignment is shown as the result of the print-horizontal= command. The number of residues printed per line can be varied by including the max_horiz= command. For example, if 60 residues per line were required, then putting max_horiz=60 in the MULTALIGN command file would have the desired effect.

5.2. Tree Alignment

Although the single-order alignment technique is very effective, a more general approach is to follow the branches of a phylogenetic tree (e.g., *see* Feng and Doolittle *[8]*). This method has the advantage of grouping the sequences into clusters, so that the most similar sequences are close together in the alignment. The AMPS package provides a straightforward procedure for generating alignments of this type.

As before, the ORDER program must be run.

order

Program O R D E R

Processes MULTALIGN pairwise output

Author: Geoff Barton

!!!
This program is supplied for use on a single machine and should not be
copied or redistributed without prior written consent from the author.
!!
Maximum Number of sequences allowed: 250
Enter pairwise filename: **ig_pairs.out**
Does the input file have timings? (Def=N): **return**
----
number of lines read in = 28

ENTER FILENAME FOR COMPARISONS TO EXCLUDE [none] **return**
Write out prolog format file [N]: **return**
Options: NGAPS (1), PIDENT (2), NAS (3), NASAL (4), RSCORE (5) ?
Enter choice: **5**

Number items compared 8

Number of identifiers 8

Produce an order file? [Y] **N**
Perform cluster analysis? [Y] **return**
Save full cluster details? [Y] **return**
Enter file to save cluster order: **ig_pairs.tord**
Enter file to save tree details: **ig_pairs.tree**
7 CLUSTERS GENERATED
8 ELEMENTS CONSIDERED
Enter plot filename: **ig_pairs.plot**
Please enter the axis title: **S. D. Score**
Please enter the plot title: **Ig Domains**
Enter page width required: **80**

Cluster analysis complete

Reanalyze the data? [Y] **N**
Program ORDER ends

This time NO is the answer to the option to produce an order file
and YES to cluster analysis. The file ig_pairs.tord will be similar, but
not usually identical to the ig_pairs.ord file generated in single-order
mode. This file tells MULTALIGN the order in which to output the
sequences once they are aligned. The new file required by MULTALIGN

```
Maximum allowed sequence length =    500
Maximum pattern length           =    500
Maximum number of sequences      =    500
-----------------
matrix file read
Mutation Data Matrix (250 PAMs)
-----------------------------
Gap penalty :     8.00
-----------------------------
Constant :     8
-----------------
Sequence file : ig.seq
-----------------------------

Order file : ig_pairs.ord
-----------------------------
Vertical format output selected
-----------------------------
Horizontal format output selected
-----------------------------
Conservation profile enabled
-----------------------------
MZ type conservation
-----------------------------

MULTIPLE mode has been defined

Checking minimal command set
-----------------------------
No iteration value defined - assume a value of 1
-----------------------------------------------
No number specified for horizontal width - 100 assumed
-----------------------------------------------

Commands satisfy this mode..
-----------------------------

Sequences defined in SEQ file:
-----------------------------
Index   Identifier   Length                Name                    File order   Score    with
    1    >1fb4hc     1191 fb4h constant domain                          2         0.00     0
    2    >3fabhc     1103 fabh constant domain Sequence from 3fab       6        30.24     2
    3    >1fb4lc     1061 fb4lv constant domain                         4         9.93     2
    4    >3fablc      973 fabl constant domain from 3fab                8        27.23     4
    5    >1fb4lv     1101 fb4l variable domain                          3         4.13     8
    6    >3fablv     1103 fabl variable domain from 3fab                7        19.88     3
    7    >1fb4hv     1101 fb4h variable domain                          1         8.15     3
    8    >3fabhv      109 Variable domain Sequence from 3fab            5        13.95     1
1*   ITERATION     1
. . . . .
. . . . .
PPPPPPGG     0.60  |+++++++  |
SSSSSSGP     0.50  |++++++   |
SSSSAVGG     0.50  |++++++   |
KKEESSVL     0.10  |++       |
SSEEGGVV     0.20  |+++      |
TTLLTAQR     0.20  |+++      |
SSQQPPPP     0.50  |++++++   |
GGAAGGGS     0.70  |++++++++ |
GGNNQQRQ     0.30  |+++++    |
TTKKRRST     0.40  |+++++    |
AAAAVVLL     0.60  |+++++++  |
AATTTTRS     0.30  |++++     |
LLLLIILL     0.90  |+++++++++|
GGVVSSST     0.50  |++++++   |
CCCCCCCC     1.00  |+++++++++|
LLLLTTST     0.40  |+++++    |
VVIIGGSV     0.40  |+++++    |
KKSSTSSS     0.40  |+++++    |
DDDDSSGG     0.40  |+++++    |
YYFFSSFT     0.40  |+++++    |

. . . . .
. . . . . Display shortened here
. . . . .

VVVVVL I     0.10  |++       |
DDEEFTHA     0.10  |++       |
```

Fig. 4. Result of single-order multiple alignment on data from Fig. 1. Note: The program header has been deleted to simplify the figure. The vertical format alignment has also been truncated to reduce the size of the figure.

```
KKKKGVGG    0.30  |++++        |
RKTTTLFC    0.20  |+++         |
VVVVGRCI    0.20  |+++         |
EEAATQSD    0.30  |++++        |
PPPPKPSV    0.10  |++          |
KKTTVKAW    0.20  |+++         |
SSEETASG    0.30  |++++        |
CCCCVACQ    0.40  |+++++       |
   SSLP G   0.00  |+           |
       S    0.20  |+++         |
       V    0.20  |+++         |
```

```
1  >1fb4hc    1  FGPDYWGQGTPVTVSSASTKGP SVFPLAPSSKSTSGGTAALGCLVKDYFPQPVT  VSW  NSGA  LTSGVHTFPAVLQ  SSGL   YS   LSS
2  >3fabhc    1     SLVTVSSASTKGP SVFPLAPSSKSTSGGTAALGCLVKDYFPEPVT  VSW  NSGA  LTSGVHTFPAVLQ  SSGL   YS   LSS
3  >1fb4lc    1  GQP KANP TVTLFPPSSEELQANKATLVCLISDFYPGAVT VAMKADGSP  VKAGVETTKP SKQ  SNNK   YA   ASS
4  >3fablc    1     TLFPPSSEELQANKATLVCLISDFYPGAVT VAMKADSSP  VKAGVETTTP SKQ  SNNK   YA   ASS
5  >1fb4lv    1  ESVLTQPPSASGTPGQRVTISCTGTSSNIGSIT VNWYQQLPG  MAPKLLIYRDAMR  PSGVPTRFSGSKSGTSA
6  >3fablv    1  SVLTQPPSVSGAPGQRVTISCTGSSSNIGAGNHVKNYQQLPG  TAPKLLIFHNNAR  FSVS  KSG   SSA
7  >1fb4hv    1  EVQLVQSGGGVVQPGRSLRLSCSSSGFIFSSYA MYWVRQAPGKGLEWVAI IMDDGSDQHYADSVKGRFTISRNDSKNT
8  >3fabhv    1  VQLEQSGPGLVRPSQTLSLTCTVSGTSFDDYY STWVRQPPGRGLEWIGYVFYHGTSDT DTPLRSRVTMLVNTSKNQ
```

```
1  >1fb4hc  101  VVTV PSSSLGTQTYICNVNHK      PSNTKVDKRVEPKSC
2  >3fabhc  101  VVTV PSSSLGTQTYICNVNHK      PSNTKVDKKVEPKSC
3  >1fb4lc  101  YLSLTPEQMKSHRSYSCQVTHE      GST  VEKTVAPTECS
4  >3fablc  101  YLSLTPEQMKSHKSYSCQVTHE      GST  VEKTVAPTECS
5  >1fb4lv  101  SLAISGLEAEDESDYYCASWNS      SDNSYVFGTGTKVTVL
6  >3fablv  101  TLAITGLQAEDEADYYCQSYDRSLRVFGGGTKLTVLRQPKAAPSV
7  >1fb4hv  101  LFLQMDSLRPEDTGVYFCARDGG      SGFCSSASC
8  >3fabhv  101  FSLRLSSVTAADTAVYYCARNL      IAGCIDVWGQG
```

```
Times: TOTAL:    2.390 USER :    2.130 SYSTEM:   0.260
```

Fig. 4 *(continued)*.

to perform tree alignments is ig_pairs.tree. In addition, a plot file is generated where the page width should be given as the largest number of characters that that printer will comfortably plot on one line.

The plot file ig_pairs.plot is illustrated in Fig. 5. The right-hand side of the file shows the numbers and identifying codes of the sequences aligned. From each sequence identifier, there extends a horizontal line that terminates in a cross alongside a similar cross at some point on a line from another sequence identifier. This point shows the level at which the two sequences are similar as indicated by the numbers running across the top of the figure. For example, 3fabhc and 1fb4hc are the most similar pair and are joined at a score of 30.55 SD, whereas 3fablc 1fb4lc is the next most similar pair (26.19 SD). These four sequences are then joined into a single cluster at 8.79 SD. With a little practice, the eye can readily pick out the major groupings within a set of sequences using plots of this kind. In Fig. 5, sequences 2, 4, 6, and 8 form one clear cluster (the Ig constant domains) and sequences 1, 3, 5, and 7 form another cluster (the Ig variable domains). The two clusters show only limited similarity to each other.

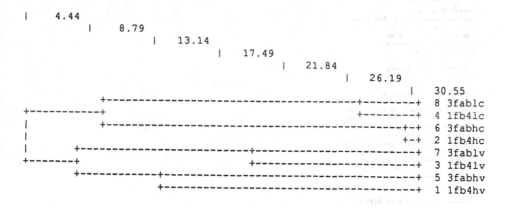

Fig. 5. Dendrogram or tree for data from Fig. 3. The identifying codes for the more similar sequences are shown clustered together.

A command file to perform tree multiple alignment follows:

Command File: ig_mult_tree.com

```
output_file=ig_mult_tree.out
mode=multiple
matrix_file=md.mat
gap_penalty=8.0
constant=8
seq_file=ig. seq
order_file=ig_pairs.tord
tree_file=ig_pairs.tree
print_vertical=
print_horizontal=
consplot=mz
```

The only difference between this file and the single-order command file ig_mult.com is the addition of the tree_file=command and the selection of ig_pairs.tord order file. The result of applying *tree* alignment to the Ig sequences is shown in Fig. 6. The principal differences between this alignment and that shown in Fig. 4 are the order in which the sequences are presented, and the details of the alignment between the Variable domains and Constant domains.

6. Notes

1. Gap penalties and sequence lengths. The gap penalty of eight used in the example is a good starting point. It should be increased (double or

```
Maximum allowed sequence length =    500
Maximum pattern length           =    500
Maximum number of sequences      =    500
-----------------
matrix file read
Mutation Data Matrix (250 PAMs)
-------------------------------
Gap penalty :     8.00
-----------------------
Constant :    8
---------------
Sequence file : ig.seq
-----------------------

Order file : ig_pairs.tord
---------------------------
Tree file defined: ig_pairs.tree
--------------------------------
Vertical format output selected
-------------------------------
Horizontal format output selected
---------------------------------
Conservation profile enabled
----------------------------
MZ type conservation
--------------------

MULTIPLE mode has been defined

Checking minimal command set
----------------------------
No iteration value defined - assume a value of 1
------------------------------------------------
No number specified for horizontal width - 100 assumed
------------------------------------------------------
Commands satisfy this mode..
----------------------------
----------------------------------
Sequences defined in SEQ file:
----------------------------------
```

Index	Identifier	Length	Name	File order	Score	with
1	>3fab1c	97	3fab1 constant domain from 3fab	8	0.00	0
2	>1fb41c	106	1fb4lv constant domain	4	0.00	0
3	>3fabhc	110	3fabh constant domain Sequence from 3fab	6	0.00	0
4	>1fb4hc	119	1fb4h constant domain	2	0.00	0
5	>3fab1v	110	3fab1 variable domain from 3fab	7	0.00	0
6	>1fb41v	110	1fb4l variable domain	3	0.00	0
7	>3fabhv	109	Variable domain Sequence from 3fab	5	0.00	0
8	>1fb4hv	110	1fb4h variable domain	1	0.00	0

```
1*  ITERATION   1
. . . . .

. . . . .
LLVVVVQQ    0.50 |++++++      |
FFFFLLLL    0.70 |++++++++    |
PPPPTTEV    0.20 |+++         |
  LLQQQQ    0.00 |+           |
  AAPPSS    0.00 |+           |
PPPPPPGG    0.60 |+++++++     |
SSSSSSPG    0.50 |++++++      |
SSSSVAGG    0.50 |++++++      |
EEKKSSLV    0.10 |++          |
EESSGGVV    0.20 |+++         |
LLTTATRQ    0.20 |+++         |
QQSSPPPP    0.50 |++++++      |
AAGGGGSG    0.70 |++++++++    |
NNGGQQQR    0.30 |++++        |
KKTTRRTS    0.40 |+++++       |
AAAAVVLL    0.60 |+++++++     |
TTAATTSR    0.30 |++++        |
LLLLIILL    0.90 |++++++++++  |
VVGGSSTS    0.50 |++++++      |
CCCCCCCC    1.00 |++++++++++++|
LLLLTTTS    0.40 |+++++       |
. . . . .

. . . . .
VVVVVVIG    0.60 |+++++++     |
EEDDFFAS    0.10 |++          |
KKKKGGGG    0.40 |+++++       |
TTKRGT F    0.00 |+           |
VVVVGGCC    0.70 |++++++++    |
AAEETTIS    0.20 |+++         |
PPPPKKDS    0.10 |++          |
TTKKLVVA    0.30 |++++        |
EESSTTMS    0.30 |++++        |
CCCCVVGC    0.70 |++++++++    |
SS LLQ      0.00 |+           |
```

Fig. 6. Example output file from tree multiple alignment using AMPS

```
R G    0.00  |+          |
Q      0.00  |+          |
P      0.10  |++         |
K      0.30  |++++       |
A      0.20  |+++        |
A      0.20  |+++        |
P      0.10  |++         |
S      0.20  |+++        |
V      0.20  |+++        |

1  >3fablc   1            T  LFP  PSSEELQANKATLVCLISDFYPGAVT VAMKADSSPVKAG V ETTTPSKQ    SNNKYAASSY
2  >1fb4lc   1       GQPKANPTVT  LFP  PSSEELQANKATLVCLISDFYPGAVT VAMKADGSPVKAG V ETTKPSKQ    SNNKYAASSY
3  >3fabhc   1            SLVTVSSASTKGPSVFPLAPSSKSTSGGTAALGCLVKDYFPEPVT VSN  NSGALTSG V ETTFPAVLQ   SSGLYSLSSV
4  >1fb4hc   1  FGPDYWGQQTPVTVSSASTKGPSVFPLAPSSKSTSGGTAALGCLVKDYFPQPVT VSN  NSGALTSG V ETTFPAVLQ   SSGLYSLSSV
5  >3fablv   1            S  VLTQPPSVSGAPGQRVTISCTGSSSNIGAGNIVKNYQQLPGTAPK L LIFHNN     ARFSVSKSGSSAT
6  >1fb4lv   1            ES VLTQPPSASGTPGQRVTISCTGTSSNIGSIT VNNYQQLPGHAPK L LIYRDAMR    PSGVPTRFSGSKSGTSAS
7  >3fabhv   1            V  QLEQSGPGLVRPSQTLSLTCTVSGTSFDDYY STHVRQPPGRGLENIGYVFYHGTDT DTPLRSRVTMLVNTSKNQ
8  >1fb4hv   1            EV QLVQSGGGVVQPGRSLRLSCSSSGFIFSSYA MYNVRQAPGKGLENVAIINDDGSDQHYADSVKGRFTISRNDSKNT


1  >3fablc  101  LSLTPEQNKSHKSYSCQVTHE GST VEKTVAPTECS
2  >1fb4lc  101  LSLTPEQNKSHRSYSCQVTHE GST VEKTVAPTECS
3  >3fabhc  101  VTV PSSSLGTQTYICNVNHKP SNTKVDKKVEPKSC
4  >1fb4hc  101  VTV PSSSLGTQTYICNVNHKP SNTKVDKRVEPKSC
5  >3fablv  101  LAITGLQAEDEADYYCQSYDRSLR  VTGGGTKLTVLRQPKAAPSV
6  >1fb4lv  101  LAISGLEAEDESDYYCASWNSSDNSYVFGTGTKVTVL
7  >3fabhv  101  FSLRLSSVTAADTAVYYCA  RNL   IAG CIDVWGQG
8  >1fb4hv  101  LFLQMDSLRPEDTGVYFCA  RDG   GHGFCSSASC

Times: TOTAL:   3.770 USER :   3.530 SYSTEM:   0.240
```

Fig. 6 *(continued)*.

treble) if sequences of very different lengths are being compared, since low gap-penalties can lead to an unrealistic "smearing" of short sequences over the length of a longer sequence. In general, this algorithm works best for sequences of about the same length that are expected to be similar over most of their length. It is therefore best to delete pro-sequences or extra domains prior to multiple alignment, if these are only present in some of the sequences to be aligned.

2. Selection of sequences for multiple alignment. Plots such as Fig. 5 can suggests which groups of sequences may be reliably aligned. For example, Barton and Sternberg *(4–6)* suggested that alignments giving pairwise SD scores > 6.0 would be correct within regions of core secondary structure. In Fig. 5, this would indicate that sequences 2, 4, 6, and 8 would be reliably aligned, and similarly 1, 3, 5, and 7, but the complete alignment of all eight Ig domains is likely to be subjet to errors. Such consideration of possible errors in alignment is vital when assessing the biological significance of any similarities found.

3. Quality of alignment. If sequences that cluster at > 6.0 SD are multiply aligned by AMPS, then the alignment is likely to be correct within the conserved core of the protein. However, it is important to remember that outside these regions, a sequence alignment may have no meaning at the residue-by-residue level. For example, a surface loop on a protein may adopt completely different conformations in two otherwise similar proteins. Clearly, any inference from a sequence alignment of

such variable regions will be very suspect. As a rule of thumb, it is best to treat parts of the multiple alignment that contain gaps with caution, while using the conservation profile as a guide to the most highly conserved, and hence structurally and functionally important parts of the sequences.

4. Time considerations. When a large number of sequences are to be aligned, performing randomizations for all pairs of sequences can be very time-consuming. A good guide to the relative similarity between the sequences is given by the normalized alignment scores NAS and NASAL. These do not require randomization in order to calculate and so considerably speed the process. Simply delete the line pairwise_random=100 from the command file to avoid performing the randomizations. When running ORDER, the user should then select option 3 rather than 5.

5. File name conventions. To avoid confusion between files and different projects, the user should adopt the following conventions for naming the various files used for alignment. For example, to start with a sequence file called PROTEIN.seq, we would use the word PROTEIN to identify the other files as follows:

Sequence file	PROTEIN.seq
Pairwise alignment command file	PROTEIN_pairs.com
Result of pairwise comparisons	PROTEIN_pairs.out
Single-order file	PROTEIN_pairs.ord
Tree order file	PROTEIN_pairs.tord
Tree file	PROTEIN_pairs.tree
Multiple alignment command file	PROTEIN_mult.com
Plot file (dendrogram)	PROTEIN_pairs.plot
Multiple alignment output	PROTEIN_mult.out

6. Single-order or tree method? In general, opt for the tree alignment method, since this groups the sequences more closely by similarity. However, this approach is slower than the single-order technique, and if the overall similarity between the sequences is very high, it offers no advantage.

7. Aligning nucleic acid sequences. Although developed for protein sequences, AMPS can also align DNA/RNA sequences. A special DNA matrix is supplied for this purpose.

8. Possible disasters. A common mistake is to leave blank lines in the command file for MULTALIGN or forget the "=" sign in a command. MULTALIGN will write an error message to the output file and stop if this occurs.

Another common mistake is to forget to put either the print_vertical= or print_horizontal= commands in the multiple alignment command file.

The program will then happily compute the alignment, but not write it out!

For alignments that are expected to take a long time, it is always best to check that all the parameters have been correctly read, by interrupting the program after a few minutes, examining the output_file so far, and then rerunning the program for the full period.

A more serious error is to attempt to align sequences that are close to the maximum length limits of the program. Sequences that are too long on input are ignored, but if during alignment the alignment grows to a length greater than that allowed, an unexpected program crash may result.

9. VAX/VMS differences. For VAX/VMS systems, the example command files shown in this chapter need two extra lines at the top:

```
$set def directory
$multalign
```

where directory is the name of the directory in which the user is working. The output_file should not be set to the same name as the VMS log file. The output_file will overwrite any file of the same name present in the working directory. This is unlike normal VAX/VMS practice where old versions of a file are retained, so if the user wants to keep an old version, he or she should remember to rename it before running MULTALIGN.

References

1. Needleman, S. B. and Wunsch, C. D. (1970) A general method applicable to the search for similarities in the amino acid sequence of two proteins. *J. Mol. Biol.* **48,** 443–453.
2. Dayhoff, M. O., Schwartz, R. M., and Orcutt, B. C. (1978) A model of evolutionary change in proteins. Matrices for detecting distant relationships, in *Atlas of Protein Sequence and Structure,* vol. 5 (Dayhoff, M. O., ed.), National Biomedical Research Foundation, Washington, DC, pp. 345–358.
3. Barton, G. J. (1990) Protein multiple sequence alignment and flexible pattern matching. *Methods Enzymol.* **183,** 403–428.
4. Barton, G. J. and Sternberg, M. J. E. (1987) A strategy for the rapid multiple alignment of protein sequences: Confidence levels from tertiary structure comparisons. *J Mol. Biol.* **198,** 327–337.
5. Zvelebil, M. J. J. M., Barton, G. J., Taylor, W. R., and Sternberg, M. J. E. (1987) Prediction of protein secondary structure and active sites using the alignment of homologous sequences. *J. Mol. Biol.* **195,** 957–961.
6. Barton, G. J. and Sternberg, M. J. E. (1987) Evalution and improvements in the automatic alignment of protein sequences. *Protein Eng.* **1,** 89–94.
7. Barton, G. J. and Sternberg, M. J. E. (1990) Flexible protein sequence patterns—a sensitive method to detect weak structural similarities. *J. Mol. Biol.* **212,** 389–402.
8. Feng, D. F. and Doolittle, R. F. (1987) Progressive sequence alignment as a prerequisite to correct phylogenetic trees. *J. Mol. Evol.* **25,** 351–360.

CHAPTER 28

TreeAlign

Jotun Hein

1. Introduction

Several approaches to the multiple alignment problem are conceivable, but virtually all are based on the Parsimony Principle: Choose the history of a set of sequences that minimizes the overall amount of change (insertion-deletions and substitutions/mutations). This formulation was first presented by Sankoff *(1)* together with a dynamic programming algorithm solving the problem. However, this was so slow that all programs in use have opted for heuristic, but much faster algorithms, and so has TreeAlign. Given the large growth in the data involving sets of homologous sequences, a good solution to this problem has high priority.

The objective for TreeAlign is to find a history of a set of sequences that minimizes the weighted sum of indels and substitutons. Indels will be weighted by $g_k = (a + b*k)$, where k is the length of the indel, k its length, and a and b are positive user-specified constants. Substitutions will be weighted by a user-specified matrix that reflects the likelihood of different types of substitutions. The more frequent the event, the less weight (penalty) it will be assigned.

In analyzing a set of homologous sequences, treating the alignment problem and the phylogeny problem is important if the frequency of insertion-deletions is not very, very low. Since phylogenetic alignment is very convenient for the user, in giving a very complete analysis from undigested sequence data, it has become very popular. Simultaneously with this development, the methods used for phylogenetic analysis have had a tendency toward statistically more well-founded methods, such

From: *Methods in Molecular Biology, Vol. 25: Computer Analysis of Sequence Data, Part II*
Edited by: A. M. Griffin and H. G. Griffin Copyright ©1994 Humana Press Inc., Totowa, NJ

as maximum likelihood *(2)* and parsimony evaluated by bootstrapping *(3)*. Unfortunately, both these methods are inapplicable to unaligned data. Maximum likelihood is inapplicable because nobody has analyzed many-sequence models of sequence evolution involving both substitutions and insertion-deletions. Parsimony with bootstrapping is inapplicable because it presumes the data are prealigned, i.e., a series of columns of nucleotides (additionally assumed independent given their common phylogeny), but the data are unaligned in the phylogenetic alignment problem. The popularity of phylogenetic alignment programs has an important lesson for the statistical methods: Prealigned data have been manipulated in a way that could seriously bias the data. If the method of prealignment was phylogenetic, it would definitely bias the data toward the tree used in the alignment.

This makes the development of statistical methods of phylogenetic alignment imperative. With the continued increase in computational power available, the development of statistical pairwise alignment *(4)* and algorithms reducing the computations in the exact phylogenetic alignment problem *(5)* gives hope for progress in this area.

Until such methods have been developed, computer simulations could provide some guide to the reliability of the phylogeny provided by phylogenetic alignment.

2. The Method

TreeAlign goes through the following steps.

1. Read Input and Input Parameters, and check if they correspond to required formats. If not, halt. If the tree is specified by the user, go directly to 4.
2. Calculate distance between pairs of sequences using pairwise alignment algorithm.
3. Construct tree from pairwise distances.
4. Given this tree, find ancestral sequences and which events have occurred on each edge in the tree connecting the sequences.
   a. Choose arbitrary root in tree.
   b. Going from tips toward the root, align sequences, reconstruct possible ancestral sequences, align these ancestral sequences, and so forth. This results in a set of possible solutions.
   c. Choose one solution only.

5. Calculate relevant quantities and write them:

Phylogeny with branch lengths

Multiple alignment with/without ancestral sequences.

Tables with substitutions and insertion-deletions.

Histogram of mutability of sites if more than 10 sequences are analyzed.

There are two facets to the phylogenetic alignment algorithm: Reconstructing the tree and the multiple alignment following that tree. The tree reconstructing method used is based on pairwise distances and does not assume a molecular clock (as does, for instance, UGPMA). The method is economical in the number of pairwise comparisons needed for a large number of sequences.

The main novelty is in the multiple alignment algorithm, which occurs in Section 2., step 4b. The traditional dynamic programming algorithm (Fig. 1) *(6)* takes two sequences, and finds one or more alignments. The alignments will imply a set of possible ancestral sequences associated with these alignments, which will be assigned to the ancestral node representing the aligned sequences. If there were only one ancestral sequence, then the traditional dynamical programming algorithm could be used up through the tree. However, frequently there are more. This problem can be circumvented by a more general algorithm, which will not be described here *(7)*. The method can only assign sequences to a node that seems minimal, knowing only the sequences at the daughter nodes. There is no guarantee that this gives a most parsimonious assignment to the complete tree, so the algorithm is heuristic.

3. An Example

TreeAlign is written in standard C *(8)* on 11 modules and is about 4500 lines. On a SUN 4, it can analyze many (>>300) and long (>10.000) sequences. It used <20 min on 18 retroviruses (10.000 bp) and <2 min to analyze 219 globins (150 amino acids).

TreeAlign always needs three input files:

"mut.dat" contains a distance matrix on amino acids or cost of mutating an amino acid to another amino acid. This file is supplied with TreeAlign.

"par.dat" (top of Fig. 2) contains user-specified parameters necessary to run TreeAlign.

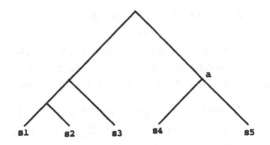

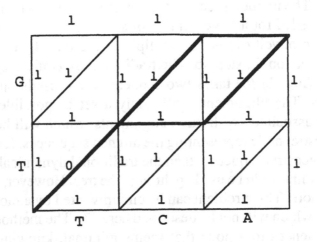

Minimal alignments:

i)   TCA   ii)   TCA
        TG-   T-G

Ancestral Sequences:

i) TC, TGA   ii)   TA, TCG

Fig. 1. Five sequences (s1–s5) are related by the tree shown. Assume that s4 and s5 are to be aligned and the ancestral sequence at "a" is to be reconstructed. One way to proceed is to align them by a pairwise alignment method and then find which ancestral sequences were possible according to the obtained alignment(s).

The best possible alignment of two sequences can be reformulated as finding the shortest path from (0,0) to (3,2) in the graph shown (1.b). The diagonal edges are assigned weight 1 if they match different nucleotides and 0 if identical nucleotides. Horizontal and vertical edges are all assigned weight 1, since they correspond to insertion-deletions. In this case, there will be two equally good alignments. For every alignment, there is a set of potential ancestral sequences.

Par.dat:

```
1  10  8  3
1 0
glob.seq
glob.ali
glob.tree
user.tree
```

Above parameters mean:
0(DNA)/1(Protein)   number of sequences (10)   gap penalty for gap k long:8+3*k
1(Ancestral sequences)/0(Not A.S.)   1(Userspecified Phylogeny)/0(Not U.P)
Filename of sequence file
Filename of output file with alignment
Filename of output file with phylogeny
Filename of file with userspecified phylogeny

Glob.seq (beginning only):

```
>
EMOGLOBIN ALPHA CHAIN - HUMAN
 V L S P A D K T N V K A A W G K V G A H A G E Y G A E A L E
 R M F L S F P T T K T Y F P H F D L S H G S A Q V K G H G K
 K V A D A L T N A V A H V D D M P N A L S A L S D L H A H K
 L R V D P V N F K L L S H C L L V T L A A H L P A E F T P A
 V H A S L D K F L A S V S T V L T S K Y R *

>
EMOGLOBIN BETA CHAIN - HUMAN
 V H L T P E E K S A V T A L W G K V N V D E V G G E A L G R
 L L V V Y P W T Q R F F E S F G D L S T P D A V M G N P K V
 K A H G K K V L G A F S D G L A H L D N L K G T F A T L S E
 L H C D K L H V D P E N F R L L G N V L V C V L A H H F G K
 E F T P P V Q A A Y Q K V V A G V A N A L A H K Y H *
```

Fig. 2. (Top) The content of the files containing parameters needed to run the program. It is called "par.dat" and must always be present in the directory. The first six lines contains the parameters, whereas the last six lines contain an explanation of the these parameters. (Bottom) The sequence file is in the NBRF format. All the sequences to be analyzed must be on the same file. A ">" at the first position indicates that a sequence entry is beginning, the next line is an entry descriptor, and the sequence then begins and is continued until a "*" is encountered. Only the first two entries of more entries are shown here.

"glob.seq" (Fig. 2 bottom) is named in "par.dat" and has the sequences to be analyzed.

The result of the analysis is written on two files (named in "par.dat"): "glob.ali" and "glob.tree".

"glob.ali" (Fig. 3) contains distance and percentwise (in multiple alignment) similarity between pairs of sequences. It contains the distance between amino acids (or nucleotides) used in comparing the

```
RELATIONSHIP BETWEEN PAIRS OF SEQUENCES:
   lower triangle distance between pairs
   upper triangle percent homology in alignment
        0     1     2     3     4     5     6     7     8     9

0             44    27    13    57    37    26    28    19    17
1      214          25    16    41    57    31    21    19    16
2      323   335          17    31    28    69    19    14    15
3      370   377   351          19    18    15    13    16    42
4      132   234   309   368          34    32    29    20    15
5      263   175   338   381   272          28    18    15    13
6      324   316    96   371   308   325          22    15    17
7      340   355   388   375   339   372   394          42    13
8      354   375   386   376   347   384   386   220          15
9      362   370   378   230   368   378   393   368   378
```

weight of insertion-deletion of length k: 8 + k*3

TABLES OF GENETIC EVENTS:

   lower triangle: symmetrisized substitutions
   upper triangle: mutation distance matrix

```
    C  S  T  P  A  G  N  D  E  Q  H  R  K  M  I  L  V  F  Y  W  B

C      2  4  4  4  3  4  5  6  5  4  4  6  4  4  4  4  3  3  3  4

S  2      1  2  1  1  1  2  3  3  3  3  3  4  4  4  2  3  3  4  1

T  1 11      2  1  2  2  3  3  3  4  3  2  3  3  4  2  4  4  5  2

P  0  3  2      1  2  4  3  3  3  3  4  4  4  3  2  3  4  4  3

A  3 32 25 19      1  3  2  2  3  4  4  3  3  4  4  1  3  4  4  2

G  0 15  6  3 26      3  2  2  4  5  3  4  5  4  4  2  4  4  3  2

N  0  5  6  0  5  4      1  2  3  2  3  2  5  4  5  4  4  3  6  0

D  0  9  1  2  5  5 16      1  2  3  4  3  4  5  5  3  5  4  6  1

E  0  4  5  3 19  3  2 24      2  4  3  2  4  5  5  2  4  5  5  0

Q  0  3  2  2  3  1  1  3 11      2  3  2  4  5  4  4  5  4  5  2

H  1  5  2  1  2  1  5  3  0  7      2  3  4  4  3  5  4  3  5  2

R  0  4  2  0  0  1  1  0  1  1  4      1  4  4  4  4  5  5  4  3

K  0 12 16  1 15  3  8  3 11  9  4 13      4  4  4  3  5  5  5  2

M  1  0  2  0  1  0  0  0  0  0  1  2      2  1  2  3  4  3  4

I  2  2  3  0  6  0  1  0  1  0  0  1  0  1      1  4  2  3  3  4

L  0  5  1  2  8  1  1  0  0  2  7  0  6 11 22      4  2  3  2  5

V  1  5  7  3 10  3  0  0  5  0  1  1  7  4 16  9      2  3  3  2

F  0  3  1  0  0  0  0  0  1  0  1  0  1  2  5 17  9      1  3  4

Y  0  0  1  0  0  0  2  0  0  0  2  1  1  1  0  6  1  8      3  3

W  0  0  0  0  0  0  0  0  1  0  0  1  0  0  3  1  2  0      5

B  0  0  0  0  0  0  0  0  0  0  0  0  0  0  0  0  0  0  0  0
```

INSERTION-DELETIONS :

```
lengths:  1  2  3  4  5  6
numbers:  9  5  3  1  1  3
```

total weight of history: 1771

Fig. 3. The first part of the alignment output file contains statistics of the analysis. The first matrix shows pairwise distances and percent homology between two sequences. Typically, a small distance will imply a large percentage homology, which is readily confirmed. For instance, the smallest distance, 96, is between the

proteins (DNA sequences) and how many of each type of mutations has occurred according to the analysis.

The most important information on "glob.ali" is the alignment of the sequences, the ancestral sequences, and the mutability histogram (Fig. 4). The more conserved a position in the molecule is, the more "#"s will appear above it, with nine as a maximum, five as an average, and zero "#"s for the highly mutable site.

After the last block in the alignment, the names of the input sequences are written. The sequences with a number, but without a name are the ancestral sequences. There is always one ancestral sequence less than the number of input sequences. Since ancestral sequences are not always of interest, it is optional if they are to be part of the alignment or not. To interpret the alignment in evolutionary terms, the phylogeny on "glob.tree" (Fig. 5) (named on "par.dat") must be consulted. The ancestral sequences correspond to internal nodes in the phylogeny. The events that have occurred on an edge can be found by comparing the sequences labeling the nodes at the ends of the edge. The weighted sum of all the events obtained in this way is proportional to the length of the edge. The phylogeny on "glob.tree" is written both in a standard form that is readable by phylogeny drawing programs in PHYLIP *(9)* and as a rooted tree with edge lengths proportional to vertical edge lengths. The phylogeny has been arbitrarily rooted at the middle point of the longest path in the tree. Parsimony trees seriously underestimate the length of longer branches. The true root of this tree is most likely at the edge leading to the plant globins. The unrootedness of parsimony trees is often circumvented by adding an outgroup sequence to the data set, so the attachment site of this sequence to the original data set roots the original data set. While running, it will give a series of messages to the screen (Fig. 6) describing the analysis performed.

---

two myoglobins, which also has the highest percent homology, 69. The second matrix contains the distance function on amino acids (upper half) and the number of mutations observed. Here it is seen that mutations penalized heavily are rare. For instance, the most frequent mutation (32 times) is an Alanine mutating to a Serine or visa versa, and has the lowest possible penalty, 1. The total weight of the proposed history is 1771, which is both the total length of the branches in the phylogeny and the weighted sum of the mutations (insertion-deletions included).

ALIGNMENT OF SEQUENCES:

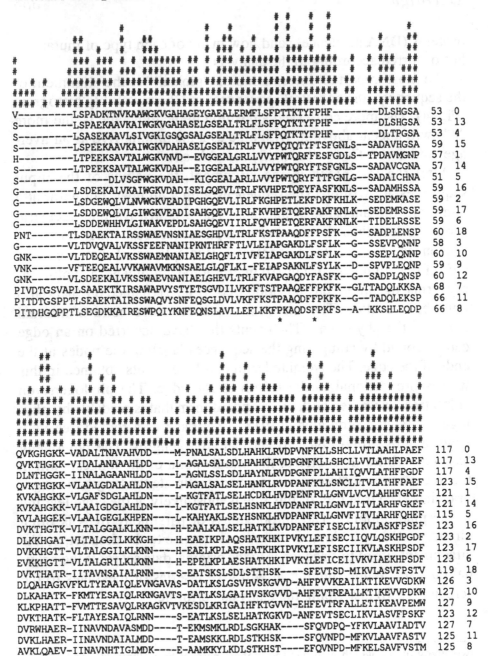

```
                          ##                #  # #
                #         #                 ###
                ##       #  ##      # #  #  #  #  # #            #
             ## ###  ### #  ### #  ## ##  #  # ### #  #  # ###   ##
        #    # #  #  ## # ### #####  ## ## ### #### ####   # #  ####
        #   ##  ############# ## ### ############# ##### ######### ####
        # #  ######  ## ####### ## ###### ########## ####### ######### ###
       #### # ########################################################### ###
       #### # ##########################################################################
V---------LSPADKTNVKAAWGKVGAHAGEYGAEALERMFLSFPTTKTYFPHF---------DLSHGSA   53   0
S---------LSPAEKAAVKAIWGKVGAHASELGSEALTRLFLSFPQTKTYFPHF---------DLSHGSA   53  13
S---------LSASEKAAVLSIVGKIGSQGSALGSEALTRLFLSFPQTKTYFPHF---------DLTPGSA   53   4
S---------LSPEEKAAVKAIWGKVDAHASELGSEALTRLFVVYPQTQTYFTSFGNLS--SADAVHGSA   59  15
H---------LTPEEKSAVTALWGKVNVD--EVGGEALGRLLVVYPWTQRFFESFGDLS--TPDAVMGNP   57   1
S---------LTPEEKSAVTALWGKVDAH--EIGGEALARLLVVYPWTQRYFTSFGNLS--SADAVCGNA   57  14
S------------DLVSGFWGKVDAH--KIGGEALARLLVVYPWTQRYFTTFGNLG--SADAICHNA   51   5
G---------LSDEEKALVKAIWGKVDADISELGQEVLTRLFKVHPETQEYFASFKNLS--SADAMHSSA   59  16
G---------LSDGEWQLVLNVWGKVEADIPGHGQEVLIRLFKGHPETLEKFDKFKHLK--SEDEMKASE   59   2
G---------LSDDEWQLVLGIWGKVEADISAHGQEVLIRLFKVHPETQERFAKFKNLK--SEDEMRSSE   59  17
G---------LSDDEWHHVLGIWAKVEPDLSAHGQEVIIRLFQVHPETQERFAKFKNLK--TIDELRSSE   59   6
PNT-------TLSDAEKTAIRSSWAEVNSNIAESGHDVLTRLFKSTPAAQDFFPSFK--G--SADPLENSP  60  18
G---------VLTDVQVALVKSSFEEFNANIPKNTHRFFTLVLEIAPGAKDLFSFLK--G--SSEVPQNNP  58   3
GNK-------VLTDEQEALVKSSWAEMNANIAELGHQFLTIVFEIAPGAKDLFSFLK--G--SSEPLQNNP  60  10
VNK-------VFTEEQEALVVKAWAVMKKNSAELGLQFLKI-FEIAPSAKNLFSYLK--D--SPVPLEQNP  59   9
GNK-------VLSDEEKALVKSSWAEVNANIAELGHEVLTRLFKVAPGAQDYFASFK--G--SADPLQNSP  60  12
PIVDTGSVAPLSAAEKTKIRSAWAPVYSTYETSGVDILVKFFTSTPAAQEFFPKFK--GLTTADQLKKSA  68   7
PITDTGSPPTLSEAEKTAIRSSWAQVYSNFEQSGLDVLVKFFKSTPAAQDFFPKFK--G--TADQLEKSP  66  11
PITDHGQPPTLSEGDKKAIRESWPQIYKNFEQNSLAVLLEFLKKFPKAQDSFPKFS--A--KKSHLEQDP  66   8
                                                           *     *
```

```
         ##                     #              #             #
         ##                     # #           # #          ### #
         ##   #  #              ### # ###     ## #          ### #  #
       # ## # #  #         #  # ### ## ## ##  ## ##        ### #  #
       ######## ###### #   ### ## ############# ###### ##  ########### #
       ######## ####### ## ### ## ############### ####### # ###########
       ######## ####### ## ### ### ############################# ########
       ######## ####### ## ### ### ##################################### #
       ######## #################### ################################# ###
       ####################### ### #### ##### ######################### ##
QVKGHGKK-VADALTNAVAHVDD----M-PNALSALSDLHAHKLRVDPVNFKLLSHCLLVTLAAHLPAEF   117   0
QVKTHGKK-VIDALANAAAHLDD----L-AGALSALSDLHAHKLRVDPGNFKLLSHCLLVVLATHFPAEF   117  13
DLNTHGGK-IINALAGAANHLDD----L-AGNLSSLSDLHAYNLRVDPGNFPLLAHIIQVVLATHFPGDF   117   4
QVKTHGKK-VLAALGDALAHLDN----L-AGALSALSELHANKLRVDPANFKLLSNCLITVLATHFPAEF   123  15
KVKAHGKK-VLGAFSDGLAHLDN----L-KGTFATLSELHCDKLHVDPENFRLLGNVLVCVLAHHFGKEF   121   1
KVKAHGKK-VLAAIGDGLAHLDN----L-KGTFATLSELHSNKLHVDPANFRLLGNVLITVLARHFGKEF   121  14
KVLAHGEK-VLAAIGEGLKHPEN----L-KAHYAKLSEYHSNKLHVDPANFRLLGNVFITVLARHFQHEF   115   5
DVKTHGTK-VLTALGGALKLKNN----H-EAALKALSELHATKLRVDPANFEFISECLIKVLASKFPSEF   123  16
DLKKHGAT-VLTALGGILKKKGH----H-EAEIKPLAQSHATKHKIPVKYLEFISECIIQVLQSKHPGDF   123   2
DVKKHGTT-VLTALGGILKLKNN----H-EAELKPLAESHATKHKIPVKYLEFISECIIKVLASKHPSDF   123  17
EVKKHGTT-VLTALGRILKLKNN----H-EPELKPLAESHATKHKIPVKYLEFICEIIVKVIAEKHPSDF   123   6
DVKTHATR-IITAVNSAIALRNN----S-EATSKSLSDLSTTHSK----SFEVTSD-MIKVLASVFPSTV   119  18
DLQAHAGKVFKLTYEAAIQLEVNGAVAS-DATLKSLGSVHVSKGVVD-AHFPVVKEAILKTIKEVVGDKW   126   3
DLKAHATK-FKMTYESAIQLRKNGAVTS-EATLKSLGAIHVSKGVVD-AHFEVTREALLKTIKEVVPDKW   127  10
KLKPHATT-FVMTTESAVQLRKAGKVTVKESDLKRIGAIHFKTGVVN-EHFEVTRFALLETIKEAVPEMW   127   9
DVKPHATK-FLTAYESAIQLRNN----S-EATLKSLSELHATKGKVD-ANFEVTSECLIKVLASVFPSKF   123  12
DVRWHAER-IINAVNDAVASMDD----T-EKMSMKLRDLSGKHAK----SFQVDPQ-YFKVLAAVIADTV   127   7
DVKLHAER-IINAVNDAIALMDD----T-EAMSKKLRDLSTKHSK----SFQVNPD-MFKVLAAVFASTV   125  11
AVKLQAEV-IINAVNHTIGLMDK----E-AAMKKYLKDLSTKHST----EFQVNPD-MFKELSAVFVSTM   125   8
```

Fig. 4. Second part of the alignment output file containing the actual alignment. The second from last column of integers contains the number of the last residue in the alignment of that sequence. The last columns are the number of the sequences in the phylogeny. Ancestral sequences are included in this alignment. Fully conserved positions are marked with "*", and above the alignment is a histogram that indicates how mutable the position is.

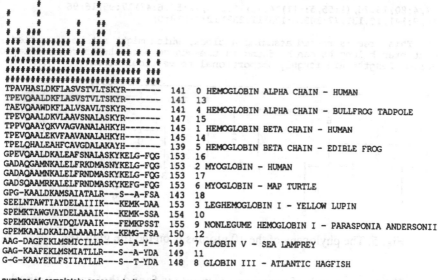

```
TPAVHASLDKFLASVSTVLTSKYR-------   141   0 HEMOGLOBIN ALPHA CHAIN - HUMAN
TPEVQAALDKFLASVSTVLTSKYR-------   141   13
TAEVQAAWDKFLALVSAVLTSKYR-------   141   4 HEMOGLOBIN ALPHA CHAIN - BULLFROG TADPOLE
TPEVQAALDKVLAAVSNALASKYR-------   147   15
TPPVQAAYQKVVAGVANALAHKYH-------   145   1 HEMOGLOBIN BETA CHAIN - HUMAN
TPEVQAALEKVFAAVANALAHKYH-------   145   14
TPELQHALEAHFCAVGDALAKAYH-------   139   5 HEMOGLOBIN BETA CHAIN - EDIBLE FROG
GPEVQAALDKALEAFSNALASKYKELG-FQG   153   16
GADAQGAMNKALELFRKDMASNYKELG-FQG   153   2 MYOGLOBIN - HUMAN
GADAQAAMNKALELFRNDMASKYKELG-FQG   153   17
GADSQAAMRKALELFRNDMASKYKEFG-FQG   153   6 MYOGLOBIN - MAP TURTLE
GPG-KAALDKAMSAIATALR---S--A-FSA   143   18
SEELNTAWTIAYDELAIIIK---KEMK-DAA   153   3 LEGHEMOGLOBIN I - YELLOW LUPIN
SPEMKTAWGVAYDELAAAIK---KEMK-SSA   154   10
SPEMKNAWGVAYDQLVAAIK---FEMKPSST   155   9 NONLEGUME HEMOGLOBIN I - PARASPONIA ANDERSONII
GPEMKAALDKALDALAAALK---KEMG-FSA   150   12
AAG-DAGFEKLMSMICILLR---S--A-Y--   149   7 GLOBIN V - SEA LAMPREY
GAG-KAAFEKLMSMIATLLR---S--A-YDA   149   11
G-G-KAAYEKLFSIIATLLR---S--T-YDA   148   8 GLOBIN III - ATLANTIC HAGFISH
```

number of completely conserved sites: 2

Fig. 4 *(continued)*.

4. Recent Improvements

Since the last description of TreeAlign *(10)*, four main improvements have been implemented.

1. Decomposition of total alignment: If the traditional alignment algorithm was used unmodified on two sequences 20.000 nucleotides long, a $4*10^8$ entry matrix would have its entries calculated, which could be done, but would be slow. Luckily, there are ways to reduce this computation considerably. There are methods that can find common segments shared by two sequences very fast. The three most common ones are hashing, suffix-trees *(11)*, and DAWG (Directed Acyclic Word Graph; Blumer et al.) *(12)*, and presently hashing is used. If this common segment is sufficiently large, it will most likely have to be embedded in the true alignment, but each time such a segment is found, the alignment can be decomposed into the sequences to the left of the common segment and to the right of the common segment. If many such segments are found, this causes a major reduction in memory requirements and in computations needed.

The segment lengths searched for are sequence length and sequence type specific. Proteins will use lengths 2 or 3, and DNA from 6 to 8. The longer the sequence, the longer the segment. The following algorithm is used to find such segments *(see* Fig. 7).

```
((((0:52,4:80)13:76,(1:55,5:111)14:76)15:113,(2:53,6:43)17:99)16:96
,((3:111,9:141)10:131,(7:105,8:123)11:232)12:41)18:0;
```

CAUTION: This tree is rooted assuming a clock, which might not be justified.
The exact branch lengths can be found in the nested parenthesis representation
just above. Lengths are roughly proportional to vertical lines.

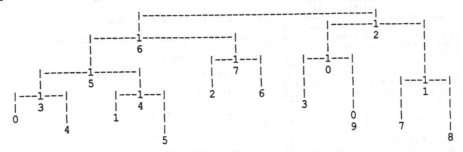

Fig. 5. The phylogeny of the 10 globins is on the file glob.tree.

The longest path of common segments is found that can be embedded in an alignment. This longest path is investigated for consecutive
segments on the same diagonal in the alignment. This defines an island
of homology with a certain number of matches and mismatches. Using
the large deviation theory of Karlin and Ost *(13)* and Arratia and
Waterman *(14)*, it is estimated whether or not a homology island is
statistically significant. If significant, it will be on the final alignment,
and it will also decompose the alignment problem into two subproblems. The introduction of this decomposition has improved the computational qualities of TreeAlign. Earlier it could align sequences at most
1000 bp, whereas now it has no problem aligning 20 complete retroviruses (10.000 bp each). It has also sped up computations considerably.

2. Extended alphabet: Previous versions only allowed the four standard
   nucleotides and 20 standard amino acids in the input sequence, and
   simply ignored nonstandard elements, which would then appear as single
   element deletions in the resulting alignment. The sequences accepted
   will be over the following alphabets:

	Protein		DNA
A	Alanine	A	Adenine
R	Arginine	C	Cytosine
N	Aspargine	G	Guanine
D	Aspartic Acid	T/U	Thymine/Uracil
C	Cysteine	M	A,C
E	Glutamic Acid	R	A,G
Q	Glutamine	W	A,T

```
% align
1: analysis of 10 protein sequences
2: gap penalty: 8 + k* 3
3: The program will reconstruct the phylogeny
4: The alignment will be written on glob.tree
5: The phylogeny will be written on glob.ali
6: finished reading parameters
7: finished reading sequences first time
8: finished reading sequences second time
9: calculated 18 pairwise distances of 45 possible
10: finished constructing distance tree using 27 comparisons
11:finished improving distance tree using 0 comparisons
12:finished making and improving parsimony tree using 9 comparisons
13: finished analysis
```

Fig. 6. After compiling and typing align, the program will run and give messages to the screen about how far in the computations it is and which kind of analysis will be performed. TreeAlign reads the data twice, first to know how large the data set is and then actually to read the content of the data set. It does not always calculate all possible pairwise distances, since this is not necessary if there are many (>30) sequences. In this case, it calculates them all (45 = number of unordered pairs choosing from a set of 10), first 18 before reconstructing the tree, but then the rest (27), when reconstructing the tree. Since there are (n−1) internal nodes in a rooted binary tree with *n* leaves, there will be nine alignments (comparisons) to reconstruct all the ancestral sequences.

G	Glycine	S	C,G	
H	Histidine	Y	C,T	
I	Isoleucine	K	G,T	
L	Leucine	V	Not	T
M	Methionine	H	Not	G
F	Phenylalanine	D	Not	C
P	Proline	B	Not	A
S	Serine	X/N	Any	
T	Threonine			
W	Tryptophane			
V	Valine			
B	(N or D)			
Z	(E or Q)			
X	Any			

The distance function used on nucleotides are: 2 for a transition and 5 for a transversion. For amino acids, the distance function is written on the file mut.dat. The distance between two sets is the shortest distance between any of its members, so if the sets overlap, the distance is zero.

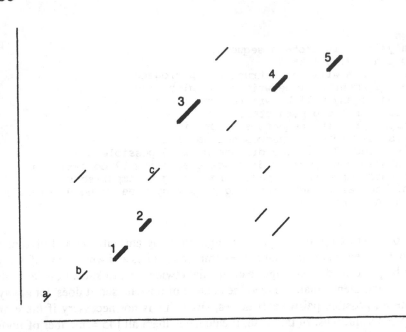

Fig. 7. Decomposition. Many long segments are found, but only five islands of homology are significant. The longest path contains segments that were not statistically significant and are ignored. The alignment problem can then be decomposed in six subproblems: The alignment of subsequences from (0,0) (lower-left corner) up island 1, the alignment of subsequences between island 1 and island 2, and so forth. The subsequences matched by the islands are aligned without insertion-deletions.

    The introduction of the larger data was a must, since it is standard in existing databases and sequences obtained by PCR techniques frequently have nucleotides that are not fully determined.

3. Incomplete sequencing (*see* Fig. 8): The distance between two sequences is the minimal number (weighted) of events to transform s1 into s2. This is a good measure, when comparing myoglobin with α globin, since they are well-defined units. With long (DNA, especially) sequences, however, they might not be well-defined units (where does a gene begin and end?) or might only be partially sequenced. This means that terminal gaps might not be genetic events, but rather arbitrary points, where sequencing stopped. Naturally, such gaps should not be included in the definition of distance between sequences, and terminal gaps should be free. This creates the problem that any two sequences can then be aligned at zero cost, by regarding the first as a deletion in the end of the other. Thus, this must be precluded by only letting terminal gaps be

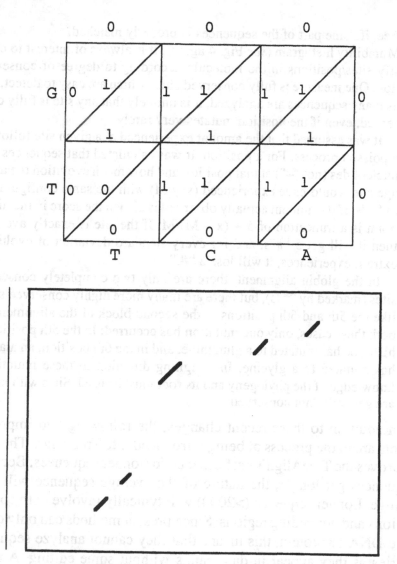

Fig. 8. Incomplete sequences. If terminal insertion-deletions do not correspond to genetic events, they should have weight 0. However, this allows for a trivial 0-weight alignment of any two sequences, by following the boundaries of the matrix, that is, lined with edges with only 0s, corresponding to the complete sequences as terminal insertion-deletions. However, if the alignment is decomposed into a series of subalignments, this artifact is fully circumvented. The decomposed alignment must pass through the statistically significant islands 1–5 and cannot go from lower-left corner to higher-right corner, only through the zero-weight edges (shown in dark) lining the matrix. If there are no statistically significant islands, this will not work, and terminal insertion-deletions must be counted as true genetic events.

free, if some part of the sequences is properly matched.

4. Mutability histogram (*see* Fig. 4 again). It is always of interest to classify sites/positions in the molecule according to degree of conservation. One measure is fully conserved sites, which are easy to detect, but as many sequences are analyzed, it is unlikely that any site is fully conserved, even if the position mutates very rarely.

It was assumed that the amount experienced at a given site followed a poisson process. For a position, it was calculated that sequences had nucleotides (not "–") at that position and how much evolution the average site would have experienced (say M) with the same configuration of "–"s. If the amount actually observed is x, then the score in the histogram is a truncation of $5 + (x - M)/M$. If the site is exactly average, then it will get 5 "#"s, and for every standard deviation of evolution extra it experiences, it will lose a "#."

In the globin alignment, there are only two completely conserved sites (marked by "*"s), but there are many more highly conserved sites, like the 5th and 6th positions in the second block of the alignment. In both these cases, only one mutation has occurred: In the 5th position, a histidine has mutated to a glutamine, and in the 6th position, an alanine has mutated to a glycine. In assigning direction to these mutations, knowledge of the phylogeny and its root must be used. Sites with indels are generally not conserved.

In addition to these recent changes, the following two improvements are in the process of being introduced into TreeAlign. The first improves the TreeAlign's performance for longer sequences. Because sequences get longer, the nature of the average sequence will also change. Longer sequence (>2000) will typically involve both coding regions and noncoding regions. Since present methods can only compare DNA or protein, this means that they cannot analyze sequence entries as they appear in data banks without some editing. A more general algorithm that has the DNA and the protein comparing algorithm as special cases is in preparation and might be incorporated into TreeAlign.

The second improves TreeAlign's performance for more distant sequences. It is fairly reliable to compare sequences with a percentage homology above 25%, but the region 15–25% is what R. F. Doolittle calls the Gray Zone, and below that is utter unreliability. Pushing these borders downward is a major concern for sequence comparing methods. The first push downward was achieved by distinguishing different types of mutations according to their frequency. A second push downward could probably be achieved by distinguishing different positions according to their mutability or importance. The string comparing algorithm used here and in most techniques does not use any information about the importance of positions, which would be crucial in a biochemist's discussion of a molecule. In the case of two sequences and no external source of information, there is no basis for treating some positions as more important than others, but as can be seen from the mutability histogram, this changes as more sequences are analyzed. Taking the information of conservedness into account in the algorithm would make the algorithm more sensitive. This could be done by assigning a weight to every position in the molecule. The mutability histogram calculates the frequency of change in every position, and Felsenstein *(15)* has already given a relationship between mutability and what would be a natural weighting for a position.

Acknowledgments

The author is very grateful to users who have provided cases where the program crashed, which helped in debugging and coming up with relevant suggestions for improvements. The author has not been able to support this program as much as should have been done because of moving and not always having the necessary computer facilities for immediate corrections.

References

1. Sankoff, D., Morel, C., and Cedergren, R. J. (1973) Evolution of 5S RNA and the non-randomness of base replacements. *Nature New Biology* **245**, 232–234.
2. Felsenstein, J. (1981) Evolutionary trees from DNA sequences: a maximum likelihood approach. *J. Mol. Evol.* **17**, 368–376.
3. Felsenstein, J. (1985) Confidence limits on phylogenies: an approach using the bootstrap. *Evolution* **39**, 16–24.
4. Thorne, J. L., Kishino, H., and Felsenstein, J. (1991) An evolutionary model for maximum likelihood alignment of DNA sequences. *J. Mol. Evol.* **33**, 114–124.
5. Kececioglu, J. (1991) Exact and approximate algorithms for the DNA sequence reconstruction problem. Ph.D. Computer Science. Tucson, AZ.
6. Sankoff, D. (1972) Matching sequences under deletion/insertion constraints. *Proc. Natl. Acad. Sci. USA* **69**, 4–6.

7. Hein, J. J. (1989a) A new method that simultaneously aligns and reconstructs ancestral sequences for any number of homologous sequences, when the phylogeny is given. *Mol. Biol. Evol.* **6.6,** 669–684.
8. Hardison, S. P. and Steele, G.L. (1987) *C: A Reference Manual* (2nd ed.) Prentice-Hall, Englewood Cliffs, NJ.
9. Felsenstein, J. (1988) PHYLIP Phylogeny Inference Package version 3.2 *Cladistics* **5,** 164–166.
10. Hein, J. J. (1990) A unified approach to alignment and phylogenies. *Methods Enzymol.* **183,** 626–645.
11. McCreight, E. M. (1976) A space-economical suffix tree construction algorithm. *J. Ass. Comp. Mach.* **23.2,** 262–272.
12. Blumer, A., Blumer, J., Haussler, D., and McConnell, R., (1987) Complete inverted files for efficient text retrieval and analysis. *J. Ass. Comp. Mach.* **34.3,** 578–595.
13. Karlin, S. and Ost, F. (1985) Maximal segmental match among random sequences from a finite alphabet. *Proceedings of the Berkeley Conference in Honor of J. Neyman and J. Kiefer,* 225–243.
14. Arratia, R. and Waterman, M. (1985) An Erdos-Renyi law with shifts. *Adv. Math.* **55,** 13–23.
15. Felsenstein, J. (1981) A likelihood approach to character weighting and what it tells us about parsimony and compatibility. *Biol. J. Linn. Soc.* **16,** 183–196.

CHAPTER 29

Using the FASTA Program to Search Protein and DNA Sequence Databases

William R. Pearson

1. Introduction

As this volume illustrates, computers have become an integral tool in the analysis of DNA and protein sequence data. One of the most popular applications of computers in modern molecular biology is to characterize newly determined sequences by searching DNA and protein sequence databases. The FASTA* program *(1,2)* is widely used for such searches, because it is fast, sensitive, and readily available. FASTA is available as part of a package of programs that construct local and global sequence alignments. This chapter will describe a number of simple applications of FASTA and other programs in the FASTA package. This chapter focuses on the steps required to run the programs, rather than on the interpretation of the results of a FASTA search. For a more complete description of FASTA and related programs for identifying distantly related DNA and protein sequences, for evaluating the statistical significance of sequence similarities, and for identifying similar structures in DNA and protein sequences (*see* ref. *2*).

All the examples below are of protein sequence comparisons. Although FASTA can be used for either DNA or protein sequence comparisons, the user should always compare sequences at the protein

*FASTA is pronounced "FAST-AYE," a name that refers to "FAST-All," and indicates its lineage from "FAST-P" (fast protein sequence comparison) and "FAST-N" (fast nucleic acid sequence comparison). FASTA can compare either protein sequences or DNA sequences.

From: *Methods in Molecular Biology, Vol. 25: Computer Analysis of Sequence Data, Part II*
Edited by: A. M. Griffin and H. G. Griffin Copyright ©1994 Humana Press Inc., Totowa, NJ

sequence level to identify sequences that share distant evolutionary ancestors. If uncertain of the open reading frame in a cDNA clone, translate the clone in all six frames, and use each of those sequences to search a protein sequence database or a translated DNA database (TFASTA). In general, protein sequence comparison allows exploration of evolutionary relationships that are 10-fold more ancient (1–2 billion yr) than DNA sequence comparison (100–200 million yr). DNA sequence comparison is most appropriate when comparing repeated sequence elements, structural RNAs, or transcription factor binding sites.

2. Materials

To use the programs described in this chapter, the user must obtain the FASTA package of programs and one or more sequence databases, and install the programs and databases. Appendix A at the end of this chapter describes how to obtain the FASTA package of programs for UNIX, IBM-PC/DOS, Macintosh, and VAX/VMS computers, and how to install it. Appendix B lists several sources for protein and DNA sequence databases. Appendix C outlines the steps required to install the sequence databases.

2.1. Computer

The FASTA package will run on UNIX machines, VAX/VMS machines, IBM-PCs, and Macintoshes, as well as any other computer that supports a "C" compiler.

2.2. Obtaining the Databases

The FASTA package does not include any sequence databases, but versions of the program work with many generally available library formats, including NBRF/PIR (National Biomedical Research Foundation/Protein Identification Resource—two formats, one for VAX/VMS and one for other machines), EMBL/SWISSPROT, GENBANK full-tape format and the simpler "FASTA" format. The VAX/VMS version of FASTA can read the NBRF/PIR protein and DNA sequence databases on VAX/VMS computers and libraries in the Genetics Computer Group format. The IBM/PC and Mac versions of FASTA can read the EMBL CD-ROM. Addresses for several of the protein and DNA sequence database distributors are listed in Appendix B.

3. Methods

The examples below assume that the user has obtained the FASTA package directly from the author, as described in Appendix A. Although this will usually be the case with IBM-PC or Macintosh versions, it may not be true on UNIX or VAX/VMS machines. Some of the FASTA programs are included with commercial software packages such as the Genetics Computer Group programs; however, the GCG FASTA program looks quite different from the program described below, although similar capabilities are available. In addition, few commercial implementations include all of the programs in the FASTA package (some include them in an "unsupported" directory); RDF2, a program for evaluating the statistical significance of an alignment, is frequently left out.

3.1. Comparing Two Sequences

To demonstrate that the FASTA program is working properly, several test sequences are included with the distribution package. Once FASTA has been been installed in a directory (and compiled if using a UNIX system):

1. Confirm that the two test sequences musplfm.aa and lcbo.aa are present in the directory.
2. Then type:

$$\text{FASTA musplfm.aa lcbo.aa}$$

Here, musplfm.aa, the first entry after the command, is the query sequence file, and lcbo.aa, the second entry, is the library sequence file. FASTA accepts a third entry on the command line, the *ktup* parameter. For protein sequences, if a *ktup* entry is not given, *ktup=2* is used.
3. After typing the command above, the user should see the following:

```
fasta musplfm.aa lcbo.aa
>musplfm mouse proliferin : 224 aa
vs library
searchlng lcbo.aa library

229 residues in      1 sequences
1 scores better than 1 saved, ktup: 2
Enter filename for results : <RET>
```

At this point, it is clear that FASTA was able to find the musplfm.aa
sequence, that it read 224 amino acids from musplfm.aa, and that it also
found a 229-amino acid sequence in the lcbo.aa file. In addition, this
search was done with the *ktup* parameter set to 2 (the default). The *ktup*
parameter sets the sensitivity of the search (*see below*, and refs. *[1–4]*).

4. Type two carriage returns (<RET>) to see:

How many scores would you like to see? [1] **<RET>**
The best scores are: initn   init1   opt
LCBO—Prolactin precursor—Bovine   381   273   432

Here, FASTA is reporting three scores that characterize the similarity
between the musplfm.aa sequence and bovine prolactin. The *init1* ("*init-
one*") score is calculated using the PAM250 matrix *(5)* from the most
similar region without gaps; this is the region bounded by the Xs in the
alignment shown below. When gaps are required to align two sequences,
there will often be several similar regions without gaps that can be combined to improve the similarity score; the score of this combination is
reported as the *initn* ("*init-n*") score. In addition, FASTA calculates an
optimal local similarity score within a 32-residue-wide band around
the best initial region; this is reported as the *opt* score. A more complete description of the calculation of these three scores and their uses
in evaluating sequence similarities are presented in refs. *(1,2)*.

5. After the similarity scores are reported, the alignment(s) may be shown
(Fig. 1). Here, the three similarity scores are reported again, and an
alignment is shown between the query sequence and the library sequence
(in this case, there is only one). On the alignment, the ":" symbol denotes
aligned identities, whereas the "." symbol indicates aligned amino acid
residues with scores ≥0. The "X"s at residues 13 and 140 in the musplfm
sequence correspond to the beginning and end of the best initial region
(the one that gives rise to the *init1* score). Note that the best initial
(*init1*) region is bounded by pairs of identical amino acids because the
search was performed with *ktup=2*. Had the search used *ktup=1*, the
"X"s would be found at residues 7 (a single aligned Q) and 179 (W).[*]

---

[*]Sometimes the introduction of gaps in the optimized alignment causes residues that
were aligned in the *init1* region to be shifted out of alignment. When this happens, the "X"
is "split" into "^" and "v." For example, one might see

```
PCSWILLLLLVNSSLLWKN
:..X:::.:.. :^v .
KGSRLLLLLVVSN-LLLCQ
```

because the gap was not present in the initial region.

```
More scores? [0] <RET>
Display alignments also? y <RET>
number of alignments [1]? <RET>
LCBO - Prolactin precursor - Bovine                    381   273   432
   36.5% identity in 219 aa overlap
```

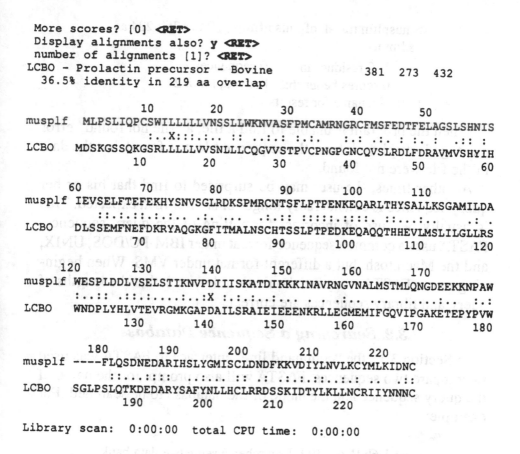

```
                  10        20        30        40        50
musplf  MLPSLIQPCSWILLLLLLVNSSLLWKNVASFPMCAMRNGRCFMSFEDTFELAGSLSHNIS
        :..X::::.:.. :: ..:.:.  :.:.    :.: .:. : :.: .:: :
LCBO    MDSKGSSQKGSRLLLLLVVSNLLLCQGVVSTPVCPNGPGNCQVSLRDLFDRAVMVSHYIH
               10        20        30        40        50        60

          60        70        80        90       100       110
musplf  IEVSELFTEFEKHYSNVSGLRDKSPMRCNTSFLPTPENKEQARLTHYSALLKSGAMILDA
        ::.:.::.:.:...  .: ...  .:.:: :::::;.::.  ::..:..   .: .
LCBO    DLSSEMFNEFDKRYAQGKGFITMALNSCHTSSLPTPEDKEQAQQTHHEVLMSLILGLLRS
               70        80        90       100       110       120

         120       130       140       150       160       170
musplf  WESPLDDLVSELSTIKNVPDIIISKATDIKKKINAVRNGVNALMSTMLQNGDEEKKNPAW
        :..:: .::.:.....:..:X :.:.:..   .  . ... ...   . :.:
LCBO    WNDPLYHLVTEVRGMKGAPDAILSRAIEIEEENKRLLEGMEMIFGQVIPGAKETEPYPVW
               130       140       150       160       170       180

          180       190       200       210       220
musplf  ----FLQSDNEDARIHSLYGMISCLDNDFKKVDIYLNVLKCYMLKIDNC
        ::...:::: ..:..  :: .: .:.:.::..:.: ..  .::
LCBO    SGLPSLQTKDEDARYSAFYNLLHCLRRDSSKIDTYLKLLNCRIIYNNNC
               190       200       210       220
```

```
Library scan:  0:00:00  total CPU time:  0:00:00
```

Fig. 1. A typical FASTA alignment.

The user has now confirmed that the FASTA program is working. One can use a similar syntax to test many of the other programs in the FASTA package. For example:

align musplfm.aa lcbo.aaa

will generate an optimal global alignment of the two sequences.

The user should always go back to this example if the FASTA program appears to have difficulty reading a query sequence or a database. For example, if the "library" sequence is not formatted correctly, or if the database is in one format, but FASTA thinks it is in another format, the user may get the results:

```
>musplfm transl. of musplfm.seq, 2 to 676 : 224 aa
vs library
        0 residues in        0 sequences
        0 scores better than 1 saved, ktup: 2
Enter filename for results :
```

When the program is unable to find a file, a "file not found" error message is displayed; here, the sequence file was found, but the data in the file were not found.

At other times, the user may be surprised to find that his or her query sequence is considerably longer or shorter than expected. This is usually because of incorrect formatting of the query sequence. FASTA uses a common sequence format under IBM-PC/DOS, UNIX, and the Macintosh, but a different format under VMS. When beginning to use FASTA, the user should double-check that the lengths of query and library sequences are correct.

3.2. Searching a Sequence Database

In Section 3.1., the "command line" interface to FASTA was used to compare two sequences. FASTA will also prompt for the name of the query sequence and the name of the database to be searched. For example:

```
% fasta
        fasta 1.6b [Nov, 1991] searches a sequence data bank
Please cite:
        W.R. Pearson & D. J. Lipman PNAS (1988) 85:2444-2448
        test sequence filename: musplfm.aa <RET>
```

Choose sequence library:

```
        P: NBRF complete database (rel 30)
        G: GENBANK Translated Protein Database (rel 70)
        D: NRL_3d structure database
        S: Swiss-Prot (rel 20)
```

```
Enter library filename (e.g., prot.lib), letter (e.g., P) or a %
followed by a list of letters (e.g., %PN): P<RET
>ktup? (1 to 2) [2] <RET>
```

In this example, if lcbo.aa were entered as the library filename, the results would have been the same as in Section 3.1.

The list of potential sequence libraries is displayed only if the libraries have been installed properly and the FASTLIBS "environment" variable has been set (*see* Appendix C). After a successful search, the program will display the sequences with the top scores. Figure 2 shows the bottom of the histogram of similarity scores and the top of the list of highest-scoring sequences. The histogram indicates the number of library sequences that obtained similarity scores in the range indicated. For example, 12 library sequences obtained *initn* similarity scores of 72 or 73; only two library sequences obtained *init1* scores in this range. This difference reflects the fact that the *initn* score is more sensitive, but the *init1* score is more selective. The means and standard errors of the distributions of *init1* and *initn* scores are also shown. This statistical calculation excludes those sequences that have high scores (>73 in this example), under the assumption that these library sequences are likely to be related to the query sequence.

3.3. Interpreting FASTA Results— What Do All of the Numbers Mean?

Most similarity searches seek to identify distantly related protein or DNA sequences that are homologous to, i.e., share a common ancestor with, an entry in a sequence database. One can imagine a perfect sequence comparison program that, after performing a search, would report infallibly: "These library sequences are homologous to the query sequence." Unfortunately, no such program exists, nor is one likely to, because many protein sequence families are so divergent that traces of common ancestry have been erased from the sequences of some of the family members *(4)*. Because of the wide range of sequence diversity present in some protein families and the large number of unrelated sequences in the databases, for diverse families there will always be some unrelated sequences that obtain similarity scores that are higher than those of related sequences. The question then becomes: "When does a high similarity score indicate homology?"

Although the three similarity scores calculated by the FASTA algorithm can be confusing, the relationships among the three scores can be used to help infer sequence homology. Consider four examples from a search of the mouse proliferin protein sequence (musplfm.aa above). (The numbers in parenthesis indicate the rank of the library sequence in the list of top-scoring sequences.)

```
        initn   initl
...
   60    38      4:--+++++++++++++++++
   62    25      2:-+++++++++++
   64    36      0:++++++++++++++++++
   66    21      2:-++++++++++
   68    20      5:---+++++++
   70    10      3:--+++
   72    12      2:-+++++
   74     5      5:===
   76     3      1:-+
   78     6      1:-++
   80     1      1:=
 > 80    94     69:------------------------------------+++++++++++++
10360161 residues in 36150 sequences
 statistics exclude scores greater than 73
 mean initn score:  24.9 (8.15)
 mean initl score:  24.3 (6.77)
 5591 scores better than 31 saved, ktup: 2, variable pamfact
 joining threshold: 28   scan time:  0:01:40
 Enter filename for results : musplfm.k2
 How many scores would you like to see? [20] 100
 The best scores are:                          initn initl   opt
 A05086 Proliferin - Mouse                      1121  1121  1121
 S05648 Proliferin 3 - Mouse                    1108  1108  1108
 A23159 Proliferin 2 - Mouse                   ·1100  1100  1100
 LCHU Prolactin precursor - Human                405   296   444
 A28867 Prolactin precursor - Human              402   296   441
 S04077 Prolactin precursor - Pig                398   292   435
 S02104 Prolactin precursor - Sheep              384   277   435
 JS0200 Prolactin precursor - Sheep              383   276   434
 LCBO Prolactin precursor - Bovine               381   273   432
 A36284 *Prolactin-like protein I, placental - Bovine  337  227  363
 JS0430 Prolactin - Elephant                     336   223   391
 ...
 JT0480 *Growth hormone - Goat                    92    92   192
 MMMSA Laminin chain A precursor - Mouse          91    43    53
 A32424 *Somatotropin precursor - Grass carp     89    60   160
 ...
```

Fig. 2. Results from a "successful" search.

	initn	initl	opt
(9) LCBO Prolactin precursor-Bovine	381	273	432
(84) STGT Somatotropin precursor-Goat	92	92	192
(85) MMMSA Laminin chain A precursor-Mouse	91	43	53
(86) A32424 *Somatotropin precursor-Grass carp	89	60	160

Bovine prolactin (LCBO), ranked 9 in the list of top-scoring sequences, is clearly homologous to the query sequence. The *initn* and *initl* similarity scores are more than 40 SD above the mean of all the sequences in the library, and the *opt* score is about 40% of that calculated when the sequence is compared with itself, implying that proliferin and prolactin are about 40% identical (the actual percent identity, 36.5%, is shown in the alignment). This is well within the limit (20–25%, ref. *[6]*) where homology can be clearly demonstrated from similarity. Thus, proliferin and prolactin share a common ancestor.

Both goat somatotpin precursor (SIGT), ranked 84, and the grass carp somatotropin (A32424), ranked 86, are also clearly related to proliferin. Here, the inference is based on the substantial increase from the *init1* score, which does not allow gaps, to the *opt* score, which does. Although the *initn* and *init1* scores only suggest that these sequences are related to proliferin (six of eight scores between 81 and 91 are from sequences that are not related to proliferin), the two- to threefold increase in the *opt* score is often found with homologous proteins (for additional examples, *see* refs. *[1–3]*). The carp somatotropin sequence provides a very typical example of a more distant, but clearly related sequence. Here, the *init1* score is much lower than the *initn* score, but it increases almost threefold when gaps are introduced in the alignment to produce the *opt* score. The lowest-ranked related sequence in the two 200 scores, a carp prolactin (ranked 121), has *initn* and *init1* scores of 69, which increase to 229 with optimization. When evaluating marginal similarity scores, look for intermediate (40–60) *init1* scores that increase to more than 150 with optimization.

The laminin scores confirm this rule. Although laminin obtains a high *initn* score and a low *init1* score, much like the carp somatotropin, the *init1* score increases only about 25%, to a value that is much lower than the *initn* score, after optimization. Laminin is unlikely to share a common ancestor with proliferin or the other members of the growth hormone family.

3.4. Increasing the Sensitivity of FASTA

Unfortunately, not all searches provide the definitive results shown in Section 3.2. Sometimes, the results look more like Fig. 3. In this example, none of the top-ranked sequences are likely to share common ancestry with the query sequence, a microsomal glutathione transferase (PIR code A28083). This inability to detect related sequences may reflect their absence from the sequence database, or it may reflect a limitation of the FASTA search. This search was done with *ktup=2* and, thus, required that initial regions be bounded by pairs of identi-

```
 70      3      0:++
 72      3      0:++
 74      1      0:+
 76      2      0:+
 78      1      0:+
 80      0      0:
> 80      5      3:--+
9697617 residues in 33989 sequences
 statistics exclude scores greater than 72
 mean initn score:  23.2 (7.37)
 mean initl score:  22.8 (6.46)
 5025 scores better than 29 saved, ktup: 2, variable pamfact
 joining threshold: 27  scan time:  0:00:34
The best scores are:                            initn initl   opt
HMNZED Hemagglutinin - Measles virus (strain Edmonston   83   46   46
HMNZHA Hemagglutinin - Measles virus (strain Halle)      83   46   46
HMNZKA Hemagglutinin - Rinderpest virus (strain Kabete   78   48   48
A25856 Neuron cytoplasmic protein 9.5 - Human            75   54   54
S04724 *NADH dehydrogenase chain 5 - Emericella nidula   75   38   40
A35694 *cut1 protein - Yeast (Schizosaccharomyces pomb   74   56   60
S06188 RNA1 polyprotein - Grapevine chrome mosaic viru   72   32   33
JQ0274 Hypothetical 29K protein (trnH-trnV intergenic   71   41   50
S13595 *6-deoxyerythronolide B synthase - Saccharopoly   71   44   47
NICLMB Nitrogenase (EC 1.18.6.1), molybdenum-iron prot   70   56   82
```

Fig. 3. An "unsuccessful" search with *ktup=2*.

cal residues. The sensitivity of the search can be increased by setting *ktup=1* or by making FASTA optimize scores for all of the sequences in the database.

If a search with *ktup=2* fails to find sequences that are likely to share common ancestry, a search should be performed with *ktup=1*.

```
                    fasta a28083.aa P 1
                    ^query-sequence file
                              ^library selection
                                   ^ktup
```

Alternatively, FASTA prompts for the the *ktup* if the query-sequence file and library file are not entered on the command line. Searches with *ktup=1* take about five times as long as searches with *ktup=2*.

There is no guarantee, of course, that sequences related to the query can be found in the database. In this example, the results of the search would be quite similar with *ktup=1*, except that the scores would be higher. Current versions of FASTA (1.5 and later) provide a second option for increasing sensitivity: calculating an optimized score for every sequence in the database. Searches performed with this option

Table 1
FASTA and SSEARCH Execution Times[a]

Algorithm	Sensitivity	Time (min)
FASTA	*ktup=2*	0.22
	ktup=1	1.07
	ktup=1	
	All scores optimized	2.35
SSEARCH	(Smith-Waterman)	22.00

[a]Execution times (minutes) required to scan the PIR2.SEQ file (preliminary entries, 12,837 sequences containing 3,384,087 amino acids) of the NBRF Protein Identification Resource protein sequence database (release September 30, 1991) using bovine prolactin (LCBO, 229 aa) as the query sequence.

are about as sensitive as searches with the rigorous Smith-Waterman algorithm *(4,7)*, but about seven times faster. (Table 1 shows the time required to search the part of the PIR protein sequence database using bovine prolactin [LCBO] as the query.) To optimize the similarity scores for every sequence in the library, two command line options are required: -o (optimize) and -c 1 (the threshold for optimization is set to 1). For example:

fasta -o -c 1 lcbo.aa P

If the -o option is used without the -c 1 option, the optimized scores are calculated only for library sequences with *initn* scores greater than a threshold (33 for a 200-residue query-sequence and *ktup=1*) that is about the mean of the unrelated sequence scores.*

Increasing the sensitivity with *ktup=1* or additional optimization not only increases the amount of time required to perform the search, but it also decreases the "signal-to-noise" ratio by increasing the scores of unrelated sequences more than the scores of related sequences *(2)*. Thus, in Fig. 4, there are many more high-scoring sequences in the histogram, but none of the high-scoring sequences are likely to be homologous to the microsomal glutathione transferase.

*The FASTA program has a large number of command-line options that can be used to modify the similarity scoring matrix, the format of the alignment, and other parameters that are used in the algorithm. These options are discussed more fully in the documentation distributed with the program.

```
   70      69      2:-++++++++++++++++++++++++++++++++++++
   72      49      0:+++++++++++++++++++++++++++
   74      38      2:-++++++++++++++++++
   76      30      2:-+++++++++++++
   78      24      0:+++++++++++++
   80      22      0:+++++++++++
 > 80      83      5:---++++++++++++++++++++++++++++++++++++++++++
 9697617 residues in 33989 sequences
  statistics exclude scores greater than 77
  mean initn score:  31.5 (10.21)
  mean init1 score:  30.4 (7.91)
  5341 scores better than 43 saved, ktup: 1, variable pamfact
  joining threshold: 32, optimization threshold: 1  scan time:  0:05:08
 The best scores are:                              initn init1   opt
 S08206 5-lipoxygenase-activating protein - Rat       34    34    84
 S11961 *Hypothetical protein - Red alga (Gracilaria ch 51   51    83
 S11150 *amiC protein - Streptococcus pneumoniae      49    49    82
 NICLMB Nitrogenase (EC 1.18.6.1), molybdenum-iron prot 79   58    82
 B28269 Protein kinase (EC 2.7.1.37), cGMP-dependent -  81    81    81
 S08164 5-lipoxygenase-activating protein - Human      34    34    81
 MNXRRW Nonstructural protein Pns9 - Rice dwarf virus   75    75    81
 A34106 *Protein kinase (EC 2.7.1.37) cGMP-dependent 1  81    81    81
 LNHUP6 Pulmonary surfactant protein A precursor (clone 61   43    80
 A35049 *Ankyrin - Human                               80    57    80
 IMECN4 Colicin N immunity protein - Escherichia coli p 81   58    80
```

Fig. 4. An "unsuccessful" search with *ktup=1*; all scores optimized.

3.5. Changing Search Parameters

Built into the FASTA program are two scoring value matrices, the PAM250 matrix for proteins *(5)* and an identity matrix for DNA sequences (+4 for a match, -3 for an unambiguous mismatch, +2 for an ambiguous match [G-R], -1 for an ambiguous mismatch [G-Y]; *see* the file upam.gbl for the precise definition). In addition, by default, all of the sequence comparison programs in the FASTA package penalize -12 for the first residue in a gap and -4 for each additional residue. Both the scoring matrix and the gap penalties can be changed by specifying an alternative SMATRIX file. This can be done on the command line:

fasta -s pam120.mat

or by changing an *environment* variable:

set SMATRIX=pam120.mat

Several alternative SMATRIX files are included with the FASTA distribution.

pam120.mat	A version of the PAM250 matrix calculated for 120 PAMs (point accepted mutations). The PAM120 matrix is used by the BLAST program *(8)* and is more selective (unrelated sequences receive lower scores).
codaa.mat	The genetic code (minumum mutation distance) scoring matrix. +6 for an identity, +2 for a single base change, -2 for two base changes, -6 for three base changes.
idnaa.mat	An identity matrix for proteins. +6 for an identity, -3 for any nonidentity.
idpaa.mat	A protein matrix that uses the PAM250 values for identical matches, and -3 for any nonidentical alignment.

All of the scoring matrices charge -12 for the first residue in a gap and 4 for each additional residue. Some algorithms refer to a gap penalty of the form:

$$q + rk$$

where k is the number of residues in the gap. The -12, -4 used by FASTA are equivalent to $q = -8, r = -4$.

3.6. Identifying Sequences in DNA Sequence Databases

The TFASTA program included in the FASTA package can be used to compare a protein sequence to a DNA sequence database, translating the DNA sequence database in all six frames. Searching a translated DNA sequence database is far more informative than searching the DNA sequence database directly; DNA–DNA sequence comparisons forgo the biochemical information encoded in the PAM250 amino acid replacement matrix.

TFASTA has the same options as FASTA; thus the command

tfasta lcbo.aa gpri.seq

might be used to search the primate portion of the GenBank DNA sequence database with *ktup=2*. Because of the size of the DNA sequence databases, translating and searching with TFASTA can be slow. However, both FASTA and TFASTA provide an optional method for selecting the libraries to be searched that makes it easier to search only the libraries of interest. For example, if the FASTLIBS file has specified that the following letters can be used to select a library

P   primate
R   rodent
M   mammal
V   other vertebrates
I   invertebrates
L   plants
B   bacteria
U   unannotated

and the user wishes to search all animal sequences, but not plants, bacteria, viruses, phage, or structural RNA sequences, then the command:

tfasta lcbo.aa %PRMVIU

would search only the primate, rodent, other mammal, vertebrate, invertebrate, and unannotated sequences. Here, the % indicates that a list of database abbreviations, rather than a file name, has been entered.

3.7. Evaluating FASTA Results with RDF2 and RSS

The FASTA package includes several other programs that can be used to evaluate the results of FASTA searches. The RDF2 and RSS programs can test the statistical significance of a sequence alignment score. Both programs compare two sequences, calculate the similarity score(s), shuffle the second sequence to generate a set of new sequences with exactly the same length and amino acid composition, and calculate similarity score(s) for each shuffled sequence. RDF2 uses the FASTA algorithm to calculate three similarity scores for each shuffled sequence; RSS uses the Smith-Waterman algorithm (7) to calculate a single score. Both shuffling programs have several options: The number of shuffles can be varied (100–200 are recommended), the shuffling strategy can be either "uniform" (each residue can be moved to any position along the sequence) or "local" (residues are shuffled in blocks; i.e., residue 1 will remain in the first 10 residues, residue 11 in the second 10, and so on), and the size of the shuffling window can be changed. Figure 5 shows a typical comparison of the microsomal glutathione transferase with the highest-scoring library sequence from Fig. 4.

```
rdf2
rdf2 1.6b [Nov, 1991] compares a test sequence to a shuffled sequence
test sequence file name: gstmicr.aa
sequence to be shuffled: s08206.aa
ktup? (1 to 2) [2] 1
number of random shuffles? [20] 200
local (window) (w) or uniform (u) shuffle [u]? w
local shuffle window size [10]  <RET>
a28083.aa :  155 aa
s08206.aa :  161 aa

      initn   init0   opt
 <  2    0       0     0:
    4    0       0     0:
    6    0       0     0:
    8    0       0     0:
   10    0       0     0:
   12    0       0     0:
   14    0       0     0:
   16    0       0     0:
   18    0       0     0:
   20    0       0     0:
   22    0       0     0:
   24    2       2     0:ii
   26    3       3     0:iii
   28    7       7     0:iiiiiii
   30   12      12     3:oooiiiiiiiii
   32   19      19     3:oooiiiiiiiiiiiiiiii
   34   27      30    12:ooooooooooooooiiiiiiiiiiiiiii I
   36   22      23    11:oooooooooooiiiiiiiiiii
 -----//-----
   66    1       0     2:io
   68    0       0     1:o
   70    1       0     1:b
   72    1       0     0:i
   74    2       0     2:bb
   76    0       0     0:
   78    0       0     2:oo
   80    0       0     0:
 > 80    1       0     0:i o
32200 residues in  200 sequences,
local shuffle, window size: 10
mean initn score:  39.9 (9.83); max initn score:  82
mean init0 score:  38.1 (6.89); max init0 score:  59
mean opt   score:  45.4 (9.50); max opt   score:  77
init  score:   34 is -0.60 s.d. above mean
init0 score:   34 is -0.59 s.d. above mean
opt   score:   84 is  4.06 s.d. above mean
ktup: 1, fact: 4  scan time:  0:00:01
```

Fig. 5. Evaluating similarities scores with RDF2.

Figure 5 shows that the relationship between the microsomal glutathione transferase and lipoxygenase activating protein is borderline. The z-value* of the optimized similarity score is only 4 (RSS** gives a z-value after a local shuffle of 5); unrelated sequences sometimes obtain z-values of 5 or more. As Fig. 5 shows, it is possible to shuffle the lipoxygenase sequence and still obtain similarity scores > 75. Both proteins are membrane-bound; this sequence similarity may simply reflect their shared hydropathy.

3.8. Evaluating FASTA Results— Local Similarity

In addition to evaluating a similarity score by Monte-Carlo shuffling, the user can also ask whether there are alternative alignments of the two sequences that yield high similarity scores. For example, several unrelated membrane proteins have high *initn* scores when compared with the β-adrenergic receptor sequence (4); in these cases, the proteins shared several alternative alignments that suggested that it was hydropathy, rather than common ancestry, that was the basis for the high scores. The LFASTA, PLFASTA, LALIGN, and PLALIGN programs can be used to display alternative local alignments between two sequences. LFASTA and LALIGN show the alignments, whereas PLFASTA and PLALIGN produce a two-dimensional dot-matrix-like plot. LALIGN and PLALIGN use a rigorous algorithm to calculate the best local alignments (9–11) and are preferred for protein sequences on faster machines with sufficient memory. LFASTA and PLFASTA use the FASTA algorithm rapidly to identify regions that share local similarity; they can be used for long sequences (2000 residues) on IBM-PCs. Analysis of the relationship between microsomal glutathione transferase and lipoxygenase activator shows a single high-scoring alignment (22% identity over 113 amino acids), a result that is consistent with the hypothesis that the two sequences share a common ancestor. As with the shuffling programs, the results are ambiguous for this pair of sequences.

---

*The z-value of a similarity score is calculated by substracting the mean score and dividing the difference by the standard deviation of the distribution of shuffled scores.

**The RSS program provides a more stringent test, since the RSS program uses a more sensitive (and less selective algorithm) than RDF2. RSS is expected to calculate higher scores for the shuffled sequences and thus makes it more difficult to obtain a high z-value.

3.9. Summary—How to Identify Sequences with FASTA and TFASTA

1. Search a protein-sequence library (or use TFASTA to compare a protein sequence to a DNA-sequence library) using the *ktup* parameter set to 2.
2. If high-scoring sequences are found (*initn* > 100, *init1* > 60, *opt* > 150), it is likely that a homologous library sequence has been found. Confirm the identification by running the *RDF2* and *RSS* programs; a *z*-value >10 is expected.
3. If a FASTA search with *ktup=2* does not turn up any likely candidates (at least 200 scores should be examined), repeat the search using *ktup=1*. (Again examine the 200 top-scoring sequences, looking for *init1* scores >60 that increase to 150 or more with optimization.) Candidates identified with *ktup=1* should have *RDF2* or *RSS* *z*-values >8. Sequence similarities with lower *z*-values are suspect; look for >15–20% sequence identity over the entire length of both sequences.
4. If no library sequences meet the criteria in step 3, consider repeating the *ktup=1* search and calculating an optimized score for every sequence in the library. However, this strategy is more likely to result in an ambiguous false-positive result than in the novel finding of homology. Searches should be done in the order shown (1, 3, 4). With some sequences, searches with *ktup=2* unambiguously show homology, whereas the conclusion is less clear as more sensitive methods are used (*ktup=1*, -o -c 1) and the similarity scores of unrelated sequences increase. One should always be extremely cautious when interpreting apparent relationships that cannot be detected without calculating an optimized score for every sequence in the library.
5. If a significant similarity is found and an alignment is to be published, the SSEARCH program should be used to produce a rigorous alignment. FASTA alignments are limited to 32-residue-long gaps; there is no limit on the length of the gaps produced by SSEARCH. For the same reason, the LALIGN and PLALIGN programs are preferred over the LFASTA and PLFASTA programs.

The examples in this chapter show a number of the uses and pitfalls of the FASTA program when characterizing distantly related protein sequences. All of the methods shown above can be applied to DNA sequences as well, but with the caution that DNA sequence comparisons suffer from "signal-to-noise" problems much more frequently than protein sequences. It is common to find a DNA sequence similarity with a modest score (*initn* score > 100 and an optimized score > 150 at the nucleotide) level that disappears (scores < 50) when the translated sequences are compared.

Appendix A—
Obtaining the FASTA Program Package

The FASTA package has progressed through several versions since it was first introduced in 1987. Newer versions correct bugs in the FASTA program, or allow the program to search more library files or additional library formats. As this chapter is being written, the current version for the UNIX, VAX/VMS, DOS, and the Macintosh is version 1.6. This latest version contains a number of additional new programs, including rigorous (but very slow) programs for library searching, local-sequence alignments, and statistical analysis.

Obtaining the Programs

The best way to obtain the program depends on the type of machine that will be used.

UNIX

The easiest way to get FASTA for a UNIX machine is via anonymous ftp from the host uvaarpa.Virginia.EDU or via electronic mail (wrp@virginia.EDU) If the user's institution has internet access, the user should try anonymous ftp first. From a machine that has access to the internet, type:

> ftp uvaarpa.Virginia.EDU or alternatively ftp 128.143.2.7
> and login with the user Name: anonymous
> and a Password: your_userid

The FASTA package is in a file of the form public_access/fasta16b.shar. (A compressed version is available as fasta16b.shar.Z. The 16b will change as newer versions become available.) To transfer the file:

> cd pub/fasta
> get fasta16b.shar

(Be sure to use binary mode for transfer of the compressed.Z file.) This is a UNIX shar file; after transferring this file to a UNIX machine, type:

> sh fasta16b.shar

and the file will be broken into the files required to recompile FASTA programs. A Makefile is included for Sun (4.2BSD), ATT SysV, and Xenix flavors of UNIX. Other makefiles are included for Turbo "C" on the IBM-PC and for the VMS operating system on a VAX.

If electronic mail is sent to "wrp@virginia.EDU" requesting the UNIX version of the program, the author can send back a set of files that can be concatenated to create the fasta.shar file, which can then be unpacked with "sh fasta.shar." Alternatively, the author can send the fasta.shar file on either IBM-PC (1.2 Mbyte 5.25 in. or 720 Kbyte/ 1.44 Mbyte 3.5 in.) or Macintosh (720 Kbyte 3.5 in.) disks.

VAX/VMS

If planning to use these programs on a VAX/VMS computer, the author can send the user a set of VAX/VMS files via electronic mail or put them on an IBM-PC or Macintosh diskette.

IBM-PC

The FASTA package comes on 5.25- or 3.5-in. floppy disks for the IBM-PC, and includes complete source code, executable versions of the programs, and also *.BGI graphics device driver programs from Borland's Turbo "C" package. The program costs $60.00 (plus $10.00 overseas shipping). Orders should be sent to:

> William R. Pearson
> 1611 Westwood Rd.
> Charlottesville, VA 22903
> USA

There is a $25.00 additional charge for purchase orders.

Macintosh

The FASTA package is also available for the Macintosh computer, although the program is not very "Mac-like." It does run in the background under Multifinder, so the user can search and do other work at the same time. The Mac version also costs US $60.00; please send checks to the address above.

Once the user has obtained the programs, he or she will need to:

1. Install them on the computer.
2. Configure a file (the FASTLIBS file) that tells FASTA where to find sequence libraries (the user may also need to edit other files that describe where the sequence databases are found).

3. Set an "environment" variable (FASTLIBS) to tell FASTA where the FASTLIBS file can be found.
4. Set the execution PATH to include the directory that contains the FASTA programs.

For example, under UNIX or DOS:

1. Edit the file called fastgbs, which is included in the distribution, changing the file names for the sequence libraries.
2. Under DOS, type the command: "set FASTLIBS=c:\fasta\fastgbs" (assuming the FASTA package was installed in a directory called "c:\fasta."
3. Under DOS, type the command: "set PATH=c:\fasta; . . . " (again assuming installation in "c:\fasta").

Equivalent commands are available under UNIX.

On the Macintosh, there are no "environment" variables, and it is easier simply to run the FASTA program from a "FASTA" folder. To mimic "environment" variables on the Macintosh, the Mac FASTA package uses a file called "environment," which contains lines like:

FASTLIBS=HD40:FASTA:FASTGBS

Appendix B—Obtaining Sequence Databases

If the user's computer has access to the internet, the DNA and protein sequence libraries are available via anonymous "ftp" from a number of sources, including those listed in Table 2. DNA and protein sequence databases are available on diskettes or tape from the sources listed in Table 3.

For most people, particularly on IBM-PCs and Macintoshes, the most difficult part of using the FASTA programs is installing the databases properly. An example of how to do this is given below.

Appendix C—Setting Up the FASTLIBS File

The FASTA and TFASTA programs can be configured to present the user with a list of databases that are available to be searched. After the list is presented, the user can enter a single letter (P) to select a database, or a string of letters prepended with a % (%PGS) to select several databases. For this menu option to work, a file that lists each of the databases, the FASTLIBS file, must be present. Unfortunately, most people who have difficulty installing FASTA have problems with the the FASTLIBS file.

Table 2
Sequence Database Available over the Internet

Provider	Internet address	Databases available	E-mail contact
NCBI, National Library of Medicine	ncbi.nlm.nih.GOV	GenBank™ DNA sequences Swiss-Prot protein sequences	info@ncbi.nlm.nih.GOV
University of Houston Gene Server	ftp.bchs.uh.EDU	Protein Identification Resource (PIR)	davison@uh.EDU

A typical FASTLIBS file looks like this:

NBRF Annotated Protein Database (rel 30)$0A/seqlib/pir1.seq 5
NBRF Protein Database (complete)$0P@/p0/seqlib/prot.nam
GB70.0 Primate$1p0/gblib/gbpri.seq 1
GB70.0 Rodent$1R/p0/gblib/gbrod.seq 1

. . .

Each line of the FASTLIBS file has four required fields and one optional field.

NBRF Annotated Protein Database $0A/seqlib/pir1.seq 5
 ^   field 1   ^23^   field 4   ^ 5   (optional)

Field 1 contains a description of the database and ends with a $. The next two characters are field 2—a number that indicates whether the database contains protein (0) or DNA (1) sequences—and field 3—a letter used to select the database. The fourth field contains the name of the database file, whereas the fifth, optional field indicates the database format.

Most people have problems filling out the fourth and fifth fields. The fourth field is complicated, because it can contain two types of database file descriptions. The first type is simple; it is simply the name of the file that contains the database. In this example, the annotated PIR database is in a file called /seqlib/pir1.seq. The second type of description is more complicated; it is a file that contains a list of database files. For example, the complete PIR protein database is dis-

Table 3
Database Distributors

Provider	Databases	Address	Media
Intelligenetics	GenBank™	GenBank NCBI Intelligenetics National Library of Medicine 8600 Rockville Pike Bethesda, MD 20894 USA	CD-ROM
European Molecular Biology Laboratory	EMBL DNA-sequence library Swiss-Prot protein-sequence library	EMBL DNA Library European Molecular Biology Laboratory Postfach 10.2209 MeyerhofStr.1 D-6900 Heidelberg, Germany	CD-ROM Nine-track tape VMS cartridge tape
Protein Identification Resource	PIR protein-sequence library	National Biomedical Research Foundation Georgetown University Medical Center 3900 Reservoir Rd. N.W. Washington, DC, 20007 USA	Nine-track tapes TK-50 cartridges (VMS) CD-ROM

tributed in three files: pir1.seq, pir2.seq, and pir3.seq. In the example above, the file prot.nam is an *indirect* file with the entries:

```
</seqlib/
pir1.seq 5
pir2.seq 5
pir3.seq 5
```

Here the </seqlib/ line indicates that /seqlib/should be prepended to each of the filenames; thus, pir1.seq becomes /seqlib/pir1.seq. The number "5" on the file name line indicates that this file is an NBRF/ PIR VMS format file *(see below)*. The inclusion of the format number allows the user to search sequence databases that have been distributed in different formats in one run of the program.

The FASTLIBS file recognizes that prot.nam is an indirect file of file names rather than an actual database because of the @ character. @prot.nam refers to an indirect file of database file names, whereas pir1.seq refers to a database file.

To use the FASTLIBS file with FASTA, the user must set the FASTLIBS environment variable.

```
set FASTLIBS=fastgbs   under DOS, or
setenv FASTLIBS fastgbs   under Unix(csh), or
define FASTLIBS fastgbs   under VMS, or
edit the "environment" file on the Macintosh.
```

In this example, fastgbs is the name of the file that contains the five fields described above. If the FASTLIBS file has not been setup, the user will not see a list of potential databases.

Even if the FASTLIBS file has been set up, the user may get the message:

```
vs NBRF Annotated Protein Database (rel 25) library
cannot open /p0/slib/lib/protein.seq library
enter new file name or <RET> to quit
```

This occurs when the file pointed to by the FASTLIBS variable contains incorrect information about where to find the sequence database. In this case, the file contained:

NBRF Annotated Protein Database (rel 25)$0A/p0/slib/lib/protein.seq 5

when it should have said

NBRF Annotated Protein Database (rel 25)$0A/seqlib/lib/pir1.seq 5

Database Formats

As this chapter is being written, FASTA recognizes six different database file formats:

0—FASTA (>seq-id-comment line/sequence data)
1—GenBank "flat-file" (LOCUS/DEFINITION/ORIGIN)
2—NBRF CODATA
3—EMBL/SWISS-PROT (ID/DE/SQ)
4—Intelligenetics (;comment line/SEQID/sequence)
5—NBRF/PIR VMS format (>P1;SEQID/comment/sequence)

Type 0 and type 5 files do not contain any reference data and are the fastest to search; as a result, the EMBL provides type 5 (PIR/VMS) format versions of their DNA database and the SWISS-PROT database on their CD-ROM. Searching the PIR/VMS format is several times faster than searching the same databases in EMBL format.

If the wrong format is selected for a database file, the total number of residues read will be wrong. If, for example, the user indicates that pir1.seq is a type 1 file (GenBank flat file format), but it is actually a type 5 NBRF/VMS file, FASTA will report that it found 0 residues in 0 sequences. To make certain that the values in the FASTLIBS file are correct, the user should confirm that FASTA has read the number of residues listed in release notes for the database.

References

1. Pearson, W. R. and Lipman, D. I. (1988) Improved tools for biological sequence comparison. *Proc. Natl. Acad. Sci. USA* **85,** 2444–2448.
2. Pearson, W. R. (1990) Rapid and sensitive sequence comparison with FASTP and FASTA, in *Methods in Enzymology,* vol. 183 (Doolittle, R. F., ed.), Academic, New York, pp. 63–98.
3. Lipman, D. J. and Pearson, W. R. (1985) Rapid and sensitive protein similarity searches. *Science* **227,** 1435–1441.
4. Pearson, W. R. (1991) Searching protein sequence libraries: Comparison of the sensitivity and selectivity of the Smith-Waterman and FASTA algorithms. *Genomics* **11,** 635–650.
5. Dayhoff, M., Schwartz, R. M., and Orcutt, B. C. (1978) A model of evolutionary change in proteins, in *Atlas of Protein Sequence and Structure,* vol. 5, supplement 3 (Dayhoff, M., ed.), National Biomedical Research Foundation, Silver Spring, MD, pp. 345–352.
6. Doolittle, R. F., Feng, D. F., Johnson, M. S., and McClure, M. A. (1986) Relationships of human protein sequences to those of other organisms. *Cold Spring Harb. Symp. Quant. Biol.* **51,** 447–455.

7. Smith, T. F. and Waterman, M. S. (1981) Identification of common molecular subsequences. *J. Mol. Biol.* **147,** 195–197.
8. Altschul, S. F., Gish, W., Miller, W., Myers, E. W., and Lipman, D. J. (1990) A basic local alignment search tool. *J. Mol. Biol.* **215,** 403–410.
9. Waterman, M. S. and Eggert, M. (1987) A new algorithm for best subsequences alignment with application to tRNA-rRNA comparisons. *J. Mol. Biol.* **197,** 723–728.
10. Huang, X., Hardison, R. C., and Miller, W. (1990) A space-efficient algorithm for local similarities. *CABIOS* **6,** 373–381.
11. Huang, X. and Miller, W. (1991) A time-efficient, linear-space local similarity algorithm. *Adv. Appl. Math.* **12,** 337–357.

CHAPTER 30

Converting Between Sequence Formats

Cary O'Donnell

1. Introduction

A "sequence format" is a punctuation style, or defined layout of text, within a computer file that separates a sequence from everything else. It allows computer programs that "understand" the format to distinguish between the sequence and any reference documentation also in the file. Some format definitions extend to the documentation itself (i.e., most database formats), allowing some software to locate specific reference information (e.g., authors, journals, species classification, coding regions).

Unfortunately sequence analysis programs do not recognize a single, universal format. Many different programs and packages have developed their own formatting style. At best this means that one software package does not read the sequence file that another has created. At worst a program reads a file in the wrong format and interprets annotations as sequence (with potentially disastrous results)!

Formats are typically named after the software package associated with them, or after an organization that defined the format. The most well known are EMBL, GCG, GenBank, IG (Intelligenetics or Stanford), and NBRF (also known as PIR). Some formats are named after individual software authors, e.g., Pearson, Staden, and Olsen.

This chapter describes how to create a specific format from the plain sequence text and how to convert between formats. The formal format definitions are not described here.

Methods 1–3 show how to copy a sequence file from one format into a new file of another format. Method 1 uses READSEQ, a single

From: *Methods in Molecular Biology, Vol. 25: Computer Analysis of Sequence Data, Part II*
Edited by: A. M. Griffin and H. G. Griffin Copyright ©1994 Humana Press Inc., Totowa, NJ

program that recognizes most formats, but does not form part of a general package. Methods 2 and 3 use programs from the GCG package.

Methods 4 and 5 describe how to reformat the EMBL *(1)* and GenBank *(2)* databases into NBRF format and create supplementary index files. This allows the most commonly used VAX packages (i.e., GCG *[3]*, NAQ *[4]*, PSQ *[5]*, XQS and ATLAS *[6]*, FASTx *[7]*, Staden *[8]*) to share one copy of the sequence and reference information.

2. Materials

1. Computers: VAX/VMS for all methods. VAX/VMS, VAX-Ultrix, and Apple Macintosh for Method 1.
2. Terminal: Any text-capable terminal is suitable as there is no graphical output.
3. Programs: The program READSEQ *(9)*, is available from the EMBL file server, the University of Houston (UH) gene-server and "anonymous ftp" from various INTERNET sites. Obtaining programs from these sources is described in Chapter 31 with the READSEQ program as an example.

    Version 7.2 of the Genetics Computer Group (GCG) package, available from:

    > Genetics Computer Group
    > University Research Park
    > 575 Science Drive, Suite B
    > Madison, WI 53711–1060
    > Electronic mail: HELP@GCG.COM.

    Methods 4 and 5 use FORTRAN program-source files supplied as part of the XQS and PIR database software distribution: createdbs.for, createinx.for, indexer.for, sorttmp.c These are the PIR programs, available from:

    > Protein Identification Resource (PIR)
    > National Biomedical Research Foundation
    > Georgetown University Medical Center
    > 3900 Reservoir Road, N.W.
    > Washington, DC 20007

4. Input files.
    The sequence file: For the READSEQ program (Method 1), the file may contain sequence only in one of the formats recognized, i.e., IG, GenBank, NBRF, EMBL, GCG, DNA Strider, Fitch, Zuker, Olsen, Phylip, plain format (i.e., the sequence only, no documenting text) and others.

For Method 2, the sequence file must be in GCG format. Such a file may be created using the GCG program SEQED. In the example, a file called CCHU.PIR1 has been retrieved from a local copy of the NBRF-Protein database (logical name PIR1), using the GCG program FETCH:

$ *fetch pir1:cchu*

For Method 3, the format of the input file must correspond to the program being used, i.e., NBRF format when the GCG program FROMPIR is used. In the first two examples, the files created in Method 2 are used. For the third example (the GCG program REFORMAT), a plain-text file is used. Method 4 requires an EMBL format database file(s). The example used here is EMBL35, which was supplied in 16 separate subsections. The Swissprot database is supplied as a single file. Both can be obtained on computer tape or CD ROM from:

> European Molecular Biology Laboratory
> Postfach 10.2209
> D-69012 Heidelberg
> Germany
> Electronic mail: DataLib@EMBL-Heidelberg.DE

Method 5 requires a GenBank format database file. The GenBank database is available on CD ROM in 14 separate subsections from:

> NCBI-GenBank
> National Center for Biotechnology Information
> National Library of Medicine
> 8600 Rockville Place
> Bethesda, MD 20894
> Electronic mail: info@ncbi.nlm.nih.gov

Additional files: Methods 4 and 5 require a single-line header file for each subsection of the database (e.g., EMBLPRI.HEADER or GBPRI.HEADER) for building the GCG-indices, e.g.:

Name: EMBLDIR:EMBLPRI LN: EM_PR SN: EM_PR Rel: 35.0Reldate: 06/93 Fordate:06/93 Type: N FORMAT: NBRF

Name: GENBANKDIR:GBPRI LN: GB_PR SN: GB_PR Rel: 77.0 Reldate: 06/93 Fordate:06/93 Type: N FORMAT: NBRF

5. Disk space: Disk space requirement is totally dependent on the size of the sequence files being converted. For single-file conversions (Methods 1–3), each conversion increases the file space used by approximately the same amount as the original file. For database formatting

(Methods 4 and 5), a total of 3.5 times the size of the original data file(s) should be available. When the processing is complete, however, the original file(s) and some intermediary files can be deleted.

3. Methods

3.1. Method 1—The READSEQ Program

1. If the example file to be converted is called TEST.SEQ, and the formatted file (i.e., the output) is to be called FORMATTED.SEQ, run the program by:

$ *readseq -v -a*
Name of output file: *formatted.seq*

2. The program lists the 18 different formats that it can write out:

1. IG/Stanford	10. Olsen (in-only)
2. GenBank/GB	11. Phylip v3.2
3. NBRF	12. Phylip
4. EMBL	13. Plain/Raw
5. GCG	14. PIR/CODATA
6. DNAStrider	15. MSF
7. Fitch	16. ASN.1
8. Pearson/FASTA	17. PAUP/NEXUS
9. Zuker (in-only)	18. Pretty (out-only)

Choose an output format (name or #):

Enter the number, or name, of the format that is required, i.e., between 1 and 18 and press return.

3. There is now a prompt for the name of the input file

Name an input sequence or -option: *test.seq*

4. There is then a repeat prompt for the input file, just press return to enter a blank line to finish.

Name an input sequence or -option:

5. There is now a file called FORMATTED.SEQ in the selected format in the default directory.

3.2. Method 2—Conversions from GCG Format

GCG has several programs that copy a file already in GCG format into a new file. GCG names the programs as TOx, where x is the new format created, i.e., TOFITCH, TOIG, TOSTADEN, TOPIR. The GCG programs TOPIR and TOSTADEN are used as examples.

1. Make sure the GCG package is available on the computer:

$ *gcg*

A banner and copyright message should appear. If this fails, consult the system manager—GCG may have another name on the user's system, e.g., GCG7, UWGCG.

2. Run the program TOPIR by entering its name:

$ *topir*

TOPIR writes GCG sequence(s) into a single file in PIR format.
TOPIR of what GCG sequence(s) ? *cchu.pir1*
    Begin (* 1 *) ?
    End (* 104 *) ?
What should I call the output file (* Cchu.Pir *) ?
CCHU 104 characters.

The new file CCHU.PIR is in NBRF format.

3. Run the program TOSTADEN by entering its name:

$ *tostaden*

ToStaden writes a GCG sequence into a file in Staden format. If the file contains a nucleotide sequence, the ambiguity codes are translated as shown in Appendix III of the PRO-GRAM MANUAL.

TOSTADEN of what GCG sequence ? *cchu.pir1*
    Begin (* 1 *) ?
    End (* 104 *) ?
What should I call the output file (* Cchu.Sdn *) ?

The new file CCHU.SDN is in Staden format.

3.3. Method 3—Conversions into GCG Format

GCG has several programs that copy a file not in GCG format into a new file. GCG names most of the programs as FROMx, where x is the format of the input file, i.e., FROMIG, FROMPIR, FROMEMBL, FROMGENBANK, FROMSTADEN. The GCG programs FROMPIR, FROMSTADEN, and REFORMAT are used as examples, with RE-FORMAT reading a plain-text file. In all cases, the file being written to is in GCG format.

1. Run the program FROMPIR by entering its name:

$ *frompir*

FROMPIR reformats sequences from the protein database of the Protein Identification Resource (PIR) into individual files in GCG format.
FROMPIR of what PIR sequence file ? *cchu.pir1*
Cchu.Gcg   104 aa
FROMPIR complete:
    Files written: 1
    Total length:104

A new file named CCHU.GCG, in GCG format, is written.
2. Run the program FROMSTADEN by entering its name:

$ *fromstaden*
FromStaden changes a sequence from Staden format into GCG format.
If the file contains a nucleotide sequence, the ambiguity codes are translated as shown in Appendix III of the PROGRAM MANUAL.
FROMSTADEN of what Staden sequence file ? *cchu.sdn*
What should I call the output file (* Cchu.Seq *) ?

A new file named CCHU.SEQ, in GCG format, is written.
3. Run the program REFORMAT by entering its name:

$ *reformat*
REFORMAT rewrites sequence file(s), symbol comparison table(s), or enzyme data file(s) so that they can be read by GCG programs.
REFORMAT what sequence file(s) ? *test.seq*
What should I call the output file (* Test.Seq *) ?
No ".." divider

A new file named TEST.SEQ, in GCG format, is written, despite the apparent failure message.

3.4. Method 4—Formatting EMBL-Format Databases

1. First compile and link all the PIR programs, from the FORTRAN source code, in the usual way, e.g.,

$ *for createdbs*
$ *link createdbs*

Repeat for the files named createinx and indexer. The sorttmp.c file is treated in a similar way:

$ *cc sorttmp*
$ *link sorttmp*

2. Create VMS symbols that will run the PIR programs. If the .EXE files are in a directory called User$disk:[Yourname.convert], then the symbols are:

> $ *createdbs:==$User$disk:[Yourname.convert]createdbs*
> $ *createinx:==$User$disk:[Yourname.convert]createinx*
> $ *indexer:==$User$disk:[Yourname.convert]indexer*
> $ *sorttmp:==$User$disk:[Yourname.convert]sorttmp*

3. Now run each of these programs in turn, with the default directory being empty, with a lot of disk space present. The first program, createdbs, "knows" about the 16 different EMBL subsections, and can process them all just by giving it the location of the source files, which are usually on a CD ROM. The user need only set up a logical name pointing to the CD ROM, e.g.,

> $ *assign DAD8: emblcd*

a. Split each of the EMBL subsections into two NBRF-style files, one with the sequence information (EMBL*.SEQ), the other with the reference information (EMBL*.REF), where * is the name of the subsection:

$ *embldbs*
Database [PIR,CODATA,GENBANK,GBNEW, EMBL,
  SWISSPROT]: *embl*
Directory for 16 *.DAT files: *emblcd:[embl]*

b. Create the basic index file (EMBL*.INX) used by the other PIR programs. This has to be run 16 times—once for each of the files just produced by the createembl program. The following is an example for the EMBLPRI (primates subsection):

$ *createinx*
Database (no file type): *emblpri*
Code length [6]: *10*
Database type (Text, Protein, or Nucleic) [P]: *n*
Database format (NBRF, CODATA, GenBank, EMBL,
    or Unknown) [NBRF]: *embl*
Database name: *emblpri*
Release date (yymmdd):*930615*
Release number:*35.0*
Database description: *EMBL Primate entries*

The same result can be achieved by entering all the information on one line (albeit somewhat encrypted):

$ *createinx/emblpri/10/2/3/0/emblpri/930615/35.0/EMBL primate entries*

    c. Create the author (.aux), accession (.acx), species (.spx), title (.ttx), and keyword (.wox) index files used by the XQS, NAQ, and PSQ programs. These files are used for building futher indices for the XQS and ATLAS programs (*see* PIR software distribution).

      To do this, run the two programs, indexer and sorttmp, one after the other, for each of the 16 subsets of data.

> $ *createinx*
> Database: *emblpri*
> Is this a preliminary update [Y]: *n*
> (ACX AUX FTX HOX JRX SFX SPX TTX WOX)
> Index to create (<ALL> for all, <ret> to run): *aux*
> Index to create (<ALL> for all, <ret> to run): *wox*
> Index to create (<ALL> for all, <ret> to run): *acx*
> Index to create (<ALL> for all, <ret> to run): *spx*
> Index to create (<ALL> for all, <ret> to run): *ttx*
> Index to create (<ALL> for all, <ret> to run):

Now run sorttmap on each of the five files just produced. The simplest way of doing this is to execute the DCL file sorttmp.com supplied with the PIR programs:

$ *@sorttmp*

    d. Create the index files (EMBL.OFFSET, EMBL.NAMES, EMBL.NUMBERS), which allow the GCG programs to read database entries directly. Again, do this for each of the 16 subsets.

$ *gcg*
$ *dbindex/nomonitor emblpri.seq*

    e. Create the file (EMBL*.SEQCAT) used by the GCG program STRINGSEARCH. Make sure the file EMBLPRI.HEADER (and other heading files) is present before running the GCG program SEQCAT.

$ *seqcat EMBLPRI.SEQ/default*

    f. After checking that all steps have been correctly completed, repeat steps b–e for each of the EMBL subsections. In release 35 these files are PHG, ORG, FUN, PRO, PLN, INV, VRT, PRI, ROD, MAM, SYN, VRL, UNC, EST, PATENT, BB.

3.5. Method 5—Formatting the GENBANK Database

Exactly the same programs may be used for the GenBank database as shown for EMBL. GenBank has 14 subsections. In release 77 these files are PHG, BCT, PLN, INV, VRT, PRI, ROD, MAM, SYN, RNA, VRL, UNA, EST, PAT.

The one-line example for createinx is slightly different for GenBank—the user must specify GenBank format by:

$ *createinx/gbpri/10/2/2/0/gbpri/930615/77.0/GenBank primate entries*

4. Notes

1. READSEQ reads IN most of the formats listed in step 5. The program is clever enough to identify the input format (unlike the GCG programs).
2. READSEQ has several optional parameters, only -v prompts for all the required input. More typically, use -o to denote the output file, -f# for the format required, e.g., to write a file in NBRF format:

$ *READSEQ -a TEST.SEQ -f3 -oFORMATTED.SEQ*

The -a parameter ensures that multiple sequences in a file are recognized, otherwise you have to select sequences by number.
3. The case of the sequence part of a file can be changed using the -c (lower case) or -C (upper case) option of READSEQ, for any format, e.g.:

$ *readseq -c TEST.SEQ -oFORMATTED.SEQ*

4. To list all the optional parameters that are available:

$ *readseq -h*

5. Do not use READSEQ on GCG format files with text included between '>' symbols in the sequence (as created when including files in SEQED). This is valid GCG format, but READSEQ does not recognize it and the output is incorrect.
6. PIR format is practically the same as NBRF format. Both names are used interchangeably for maximum confusion, and both are valid for protein and nucleotide sequences. The title information given by the FROMPIR program could be mistaken to mean that PIR format refers to protein sequences only.
7. All formats, except GCG and plain-text, allow for multiple sequences in a single file. When converting from a multiple sequence format, the programs READSEQ and FROMPIR create several GCG-format files (note the option in READSEQ to write to MSF files, these are multiple-sequence files acceptable by GCG), with one sequence in each file.

8. The reverse case, copying several GCG sequences into a multiple-sequence format, cannot be done with READSEQ. Instead, use the GCG program REFORMAT with a file of sequence names. First enter the names of all the GCG-format sequence files into a file, say LIST.NAM. Then:

$ reformat/msf @list.nam

The output file is in MSF format, which may subsequently be converted using READSEQ.

9. Care should be taken when converting between Staden and GCG formats. The TOSTADEN program converts all lower case letters in GCG format into Staden ambiguity codes, which are numbers. Many GCG programs create sequences with lower case letters, which do not correspond to Staden ambiguity codes. Instead these are usually consensus sequences where lower case refers to a majority symbol. (e.g.,'c' in Staden format means C or CC).

10. The GCG program FROMPIR is usually set up to convert the NBRF sequence symbol '-' into '?'. It is the author's opinion that '-' should be converted into '.'. The computer manager may be able to make this change by amending source code in the GCG package.

Add one line in the file *gensource:frompir.for* so it reads:

```
If (Calls.eq.1) then
    Call SymbolSet(Symbols)
    Symbols(IChar('.')) = Char(0)
    Symbols(IChar('*')) = Char(0)
    Symbols(IChar('-')) = '.'
    Calls = -1
end if
```

Information on how to compile and relink the program can be found in the GCG System Manual.

11. The NBRF-protein *(10)* and nucleic acid databases are supplied, on tape, with steps a–e of Method 4 already completed. The GCG indexes supplied with recent NBRF-protein releases have been incorrect, so the user has needed to repeat steps d and e to create GCG-readable files. The NBRF-protein database is in three parts, with files named PIR1.SEQ, PIR2.SEQ, and PIR3.SEQ, so steps d–e must be carried out for each of these sequence files.

12. The Swissprot database *(11)* is provided as a single file in EMBL format. The name SWISSPROT need only be substituted for EMBL

throughout the procedure. In SWISSPROT.HEADER, change the TYPE: N part so it reads TYPE: P

Name: SWISSPROTDIR:SWISSPROT LN: SWISS SN: SW Rel: 18.0
    Reldate: 09/91 Fordate:09/91 Type: P FORMAT: NBRF

13. Methods 4 and 5 process very large quantities of data and take a lot of computing time (CPU). It is unreasonable to run the programs interactively, and a batch method should be used. By far the longest period is taken up by the PIR program CREATEINX.

14. GCG format has a maximum sequence length of 350,000 bases. Any entries in the databases that might exceed this cannot be used correctly by the GCG package. Amendments can be made to CREATEDBS.FOR to identify all sequences that exceed a given maximum size, split them and create additional entries in the database. An overlap of 10,000 bases, common to both entries, should be made to ensure that pattern searching programs do not miss features that span the break-point. The amendments required are too lengthy to detail here.

15. The GCG programs EMBLToGCG and GenBankToGCG can be used to reformat the EMBL and GenBank databases, in place of Methods 4 and 5. The database is then only available to the GCG software.

16. The sequence analysis software defines the format of the sequence data, not the source of that data. For example the FETCH program in the GCG package retrieves entries from the EMBL, GenBank or NBRF databases. The retrieved sequence is written to a file in GCG-format. Similarly the COPY command, in the NAQ or XQS program, retrieves a database entry, but the file written is in NBRF format.

17. The PIR program indexer cannot build all of the indices it promises. FTX, HOX, SFX are NOT created even if the ALL option is selected. Neither is the Journal Index built, as the information in the EMBL and GenBank files is in the wrong format.

References

1. Stoehr, P. J. and Cameron, G. N. (1991) The EMBL data library. *Nucleic Acids Res.* **19,** 2227–2230.
2. Burks, C., Cassidy, M., Cinkosky, M. J., Cumella, K. E., Gilna, P., Hayden, J. E.-D., Keen, G. M., Kelley, T. A., Kelly, M., Kristofferson, D., and Ryals J. (1991) GenBank. *Nucleic Acids Res.* **19,** 2221–2225.
3. Devereux, J., Haeberli, P., and Smithies, O. (1984) A comprehensive set of sequence analysis programs for the VAX. *Nucleic Acids Res.* **12,** 387–395.
4. Orcutt, B. C., George,D. G., Fredrickson, J. A., and Dayhoff, M. O. (1982) Nucleic acid sequence database computer system. *Nucleic Acids Res.* **10,** 157–174.

5. Orcutt, B. C., George,D. G., and Dayhoff, M. O. (1983) Protein and nucleic acid sequence database computer systems. *Ann. Rev. Biophys. Bioeng.* **12,** 419-441.
6. Hunt, L. T. (1990) in *Protein Identification Resource Newsletter,* vol. 9, May. National Biomedical Research Foundation, Washington, DC.
7. Pearson, W. R. and Lipman, D. J. (1988) Improved tools for biological sequence comparison. *Proc. Natl. Acad. Sci. USA* **85,** 2444–2448.
8. Staden, R. (1986) The current status and portability of our sequence handling software. *Nucleic Acids Res.* **14**(1).
9. Gilbert, D. G. (1989) ReadSeq, C and Pascal routines for converting among nucleic acid & protein sequence file formats, suitable for various computers. Published electronically on the Internet, available via anonymous ftp to ftp.bio.indiana.edu.
10. Barker, W. C., George, D. G., Hunt, L. T., and Garavelli, J. S. (1991) The PIR protein sequence database. *Nucleic Acids Res.* **19,** 2231–2236.
11. Bairoch A. and Boeckmann B. (1991) The SWISS-PROT protein sequence data bank. *Nucleic Acids Res.* **19,** 2247–2249.

CHAPTER 31

Obtaining Software via INTERNET

Cary O'Donnell

1. Introduction

Many noncommercial software packages and individual programs are available at several computer sites around the world. Some sites make these files available via "file-servers," and some via "anonymous FTP." There are usually information files available listing everything that can be obtained at the site. Nearly all computer sites worldwide have direct or indirect access to the INTERNET computer network, and so can communicate electronically.

Anonymous FTP requires logging on to the remote computer with the user name "anonymous." This user name allows restricted use of the remote computer, enough to search for the required files and send them back. Anonymous FTP allows direct transfer of both text and binary files across the computer network. Method 2 describes this procedure using the anonymous FTP service provided at the University of Indiana.

To retrieve information and software from a file-server, an electronic mail message is sent to the file-server address, requesting particular files. The file-server is a program that responds by mailing the requested files back to the original sender. Program files, or large sets of files, are normally supplied in an encoded form; help information is supplied as readable text. Two file-servers are described here: The EMBL file-server (1) in Heidelberg, Germany; and the UH geneserver (2) at the University of Houston.

Several types of file encodings are used at FTP and file-server sites:"ZOO," "UUE," "tar," and "shar." A ZOO file contains several

From: *Methods in Molecular Biology, Vol. 25: Computer Analysis of Sequence Data, Part II*
Edited by: A. M. Griffin and H. G. Griffin Copyright ©1994 Humana Press Inc., Totowa, NJ

other files compressed into just one file, with "white space" removed, and in binary code. Any binary file can be converted into a UUE file, made up of only 94 different ASCII (i.e., text) characters. With only 60 characters/line, UUE files can be sent by mail between any computer combination without introducing errors. Most software sent by file-servers is in the UUE form and once decoded is in the ZOO form, which must then be decompressed.

Tar files are compressed binary files, which may be decompressed under the UNIX operating system, or with appropriate VMS utilities. Shar files are in ASCII form and may be self-unpacking, or may require a spacial unshar program to unpack the files. FTP sites commonly provide program files in all these formats.

Each file-server or ftp site usually provides copies of the software, which must be used to decode the files. In Method 1, described below, the UU-decoding software is retrieved at the same time as the software READSEQ.

2. Materials

1. Computers: VAX/VMS-Ultrix, and others.
2. Terminal: Any VT100 terminal is suitable since there is no graphical output.
3. Programs: C language compiler VAX/VMS (cc) and libraries, or, for the VAX-Ultrix, an ANSI-C language compiler, e.g.,(vcc), and libraries.
4. Electronic mail privileges: Method 1A requires access to INTERNET for sending and receiving e-mail. It may be necessary when mailing from the UK to send mail explicitly via the JANET address UK.AC.NSFNET-RELAY, e.g., to user%site@nsfnet-relay. VAX users in the UK should use cbs%nsfnet-relay::site::user. From the rest of Europe: to EMBL-HEIDELBERG.DE, or local variation of that address. Method 1B requires access to the US part of INTERNET, specifically the address: BCHS.UH.EDU.
5. Special equipment: For Method 2 a direct connection to INTERNET or to FTP.BIO.INDIANA.EDU. The user may need to find out from a local computer manager how this connection is made. In the UK, sites without an INTERNET connection may use the "Guest FTP service" provided by the University of London Computer NSF.SUN. The user may need to obtain special privileges at a local computer ("host site") to connect to this service.
6. Disk space: Up to 3000 blocks (1 block is 512 bytes) of file space is required for the steps described, but less space is necessary if files not

required for later steps are deleted. The final READSEQ executable file takes up to 189 blocks.

3. Methods

3.1. Method 1A—The EMBL File-Server

1. Create a text file called EMBL.SERV containing the following text:

```
GET VAX_SOFTWARE:READSEQ.UAA
GET VAX_SOFTWARE:ZOO.UAA
GET VAX_SOFTWARE:UUD.C
```

2. Mail the file to Netserv@embl-heidelberg .de, the exact syntax is very dependent on the mail system used at a local site. The following examples illustrate two likely alternatives, the second for mailing from UK sites via NSFNET-RELAY.

    $ *mail embl.serv netserv%de.embl-heidelberg*
    $ *mail embl.serv cbs%nsfnet-relay::de.embl-heidelberg::netserv*

3. The requested files arrive as seven electronic mail messages. Extract the seven mails into seven files UUD.C, READSEQ.UAA, READSEQ.UAB, READSEQ.UAC, ZOO.UAA, ZOO.UAB, and ZOO.UAC. Take care when identifying the many different file names: a clue is given in the subject line of the mail, e.g.:

Subject: Reply to: GET VAX_SOFTWARE:READSEQ.UAA (part 2 of 3)

should, obviously, be saved as READSEQ.UAB. For example, when using VMS mail, use the "extract" command:

    MAIL> *extract readseq.uab*

4. After exiting from MAIL, edit the file UUD.C using a text editor. The mail headings that have been added to the top of the file must be removed. The very first line of the file should then be:

    /* Uud -- decode a uuencoded file back to binary form.

The other files do not need to be edited, unless errors occur in the remaining steps.

5. First compile and link the UUD program:

       $ *cc uud*
       $ *link uud*

6. Users of VMS have to create a DCL symbol to run UUD. First determine the name of the current disk and directory:

$ *show default*

The operating system replies with something like:

Current: User$disk:[Yourname.subdirectory]

Substituting appropriately, define the VMS symbol:

$ *uud:==$User$disk:[Yourname.subdirectory]uud*

7. Decode the ZOO and READSEQ files:

$ *uud ZOO.UAA*
$ *uud READSEQ.UAA*

The UUD program automatically picks up the .UAB and .UAC files and gives an error if any file is not present when required. On correct completion, the files READSEQ.ZOO and ZOO.EXE are present. It is good practice to place the READSEQ.ZOO file in a directory of its own from now on.

8. The decoded ZOO program is the decompression software. Again VMS users must create a DCL symbol to run it:

$ *Zoo:==$User$disk:[Yourname.subdirectory]zoo*

9. Decompress the READSEQ.ZOO file:

$ *Zoo -ex READSEQ*

Several files are now present that comprise the source code and any documentation. The file called READSEQ.EXE is the executable program file.

10. Again make a VMS symbol to allow the program to be executed. The program may then be run as described in Chapter 30.

$ *readseq:==$User$disk:[Yourname.subdirectory]readseq*

(Steps 1–9 need to be carried out only once, after which the symbol definition in 10 should be copied into the LOGIN.COM file.)

11. If the operating system reports errors when running READSEQ, then try recompiling and relinking the source code:

$ *cc readseq, ureadseq*
$ *link readseq, ureadseq, sys$library:vaxcrtl/lib*

Alternatively, execute the MAKEFILE that also carries out some data checking:

$ *@make.com*

3.2 Method 1B—The UH Gene-Server

1. Create a completely blank file called, e.g., UH.SERV.
2. Mail the file to the following address (but check the mail syntax at the host site):

gene-server@bchs.uh.edu

or

gene-server%bchs.uh.edu@cunyvm

The commands to the server must be contained in the subject line of the mail. For example, for the READSEQ software, mailing from the UK:

$ *mail cbs%nsfnet-relay::edu.uh.bchs::gene-server/subject=*
"SEND VAX READSEQ.UUE"

$ *mail cbs%nsfnet-relay::edu.uh.bchs::gene-server/subject=*
"SEND VAX ZOO.UUE"

$ *mail cbs%nsfnet-relay::edu.uh.bchs::gene-server/subject=*
"SEND VAX UUD.C"

3. The requested files arrive as three electronic mail messages. Extract the files as READSEQ.UAA, ZOO.UAA, and UUD.C and continue from step 4 of Method 1A.

3.3. Method 2—Retrieving Files by Anonymous FTP

1. First log on to the computer, which has the IP connection to INTERNET. This is the INTERNET-HOST computer. Many users will find this facility readily available on the local computer. For sites in the UK without Internet connectivity the University of London provides an open account called guestftp. From a JANET-HOST, log on to this computer:

```
$ pad call nsf.sun
login:guestftp
password:guestftp
Enter your reference for this session: Cary
guest_ftp> dir
total blocks:0
```

The reference name creates some temporary storage space. It allows users to log on at a later time to locate any files.

2. Once on an INTERNET-HOST computer (which may be the user's own host computer), run the FTP program that will make a connection to any other INTERNET site:

guest_ftp> ftp (or, possibly $ ftp)

When the ftp prompt appears, open the connection to the **remote** site:

ftp> open ftp.bio.indiana.edu
connected to ftp.bio.indiana.edu

3. The remote computer then requests the user to log on. Log on as "anonymous," but give own electronic mail address as password, in INTERNET style, for the providers' records.

Name (ftp.bio.indiana.edu:guestftp): *anonymous*
331 guest login OK, send e-mail address as password
password: *odonnell@afrc.ac.uk* (the user will not see this on
the terminal)

4. Now look at the directories available and try to locate the software required.

ftp> dir

Change directory to the molbio directory.

ftp> cd molbio
ftp> dir

Each set of programs has its own subdirectory. The example here is to obtain the READSEQ software:

ftp> cd readseq
ftp> dir

Look for a file with a title that suggests it is readable, such as a README or FAQ file. A FAQ file usually gives answers to "Frequently Asked Questions." This file should be retrieved first and read on the home computer. It will inform a user which files are needed. The user should take note of which files are text (ASCII) files and which are binary files. If the file required is a binary file, as in this case, the following command must be given:

ftp> binary

5. Send the file across to your host computer.

ftp> get readseq.shar

6. Log off the remote computer, and then exit the ftp program.

ftp>*close*

ftp>*bye*

7. Now all the files are present on the INTERNET host computer (which may be the user's own host site). If using NSF.SUN, the user will have an additional step—sending the files back from NSF.SUN to the JANET host site.

```
guest_ftp> dir
guest_ftp> push
Okay lets push a file using NIFTP
Give local filename: readseq.shar
Give remote filename: readseq.shar
Give NRS name of remote host: afrc.arcb
Do you want binary or <default> ascii (input b or a): b
Give user name on remote host: odonnell
Give user password on remote host:
Re-type password to make sure:
```

Finally logoff NSF.SUN

guest_ftp>*lo*

8. Unpack the READSEQ.SHAR file to produce the same files as achieved at the end of Section 9 of Method 1A. At the time of writing an unpacking program for VMS was not available at ftp.bio.indiana.edu, but the files VMS.UNSHAR.1, VMS.UNSHAR.2 and VMS.UNSHAR.3 were available on the UH Gene-Server.

After unpacking carry out step 11, and then 10 of Method 1A.

4. Notes

1. To obtain information files, mail the following text to the EMBL file server:

HELP

DIR

HELP VAX_SOFTWARE

DIR VAX_SOFTWARE

2. The equivalent file for the UH file server is to put the following in the SUBJECT LINE of the mail:

> SEND HELP
> SEND INDEX
> SEND VMS HELP
> SEND VMS INDEX

3. Computer managers may restrict the use of mail and ftp from the host computer. For any failure whereby the mail does not appear to be sent, or the INTERNET connection cannot be made, contact the local computer manager.

4. The user may get the following error message when attempting to connect to the remote site:

ftp> *open ftp.bio.indiana.edu*
failed to get host information for ftp.bio.indiana.edu from database

This means that ftp.bio.indiana.edu is not in the site's telephone book! In this case, the user will have to make the connection directly using the INTERNET number of the site (**warning:** These numbers can change):

ftp> *open 129.79.224.25*

5. Many program files are supplied at ftp sites in binary form. When retrieving such files you must set ftp to work in binary:

ftp> *binary*

To switch back for text transfer:

ftp> *ascii*

6. The University of Houston is also reachable by anonymous ftp. The address is: ftp.bchs.uh.edu, and the direct number is 129.7.2.43.

Instead of steps 4 and 5 of Method 2, do the following to collect the correct files:

ftp> *cd pub/gene-server/vms*
ftp> *dir*
ftp> *get uudecode.c*
ftp> *get zoo.uue*
ftp> *get readseq.uue*

After retrieving them continue at step 4 of Method 1A.

7. EMBL provides an FTP service on ftp.embl-heidelberg.de

8. NSFNET-RELAY is a "gateway" between the JANET network and INTERNET, allowing mail and files to be transferred. Other gateways from JANET include:

> UKNET (to UUCP network)
> EARN-RELAY (to EARN/BITNET)

EARN/BITNET and INTERNET also have gateways linking them. Mail from JANET to INTERNET sites sent through uk.ac.earn-relay will be forwarded to INTERNET.

Some computer sites may not be able to send mail via one or more of these gateways. Some mail servers on INTERNET may attempt to send mail through a gateway that the user's site does not subscribe to. In that case the user will receive nothing from the file server!

9. On the EMBL file server an acknowledgment mail is usually received first. Check this to see the mail has been understood.

10. Large files tend to be mailed back at off-peak periods. Sometimes the file servers, or intermediate gateways, get very busy, and a wait of a day or more is not unusual.

11. Anonymous FTP sites usually request that they be used only at off-peak times if requesting large files, i.e., avoid the 8am–6pm period, taking account of local time at the remote site. For the same reasons, transfers across JANET from NSF.SUN can take a long time.

12. Some very large pieces of software are still large even when ZOO-ed. The UU-encoded files are then split into several files to enable them to be sent across the networks. The information listed on the EMBL file server indicates which software arrives in several parts. Such files are usually listed with the extension .UAA, .UAB, .UAC, and so forth, instead of .UUE. All the files should be present before starting the UUDECODE process.

13. On a UNIX system compiling the READSEQ software is carried out as follows:

% vcc readseq.c ureadseq.c

An alias definition for READSEQ is not usually required on Unix.

14. The READSEQ.C source code can also be compiled on the Apple Macintosh, according to its documentation. There is information in the READSEQ.C file that provides a script to do this.

15. When preparing the above examples it was noticed that the version of READSEQ available at ftp.bio.indiana.edu was a more recent version than that available at the file servers. The UH gene server also had the more recent 3-file uu-encoded files of READSEQ available (the single file version in the example was an older version still). This is not unexpected as the ftp.bio.indiana.edu site is the source site used by the file servers, so there may be a delay before a new version is available at the file servers.

Acknowledgments

I am very grateful to my former colleague Philip O'Connor for his advice on using ftp. Thanks to Frank Wright for reading an early draft of the manuscript.

References

1. Fuchs, R., Stoehr, P., and Rice, P. (1990) New services of the EMBL data library. *Nucleic Acids Res.* **18,** 4319–4323.
2. Davison, D. B. and Chappelaar, J. E. (1990) The GenBank Server at the University of Houston. *Nucleic Acids Res.* **18,** 1571,1572.

CHAPTER 32

Submission of Nucleotide Sequence Data to EMBL/GenBank/DDBJ

Catherine M. Rice and Graham N. Cameron

1. Introduction

The EMBL Data Library *(1)* was founded in 1980 as a direct consequence of the amount of sequence data appearing in the journals. Over the past 11 years, the growth in data acquisition has been exponential. With the latest developments in genome projects, we foresee no let up in the amount of data they will receive in the next few years. We do envisage, however, that a larger proportion will not be accompanied by detailed biological knowledge.

In 1982, a direct collaboration was established between GenBank *(2)* (in Los Alamos) and the EMBL Data Library to facilitate coverage of all primary nucleotide sequence data. The DNA Data Bank of Japan (DDBJ) joined more recently. The three databases are equivalent, and published data are exchanged daily. Data are incorporated either by computer-readable submissions from authors or (much more rarely) by entering the published sequences by hand. Data entry is error-prone for a number of reasons, including legibility of the original sequence-containing article. Author submission is more accurate and results in faster incorporation of the data.

2. Submission Methods

Many journals have mandatory submission policies with member databases. In these instances, acceptance of sequence-containing papers for publication requires proof of submission. Sequence data submitted to EMBL/GenBank/DDBJ receive unique identifiers in the

From: *Methods in Molecular Biology, Vol. 25: Computer Analysis of Sequence Data, Part II*
Edited by: A. M. Griffin and H. G. Griffin Copyright ©1994 Humana Press Inc., Totowa, NJ

form of an accession number that provides such proof. The number identifies an individual sequence, so many accession numbers could be cited in any given publication. Details for each journal are given in their individual "notes to submitting authors." The databases are, of course, happy to receive all sequence data, including sequences that will never appear in any publication. Relevant addresses for all three collaborating databases are given in Appendix 1, but specifics are given only for EMBL.

Much of the communication in the databases is done by electronic mail. Not only do the databases often use this to communicate among themselves, but it is also a very quick and efficient way to contact submitters. The EMBL Data Library receives approx 60% of total submissions by electronic mail. These come either as sequences appended to a copy of the submission form or as Authorin output. Any queries resulting from a submission can be readily answered. Subsequent notification of newly assigned accession numbers is preferably given this way. Apart from the submission address, we also have an address for general queries (*see* Appendix 1). Our internet address can be reached via various gateways, including Bitnet, Usenet, and JANET. Advice on how to contact us can be obtained either from a local network expert or by contacting the Data Library itself.

A part of our direct submission data comes as Authorin output sent by electronic mail.

The Macintosh version of Authorin (currently version 2.1) is available from the fileserver. To access the program from EMBL, send electronic mail to the EMBL fileserver *(3)* with the following commands:

HELP software
GET Mac_software:authorin.hqx

The IBM-PC version of Authorin (currently version 1.1, but 1.2 is due soon) is available from Intelligenetics at no extra cost (*see* Appendix 1). Full instructions for use are provided with the software.

A large proportion of our submissions arrive by electronic mail as filled out submission forms with appended sequence data. A computer-readable copy of the submission form is available and is included in each release (for all the collaborating databases). This computer-readable form can also be accessed as follows: Send a mail message to the EMBL fileserver *(3)* (Appendix 2), and include the following command either in the subject line or in the body of the message:

GET DOC:datasub.txt

One submission form should be filled out for each nucleotide and protein sequence (where applicable). The EMBL Data Library also provides software for VAX/VMS users, which simplifies the process of filling out and mailing the submission form. To retrieve that instead, include the commands:

HELP software
GET Vax_software:subform.uaa

on separate lines in the body of the message.

Printed copies of the submission form are available in the first issue each year of *Nucleic Acids Research* (*see* Fig. 1). They are also available on request from the EMBL Data Library.

Submissions arrive also on diskette by mail, usually accompanied by a paper copy of the submission form. We rarely contact the author by post, except in the absence of any electronic mail or fax address.

The EMBL Data Library supports Macintosh or IBM-PC compatible (3.25 or 5.25 in.) diskettes. When sending a submission by Macintosh or IBM-PC diskette, one can either use the relevant Authorin program (preferred) or simply send the sequence as text format with an accompanying hard copy or machine-readable submission form.

The large variety of wordprocessing applications now available for Macintosh and IBM-PC machines make it difficult for us to guarantee readability for all given formats. For this reason, we request that all sequence data be saved as simple text on the diskette. The data are easier to handle, and therefore, the submitter benefits by quicker receipt of accession numbers.

We do not accept sequence data as such by fax, but in the absence of an electronic mail address, we use the fax address for any further communications with the author. This has the benefit of speed for cases when the journal publication deadline is at hand and accession numbers are urgently required. (*See* Appendix 1 for Data Library addresses.)

Sequence Data Submission Form

This form solicits the information needed for a nucleotide or amino acid sequence database entry. By completing and returning it to us promptly you help us to enter your data in the database accurately and rapidly. These data will be shared among the following databases: DNA Data Bank of Japan (DDBJ; Mishima, Japan); EMBL Nucleotide Sequence Database (Heidelberg, FRG; GenBank (Los Alamos, NM, USA and Mountain View, CA, USA); International Protein Information Database in Japan (JIPID; Noda, Japan); Martinsried Institute for Protein Sequence Data (MIPS; Martinsried, FRG); National Biomedical Research Foundation Protein Identification Resource (NBRF-PIR; Washington, D.C., USA.); and Swiss-Prot Protein Sequence Database (Geneva, Switzerland and Heidelberg, FRG).

Please answer all questions which apply to your data. If you submit 2 or more non-contiguous sequences, copy and fill out this form for each additional sequence. Please include in your submission any additional sequence data which is not reported in your manuscript but which has been reliably determined (for example, introns or flanking sequences). When submitting nucleic acid sequences containing protein coding regions, also include a translation (SEPARATELY from the nucleic acid sequence). Then send (1) this form, (2) a copy of your manuscript (if available) and (3) your sequence data (in machine readable form) to the address shown below. Information about the various ways you can send us your data and about formats for the sequence data is given in the following two sections.

Thank you.

SUBMITTING DATA TO THE EMBL DATA LIBRARY

We are happy to accept data submitted in any of the following ways: **(1) Electronic file transfer:** files can be sent via computer network to: DATASUBS@EMBL-Heidelberg.DE. This INTERNET address can be reached via various gateways from Bitnet, JANET, etc. Ask your local network expert for help or phone us. Please ensure that each line in your file is not longer than 80 characters; longer lines often get truncated when they are sent. **(2) Floppy disks:** we can read Macintosh and IBM-compatible diskettes. Please use the 'save as text only' feature of your editor to save your sequence file, as otherwise we might have difficulty processing it. **(3) Magnetic tapes:** 9-track only (fixed-length records preferred); 800, 1600 or 6250 bpi (any blocksize); ASCII or EBCDIC character codes; any label type or unlabelled. Our address is:

EMBL Data Library Submissions	Computer network DATASUBS@EMBL-Heidelberg.DE
Postfach 10.2209	Telefax (+49) 6221 387 519
D-6900 Heidelberg	Telephone (+49) 6221 387 258
Federal Republic of Germany	

When we receive your data we will assign them an accession number, which serves as a reference that permanently identifies them in the database. We will inform you what accession number your data have been given and we recommend that you cite this number when referring to these data in publications.

If your manuscript has already been accepted for publication, the accession number can be included at the galley proof stage as a note added in proof. So that we can process your data and inform you of your accession number before you receive the galley proofs, please return this form to us as soon as possible. We suggest that the note added in proof should read approximately as follows: "The nucleotide sequence data reported in this paper will appear in the EMBL, GenBank and DDBJ Nucleotide Sequence Databases under the accession number _____."

A computer-readable version of this form is available on the distribution tapes of the EMBL Data Library from Release 11 onwards and on GenBank Releases 48 onwards and via the EMBL and GenBank File Servers. Feel free to use the computer-readable form rather than this printed one. In this case, the form should be filled out with a text editor and sent via computer network or normal post to the address indicated above.

FORMATS FOR SUBMITTED DATA

We would appreciate receiving the sequence data in a form which conforms as closely as possible to the following standards.

Each sequence should include the names of the authors.

Each distinct sequence should be listed separately using the same number of bases/residues per line. The length of each sequence in bases/residues should be clearly indicated.

Enumeration should begin with a "1" and continue in the direction 5´ to 3´ (or amino- to carboxy- terminus).

Amino acid sequences should be listed using the one-letter code.

Translations of protein coding regions in nucleotide sequences should be submitted in a separate computer file from the nucleotide sequences themselves.

The code for representing the sequence characters should conform to the IUPAC-IUB standards, which are described in: Nucl. Acids Res. 13: 3021-3030 (1985) (for nucleic acids) and J. Biol. Chem. 243: 3557-3559 (1968) and Eur. J. Biochem 5: 151-153 (1968) (for amino acids).

Fig. 1. A sample nucleotide sequence database submission form.

Please fill out with a typewriter or write legibly

I. GENERAL INFORMATION

Your last name	First name	Middle initials
Institution		
Address		
Computer mail address	Telex number	
Telephone	Telefax number	

On what medium and in what format are you sending us your sequence data? (see instructions on front page)
[ ] electronic mail
[ ] diskette: computer _____ operating system _____
            editor _____ file name _____
[ ] magnetic tape (specify format_____)

II. CITATION INFORMATION

These data represent [ ] new submission [ ] correction (Accession number of affected sequence _____)

These data are [ ] published   [ ] in press   [ ] submitted   [ ] in preparation [ ] no plans to publish
authors

title of paper

journal   volume   first-last pages   year

Do you agree that these data can be made available in the database before they appear in print?
[ ] yes   [ ] no, they can be made available after : _____ (please fill in date)

Does the sequence which you are sending with this form include data that do not appear in the above citation?
[ ] no   [ ] yes, from position _____ to _____ [ ] bases OR [ ] amino acid residues
        (If your sequence contains 2 or more such spans, use the feature table in section IV to indicate their positions)
If so, how should these data be cited in the database?
        [ ] published   [ ] in press   [ ] submitted   [ ] in preparation [ ] no plans to publish
authors

address(if different from that given in section I)

title of paper

journal   volume   first-last pages   year

List references to papers and/or database entries which report sequences overlapping with that submitted here.

first author	journal, vol., pages, and/or database, accession numbers

Fig. 1 *(continued)*.

III. DESCRIPTION OF SEQUENCED SEGMENT

Wherever possible, please use standard nomenclature or conventions. If a question is not applicable to your sequence, answer by writing N.A.; if the information is relevant but not available, write a question mark (?).

What kind of molecule did you sequence? (check all boxes which apply)

[ ] genomic DNA   [ ] genomic RNA   [ ] cDNA to mRNA   [ ] cDNA to genomic RNA

[ ] organelle DNA   [ ] organelle RNA   please specify organelle _____

[ ] tRNA   [ ] rRNA   [ ] snRNA   [ ] scRNA

for viruses: [ ] virus or   [ ] provirus or   [ ] viroid   [ ] DNA   or   [ ] RNA

         [ ] ds   or   [ ] ss   or   [ ] circular   [ ] enveloped or   [ ] nonenveloped

[ ] other nucleic acid (please specify)_____

[ ] peptide: [ ] sequence assembled by   [ ] overlap of sequenced fragments   [ ] homology with related sequence

                               [ ] other (please specify) _____

         [ ] partial: [ ] N-terminal   or   [ ] C-terminal   or   [ ] internal fragment

---

length of sequence   [ ] bases   or   [ ] amino acid residues

---

gene name(s) (e.g., *lacZ*)

---

gene product name(s) (e.g., beta-D-galactosidase)

Enzyme Commission number (e.g., EC 3.2.1.23)

---

gene product subunit structure (e.g.; hemoglobin $\alpha 2\beta 2$)

The following items refer to the original source of the molecule you have sequenced.

     organism (species) (e.g., Mus musculus)   plant cultivar

     strain (e.g., K12: BALB/c)   substrain

     name/number of individual or isolate (e.g., patient 123; influenza virus A/PR/8/34)

     developmental stage   [ ] germ line   [ ] rearranged

     haplotype   tissue type   cell type

     allele   variant   [ ] macronuclear

The following items refer to the immediate experimental source of the submitted sequence.

     name of cell line (e.g., Hela; 3T3-L1) or plant cultivar

     clone library   clone(s), subclone(s)

The following items refer to the position of the submitted sequence in the genome.

     chromosome (or segment) name/number

     map position   units: [ ] genome % or [ ] nucleotide number or [ ] other _____

Using single words or short phrases, describe the properties of the sequence in terms of: its associated phenotype(s); the biological/enzymatic activity of its product; the general functional classification of the gene and/or gene product macromolecules to which the gene product can bind (e.g., DNA, calcium, other proteins); subcellular localisation of the gene product; any other relevant information.

Example (for viral *erbB* nucleotide sequence): transforming; EGF receptor-related; tyrosine kinase; oncogene; transmembrane protein.

Fig. 1 *(continued)*.

IV. FEATURES OF THE SEQUENCE

Please list below the types and locations of all significant features experimentally identified within the sequence. **Be sure that your sequence is numbered beginning with "1."** Use < or > if a feature extends beyond the beginning or end of the indicated sequence span.

In the column marked   fill in

feature	type of feature (see information below)
from	number of first base/amino acid in the feature
to	number of last base/amino acid in the feature
ba	x, if your numbers refer to positions of bases in a nucleotide sequence
aa	x, if your numbers refer to positions of amino acid residues in a peptide sequence
id	method by which the feature was identified. E = experimentally; S = by similarity with known sequence or to an established consensus sequence; P = by similarity to some other pattern, such as an open reading frame
comp	x, if feature is located on the nucleic acid strand complementary to that reported here

Significant features include:

regulatory signals (e.g., promoters, attenuators, enhancers)
transcribed regions (e.g., mRNA, rRNA, tRNA). (Indicate reading frame if start and stop codons are not present)
regions subject to post-translational modification (e.g., introns, modified bases)
translated regions
extent of signal peptide, prepropeptide, mature peptide
regions subject to post-translational modification (e.g., glycosylated or phosphorylated sites)
other domains/sites of interest (e.g., extracellular domain, DNA-binding domain, active site, inhibitory site)
sites involved in bonding (disulfide, thiolester, intrachain, interchain)
regions of protein secondary structure (e.g., alpha helix or beta sheet)
conflicts with sequence data reported by other authors
variations and polymorphisms

The first 2 lines of the table are filled in with examples.

Numbering for features on the sequence submitted here [ ] matches paper [ ] does not match paper

	feature	from	to	ba	aa	id	comp
EXAMPLE	TATA box	1	8	x			
EXAMPLE	exon 1	9	>264	x			

Fig. 1 *(continued)*.

Occasionally, we receive queries concerning submissions or services by phone. Any such communications are difficult for us to record in an accurate way and are therefore not recommended except in unusual circumstances.

3. Processing Submissions

The EMBL Data Library processes data submissions within seven working days of receipt, and then either sends accession numbers to the submitting author or, if there is a problem, a request for further information. Submitters can help us a great deal in the following ways:

- Completeness—do give all the information possible;
- Check accuracy and explain any apparent inconsistency; and
- Give us a fast way to contact the submitting author—electronic mail, fax, or telex.

Once an entry has been produced from submission information, then a copy will be sent to the submitting author to review or update, if necessary. This entry may or may not be accessible to the public (*see* Fig. 2).

Entries are released to the user community initially by fileserver, where they will be available on the day they are made public. For those who do not receive daily updates through their EMBnet node, the entries will appear at the next quarterly release on tape and CD-ROM. Figure 2 is a sample finished entry. This flat file format is exactly how it appears to any user. Actual release of an entry to the public is either at the point of completion or, when specifically requested, after publication of the relevant sequence-bearing article. There are inherent problems associated with matching citations with data being held until publication. These can be alleviated by the author citing all relevant accession numbers in the publications and by informing the Data Library of publication.

Appendix 1: Nucleotide Sequence Databases' Addresses

1. EMBL Data Library:
   a. Internet electronic mail addresses:
      Data submissions: datasubs@EMBL-Heidelberg.DE
      General enquiries: datalib@EMBL-Heidelberg.DE

```
ID   ECMALZ    standard; DNA; PRO; 2345 BP.
XX
AC   X59839;
XX
DT   08-NOV-1991 (Rel. 29, Last updated, Version 6)
DT   29-OCT-1991 (Rel. 29, Created)
XX
DE   E.coli malZ gene for alpha-1,4-D-glucosidase
XX
KW   alpha-1,4-D-glucosidase; alpha-glucosidase; malZ gene.
XX
OS   Escherichia coli
OC   Prokaryota; Bacteria; Gracilicutes; Scotobacteria;
OC   Facultatively anaerobic rods; Enterobacteriaceae;
OC   Escherichia.
XX
RN   [1]
RP   1-2345
RA   Tapio S.;
RT   ;
RL   Submitted (27-MAY-1991) on tape to the EMBL Data Library by:
RL   S. Tapio, University of Konstanz, Dept of Biology AG Boos,
RL   P.O. Box 5560, W-7750 Konstanz, Germany.
XX
RN   [2]
RA   Tapio S., Yeh F., Shuman H., Boos W.;
RT   "The malZ gene of Escherichia coli, a member of the maltose
RT   regulon encodes a maltodextrin glucosidase";
RL   J. Biol. Chem. 266:19450-19458(1991).
XX
CC   *source: strain=K12;
XX
FH   Key            Location/Qualifiers
FH
FT   misc-binding   188..197
FT                  /bound_moiety="MalT"
FT   misc-binding   201..210
FT                  /bound_moiety="MalT"
FT   misc-binding   224..233
FT                  /bound_moiety="MalT"
FT   -10-signal     254..259
FT   RBS            280..282
FT   mRNA           265..>2103
FT                  /evidence=EXPERIMENTAL /gene="malZ"
FT   CDS            289..2103
FT                  /gene="malZ" /EC_number="3.2.1.20"
FT                  /product="alpha-glucosidase"
FT   stem loop      2257..2279
FT   stem loop      2301..2324
XX
SQ   Sequence 2345 BP; 529 A; 579 C; 683 G; 554 T; 0 other;
     tatcgggttg attggttatc acccggatac gcgtatctcg ctgtatgtcg
     gtttcgcgtg gattgttgtg ctgttgattg gctggatgtt taaacgccgc
     cacgatcgtc agctggctga


         .         .         .             .
         .         .         .             .
         .         .         .             .


     ccagacgtgg gcggcggctt gccatgccgt ttaacacgtt ctggatgaaa
     tccatatcgc gatagcgcac cagccactgc tctgaccaca agtaattgtt
     cagattgata aaacgtggcg
     gtgag
```

Fig. 2. A sample entry.

    b. Postal address:
        Data Submissions,
        Postfach 10.2209,
        Meyerhofstrasse 1,
        W6900 Heidelberg.
        Federal Republic of Germany
    c. Telephone:
        +49 6221 387258
    d. Telefax:
        +49 6221 387519
    e. Telex:
        461613 (embl d)

2. GenBank:
    a. Internet electronic mail address:
        gb-sub@genome.lanl.gov
    b. Postal address:
        GenBank Submissions,
        Mail Stop K710,
        Los Alamos National Laboratory,
        Los Alamos, NM 87545
    c. Telephone:
        +1 505 665 2177
    d. Telefax:
        +1 505 665 3493

3. DNA Data Bank of Japan:
    a. Internet electronic addresses:
        Data submissions: ddbjsub@ddbj.nig.ac.jp
        General enquiries: ddbj@ddbj.nig.ac.jp
    b. Postal address:
        Laboratory of Genetic Information Analysis,
        Center for Genetic Information Research,
        National Institute of Genetics, Mishima,
        Shizuoka 411, Japan
    c. Telephone:
        +81 559 750771 ×647
    d. Telefax:
        +81 559 756040

4. Intelligenetics:
    a. Internet electronic mail addresses:
        Authorin software: authorin@GenBank.bio.net
        or ftp from GenBank.bio.net

b. Postal address:
   Intelligenetics Inc.,
   700 East El Camino Rd,
   Mountain View, CA 94040
c. Telephone:
   +1 415 962 7364
   or 800 477 2459 within the US
d. Telefax:
   +1 415 962 7302

Appendix 2. EMBL Services

1. Related databases crossreferenced by EMBL:
   Drosophila genetic map database; FLYBASE *(4)*—available from EMBL
   E. coli database; ECD *(5)*—available fom EMBL
   EC nomenclature database; ENZYME (6)—available from EMBL
   Eukaryotic promoter database; EPD *(7)*—available from EMBL
   Genome database; GDB *(8)*
   Online Mendelian inheritance in man; OMIM *(9)*
   Protein pattern database PROSITE *(10)*—available from EMBL
   Protein sequence database; SWISSPROT *(11)*—available from EMBL
   3-D protein structure database; PDB *(12)*—available from EMBL
   Restriction enzyme database; REBASE *(13)*—available from EMBL
   Transcription factor database; TFD *(14)*—available from EMBL
2. Fileserver *(3):*
   To access the EMBL fileserver, send a standard electronic mail to the address Netserv@EMBL-Heidelberg.DE.

The most important command is "HELP", sent either on the subject line or in the body of the text.

References

1. Stoehr, P. S. and Cameron, G. (1991) The EMBL data library. *Nucleic Acids Res.* **19,** 2227–2230.
2. Burks, C., Cassidy, M., Cinkosky, M. J., Cumella, K. E., Gilna, P., Hayden, J. E.-D., Keen, G. M., Kelley, T.A., Kelly, M., Kristofferson, D., and Ryals, J. (1991) GenBank. *Nucleic Acids Res.* **19,** 2221–2225.
3. Stoehr, P. and Omond, R. (1989) The EMBL network fileserver. *Nucleic Acids Res.* **17,** 6763–6764.
4. Ashburner, M. (1990) University of Cambridge, Cambridge.

5. Kroeger, M., Wahl, R., and Rice, P. (1991) Compilation of DNA sequences of *Escherichia coli* (update 1991). *Nucleic Acids Res.* **19,** 2023–2043.
6. Bairoch, A. (1990) University of Geneva, Geneva.
7. Bucher, P. and Trifonov, E. N. (1986) Compilation and analysis of eukaryotic POL II promoter sequences. *Nucleic Acids Res.* **14,** 10,009–10,026.
8. Pearson, P. L. (1991) The genome database (GDB)—a human gene mapping repository. *Nucleic Acids Res.* **19,** 2237–2239.
9. McKusick, V. (1990) *Mendelian Inheritance in Man.* Johns Hopkins University Press, Baltimore, MD.
10. Bairoch, A. (1991) PROSITE: a dictionary of sites and patterns in proteins. *Nucleic Acids Res.* **19,** 2241–2245.
11. Bairoch, A. and Boeckmann, B. (1991) The SWISS-PROT protein sequence data bank. *Nucleic Acids Res.* **19,** 2247–2249.
12. Bernstein, F. C., Koetzle T. F., Williams, G. D. B., Meyer, E. F., Brice, M. D., Rodgers, J. R., Kennard, O., Shimanouchi, T., and Tasumi, M. (1977) The protein data bank: A computer-based archive file for macromolecular structures. *J. Mol. Biol.* **112,** 535–542.
13. Roberts, R. J. (1985) Restriction and modification enzymes and their recognition sequences. *Nucleic Acids Res.* **13,** r165–r200.
14. Ghosh, D. (1991) New developments of a transcription factors database. *TIBS* **16,** 445–447.

Index

425

Printed in the United States
By Bookmasters